D1085260

LOGISTICS ENGINEERING

NEW DIMENSIONS IN ENGINEERING

Editor
Rodney D. Stewart

SYSTEM ENGINEERING MANAGEMENT
Benjamin Blanchard

LOGISTICS ENGINEERING
Linda L. Green

DESIGN TO COST
Jack V. Michaels
William P. Wood

COST ESTIMATING, SECOND EDITION
Rodney D. Stewart

LOGISTICS ENGINEERING

LINDA L. GREEN

A Wiley-Interscience Publication
JOHN WILEY & SONS, INC.
New York • Chichester • Brisbane • Toronto • Singapore

Copyright © 1991 by John Wiley & Sons, Inc.

Library of Congress Cataloging in Publication Data:

Green, Linda L.
  Logistics engineering/Linda L. Green.
  p.   cm.—(New dimensions in engineering)

  "A Wiley-Interscience publication."
  Includes bibliographical references.

  ISBN 0-471-50632-X
  1. Systems engineering. 2. Logistics. I. Title. II. Series.
TA168.G75   1991
                                                90-44286
658.5—dc20
                                                CIP

To my mother, Opal Jeffries Green, whose friendship and advice I shall deeply miss, and my dog, Cookie, my friend and companion who sat by day and night as I prepared this manuscript.

FOREWORD

As the second volume in the "New Dimensions in Engineering" series, this exhaustive treatment of the discipline of logistics covers what is rapidly becoming a complex and multifaceted part of the conduct and procurement of major products, projects, processes, and services. No longer is the initial investment the sole consideration in acquiring new work outputs or initiating new work activities. Costs of services and products that support ownership often exceed the cost of the original acquisition, and the adequacy of these supporting services and products can have far-reaching effects on the long-term suitability of the systems they support. The "after-market" activities of supply, training, repair, maintenance, and field support are often deciding factors in the purchase decision. Suitability of the "logistics" function, as broadly defined, can determine the long-term usefulness of a system. Logistics support, which previously was often planned and analyzed far too late in the life cycle of the system, has achieved a predominant role in the planning of any new major project.

Therefore, we recommend this volume as a reference source for individuals and organizations engaged in the research, development, production, and support of major systems and products. Its synergy with other volumes in the "New Dimensions" series is strengthened by the interrelationships between logistics and systems engineering; quality, reliability, and safety; maintainability; and many other emerging "horizontal'" disciplines that cut across the vertical engineering facets of major programs. Logistics is a vital line that ties together the practical aspects of using and supporting new additions to the growing high-technology inventory. This book clarifies, organizes, explains, and instructs in this important new dimension in engineering. It is hoped that the readers' projects or products will be enhanced by thoughtful, systematic, methodical, and creative application of the facts, figures, and methods contained herein.

RODNEY D. STEWART

Series Editor

PREFACE

Logistics is not unique to the military world of weapons systems and field forces or even to actual government systems for it embraces a wide gamut of professional disciplines in both the public and private sectors. In recent years, systems and products have become more complex. The trend toward integration in logistics surfaced with the belief that integrated performance produces results superior to those produced by individual functions acting in relative isolation, which results in reduced costs.

The challenge for the 1990s is to recognize and act on the interdependence between the logistics elements. A systems orientation to logistics forces this recognition and requires the logistics manager to develop logistics strategies that are properly designed to support the objectives of the organization. In the military context, logistics is concerned with the various aspects of maintenance and system–product support, particularly from the point in time when systems come into operational use. Logistics is part of the total systems acquisition cycle and must be planned for early in the development process to preclude the "pay me now or pay me later" syndrome of playing "catch up" downstream. Thus, the primary objective of integrated logistics support is to develop a package of logistics resources that optimize the operation of any system.

The study of space logistics is in its infancy. Although there have been a number of conceptual studies of on-orbit maintenance in the past few years, there is no unifying literature, no "generally accepted" quantitative approach, and no inventory of space logistics models.

The rapidly expanding dependence on space systems by the military and civilian communities is rapidly placing new requirements on logistics. The requirements are driven primarily by the need or improved cost-effectiveness. In the largely research and development (R&D)-oriented space environment of the past 25 years, traditional logistics functions were designed out of systems. Replacement has been achieved by a combination of on-orbit spaces and scheduled launches. Spacecraft have been designed and built using exceptionally reliable components and redundancy. The successful retrieval and repair of the Palapa B and Westar 6 satellites as well as the repair of the SOLAR MAX have demonstrated the technical feasibility of in situ repair. A significant issue, however, still remains—the cost-effectiveness of satellite repair in space.

Current studies concerning on-orbit maintenance center on the issue of cost-effectiveness. Existing supply support models largely reflect a logistics system structured to support nonspace missions and a maintenance process dictated by the terrestrial environment in which the missions are carried out. Generally any attempt to modify or build on present depot-supply, base-supply, bench-stock model structures used by the U.S. Army or the U.S. Air Force is fraught with the danger of assuming away significant parts of the problem.

The option of on-orbit maintenance and servicing places very significant demands on standardization and commonality of design to assure compatible hardware and software interfaces. For the "strategic defense initiative" (SDI) to reduce the number of maintenance actions necessary, systems must be designed to allow for maintenance of a majority of the subsystems by modular replacement. Subsystems that cannot be modularized must be designed for disassembly, repair, and/or replacement utilizing standard manipulators and/or tool-sets for either manned or unmanned telepresence servicing missions. The optimum mix of manned versus unmanned maintenance and servicing may be influenced more by the environmental risks involved in the manned missions (radiation exposure) than by tradeoff studies of most efficient and effective performance functions. Most of the SDI satellites are in locations where the current crew radiation shielding systems severely limit the period of time that a manned mission can be exposed to the radiation environment. Where unmanned support missions are envisioned, the appropriate levels of space-qualified robotic capabilities are required to facilitate the accomplishment of maintenance and servicing actions. Required capabilities may range from intensively interactive person-machine (teleoperator) systems to semi-autonomous (telerobotic) systems. Telerobotic capabilities are currently under development with projected availability of space-qualified systems by the year 2000.

The possibility of locating manned support stations at low altitudes in or near the orbital planes of the SDI assets has been proposed, but the radiation environment is still a concern. In addition, the energy expenditures necessary to bring a large SDI satellite down to a lower-altitude orbit for maintenance and servicing and reinserting the satellite in its original orbit makes this a potentially high-cost approach. Also, the ground processing of SDI satellites must be done in a more "standardized" manner than current space system processing practices. Consolidated repair, refurbishment, warehousing, and payload processing–integration facilities must be established within reasonable transportation distances of the launch–recovery facilities. Even through all this, the government is still advocating the use of the logistics support analysis (LSA) process under Military Standards MIL-STD-1388-1A and 2A in conjunction with repair level analysis and SDI logistics support model (under development) to identify and quantify support requirements for SDI systems. The use of the systems engineering process will still be valid for space logistics.

Because of the costs and unique capabilities of future systems, it will become increasingly difficult to use on-orbit spares or rely on ground-based backups. It will not be economically feasible to fund the costs of redundant systems, including the recurring costs of production and launch; nor will it be feasible to discard a platform if it fails. As a result of high initial acquisition and support costs, plus an increasing dependence on these systems, users of future spacecraft and space platforms can be expected to desire longer mission lifetimes and greater availability in conjunction with the ability to upgrade both the hardware and software of their systems.

Scarce resources, budgeting constraints, and public scrutiny will influence government and commercial developers of space. On-orbit servicing and its related effects will have the potential to amortize fixed costs over longer system lifetimes, decrease payload turnaround time, simplify launch manifesting, and allow users to accomplish platform technology upgrades and make design corrections or change missions. On-orbit servicing will benefit commercial users by decreasing insurance premiums.

Logistics influenced one important part of the one-shot era—reliability. The cost of expendable launch vehicles and satellites forces satellite producers to make the most of reliable payloads possible. Successful logistics strategies still require highly reliable satellites.

Standardization and interchangeability provide the greatest potential gains in the acquisition strategies. Standardization probably provides the fastest and easiest manner to move toward serviceability. Standardization will decrease procurement costs and accelerate the process of inserting new technologies by simplifying the qualification of space-rated components. Standardization will allow the automated manufacture of high-labor-content products such as wiring hardnesses. Standardization can play a role in high-order programming languages and test languages so that on-board families of computers and standardized automated test equipment can become feasible. Standardization of platforms is currently receiving the greatest amount of attention.

The purpose of this book is to provide an in-depth study of logistics as it functions in the Department of Defense with emphasis on new dimensions evolving around warranty programs, computer resources support, manpower and personnel integration (MANPRINT), configuration management, and the use of nondevelopmental items in systems acquisition. In addition, the design of a system for supportability has a tremendous impact on future required resources. Proper planning for logistics early on can ensure that supportability is achieved with the lowest life-cycle cost.

All acquisition programs within the Department of Defense (DoD) require an integrated logistics support effort. This book addresses that effort and is divided into seven modules (parts). Part 1 introduces the integrated logistics support (ILS) process and its objectives, ILS planning requirements, the initial development of the work-breakdown structure (WBS), and the acquisition process and ILS involvement. Part 2 describes the ILS impacts on design and support decisions through reliability and maintainability (RAM), ILS in the systems engineering process, and logistics support analysis (LSA). Part 3 emphasizes the new dimension in logistics. Part 4 focuses on the ILS role in planning for and accomplishing the transition to production and, operational and postproduction support.

This book is designed for use by the practicing professional in the field or by the student of logistics in the classroom. Concepts and principles discussed are applicable to any type of system, and the functions can be tailored to large or small programs. Numerous figures and table are used to provide further understanding in specific areas.

I wish to thank Mr. Rod Stewart for his encouragement in the preparation of this book, and Miss Debbie Steger for her diligence in the preparation of the manuscript. I also wish to thank all of those friends and co-workers who provided day-to-day encouragement in this my first endeavor in professional publication of a book-sized manuscript.

LINDA L. GREEN

Huntsville, Alabama
January, 1991

CONTENTS

LOGISTICS ENGINEERING

PART 1

INTRODUCTION TO INTEGRATED LOGISTICS

Part 1 lays the basic foundations for integrated logistics support and its application within the Department of Defense.

Logistics managers must have a thorough understanding of ILS elements and a comprehensive knowledge of the materiel acquisition process and the role of logistics in that process. Logistics managers must ultimately determine whether a design–acquisition effort has achieved ILS objectives, and that assessment can be difficult if measurement is based on a reduced support structure. The ILS objective is to have the right item, in the right quantity, in the right place, at the right time within acceptable resource limits.

The fundamental concept of integrated logistics support is to relate support to design and to use an engineering analytical approach for designing the logistics support subsystems for hardware–product acquisition. ILS is crucial for any system to serve its intended purpose after it is acquired. To accomplish this, logistics supportability requirements must be defined early in the acquisition process and be accorded emphasis comparable to cost, schedule, and performance in acquisition decisions.

The acquisition process can be cumbersome but can be streamlined or tailored to provide greater flexibility to meet the objectives of the Department of Defense for the acquisition of cost-effective equipment and to take advantage of parallel initiation.

1
INTEGRATED LOGISTICS SUPPORT FUNDAMENTALS

1.1 INTRODUCTION

Major systems and products have been planned, designed and developed, produced, and delivered to the customer with very little consideration given to the aspect of logistics support. Sustaining life-cycle support of the system has generally been addressed after the fact. This practice has been costly because the costs of the support system often constitute a major portion of the overall life-cycle cost, and the costs of the system support are largely influenced by decisions made in the early phases of the life cycle. Because of the cause-and-effect relationships between system–product design and support and the fact that logistics costs may assume major proportions, planning for logistics support must be included in the early stages of system–product planning and design.

The fundamental concept of integrated logistics support (ILS) is to relate support to design and to use an engineering analytical approach for designing the logistics support subsystems for hardware–product acquisition. The ILS process pursues two objectives simultaneously: (1) design influence to reduce operating and support costs and simplify equipment operation and maintenance and (2) design, development, test, and acquisition of support to assure satisfactory operation and readiness of the system in the field. The effectiveness of the first objective reduces the demands on the second.

Integrated logistics support is crucial to the ability of a system–product to serve it's intended purpose after it is acquired; therefore, ILS must positively influence the development or selection of items of equipment and also ensure that the support items needed to operate and maintain a system are developed concurrently with the system. Through the integrated development of the ILS elements [and consideration of the manpower and personnel integration (MANPRINT) process], the materiel developer is able to proactively influence the acquisition strategy, system design, acquisition, and ensure the total support requirements for a system will be considered. Logistics supportability

3

requirements must be defined early in the acquisition process and be accorded emphasis comparable to cost, schedule, and performance in acquisition decisions, according to Department of Defense Directives (DoDD) 5000.1 and 5000.39. In addition, the effective utilization and consideration of ILS and MANPRINT will help ensure that the soldier–machine interface is considered; tradeoff evaluations consider the impact of operation and support requirements in addition to technical and performance needs; the system can be operated and maintained within the personnel, skill, and cost resources that will be available at fielding; and the needed support structure is in place to effectively maintain the system to its specified level of readiness.

The basic principle of ILS planning is the development of harmony and coherence among all the diverse support elements. ILS is the unified and iterative approach to the management and technical activities needed to influence operational and materiel requirements and design specifications, define support requirements best related to system design and to each other, and seek improvements in the materiel system and support systems during the operational life. Thus, a logical approach to integrated logistics support requires the study of the individual elements that form the functional building blocks; the tools, methods, and techniques used to combine them; and the master overview. These elements will be discussed further later in this and other chapters.

ILS is a management function in which the initial planning, funding, and controls are established that help assure the ultimate user will receive a system–product that will meet performance requirements, and can be expeditiously and economically supported throughout its programmed life cycle. ILS is also the integration of related technical parameters that assures the compatibility of all physical, functional, and program interfaces in a manner that optimize the total system definition and design.

1.2 BACKGROUND

Logistics is not a new discipline, yet it forms a new dimension in engineering when coupled with the high-technology systems and products of the 1990s. Logistics incorporates physical facilities at various locations, transportation, inventories, handling, and storage. These have been necessary functions since the beginning of time. What is new about logistics is that it is now considered to be a formalized support element, and its newness stems from the integrated approach that became necessary in defense programs in the mid-1950s.

Logistics came to include all support required for the introduction of a new product or significant changes to an existing product. This, in turn, would require training, the development of technical publications, and the acquisition of spare parts, special tools, and test equipment. Thus, logistics become the "process of having the right quantity of the right item in the right place at the right time." Management of the integrated logistics support activities mentioned above has become the task of ensuring that these objectives are achieved within acceptable resource limits.

The Department of Defense (DoD) defined ILS in DoD Directive 4100.35, stating that "Integrated logistics support is a composite of all support considerations necessary to assure effective and economical support of a system throughout its existing life. It is an integral part of all aspects of the system and its operation. Integrated logistics support is characterized by harmony and coherence among all the logistics elements."

The economic climate of the 1950s led to a squeeze on profits, leading to a massive search for cost-reduction measures. Prior to this, logistics was implemented only piecemeal, and it was not realized that integrated logistics could lead to improved performance and cost reduction.

Between 1956 and 1965 the ILS concept grew by leaps and bounds because of major developments in customer service, distribution networks, cost analysis techniques, and a system approach to new products. During the late 1960s, the logistics field and the logistics manager were recognized as a necessity within the DoD community. Logistics at this time was concentrated largely in the material management arena. Material management emphasized the orderly flow of materials with time-phased delivery being scheduled in accordance with the needs of the customer.

In the 1970s, the energy crisis forced logistics to face the need for improvements in productivity. The crisis was worsened by the high visibility of transportation and storage costs: basic logistics activities. Logistics costs began to soar.

The federal government and the vast military industrial complex had a significant impact on the development of an ever-growing logistics capability. Government influence in logistics became a reality when it was realized that costs associated with the logistics support throughout the useful life of a system (from concept exploration to displacement) frequently exceeded initial acquisition cost. Government contractors began implementing logistics as cost-reduction moves. Many of these same contractors also produced commercial products to which they also applied logistics to improve customer service.

The trend toward integration surfaced when it was realized that integrated performance produces results superior to those produced by individual functions acting in isolation. Integrated logistics combines all elements of logistics into a whole or composite that strikes a balance between reasonable performance levels and realistic cost expectations. ILS is a balanced discipline whose application stems from a requirement to effectively support systems throughout their life cycle.

1.3 LOGISTICS MANAGEMENT

The challenges logistics managers face are numerous. Their responsibilities include the development of requirements, the production of supplies, and the acquisition of goods and services, the transportation and distribution of materials, storage, the determination and maintenance of stock levels, and arrangements for support in coalition warfare. These are challenges for any logistics manager, whether in the DoD community or in the commercial arena.

Early in the acquisition phase an integrated logistics support program is implemented for the life-cycle management of major systems to ensure adequate logistics support. The primary objective of the integrated logistics support program is to achieve system readiness objectives at an affordable life-cycle cost. The Department of Defense establishes broad policy guidelines for all military departments in the planning and development of logistics support for new systems. The guidelines aim at a disciplined, unified, and consistent approach to management and technical activities to integrate and acquire support for systems and equipment. Logistics support is clearly established to meet the acquisition system goals of system readiness, operational performance, cost, and schedule. Commercial businesses can profit heavily from being organized to perform ILS. The system approach delineates the activities to be accomplished and the procedures to be followed in the definition, design, development, provisioning, acquisition, production, test, delivery, and operation of logistics support. Certain logistics support activities, such as maintenance concept definition and support engineering analysis, provide input into the design engineering process.

Logistics is a support function, and the product of logistics is performance. Customers are frustrated when the products they purchase fail to operate properly, especially if the product is a high-cost item such as an automobile, a dishwasher, a refrigerator, or a stereo system. Performance within the logistics element is a function of availability, capability, and quality; "availability" is a function of safety stock levels and is measured by the probability that an item will be available when requested; "capability" refers to the speed and consistency of logistics performance cycles; and "quality" relates to the number of incorrect, missing, or damaged items. Each of these measures must be taken over time to preclude erroneous conclusions regarding the level of logistics performance.

1.4   GOAL OF LOGISTICS

The goal of logistics is to reduce the burdensome cost of logistics through better management, organization, and utilization of all resources to the maximum extent possible. ILS must ensure that logistics considerations are integrated into the design effort of developmental and product-improved systems. It must also ensure that the material acquisition process includes the timely availability of all required logistics resources. When ILS is properly implemented, it produces a step-by-step development process that can be subjected to management appraisal and control techniques.

When logistics is considered independently of the product, the life-cycle perspective is lost as a result of parochialism, lack of coordination and communication, compromise of person–machine interrelationships, and misplaced emphasis.

1.5 LOGISTICS PLANNING

Logistics planning involves the determination of supply, transportation, maintenance, construction, and related logistics requirements and the existing capabilities to meet these requirements. It is necessary to understand the following influences and basic considerations of logistics planning in the development of effective planning procedures:

1. *Leadtime.* In general terms, leadtime is considered to be that time between action taken to obtain an item for use and the arrival of the item in the hands of the user. You might wonder why the part your car needs is never on hand and you are inconvenienced to bring the car back when the part arrives. Poor planning for known high failure-parts causes shortages of these parts.

2. *Limited Resources.* Resources (e.g., personnel materiel, and money) are always limited. The concept of resource management promulgated by DoD analysts recognizes this consideration by establishing a system for evaluating the essentiality of conflicting defense programs.

3. *Critical Shortages.* This has always been a problem in logistics planning. The logistics planner and the logistics system must expect that somewhere along the line a critical shortage will develop and emergency measures must be taken to correct the shortage. For example, many car dealership parts departments are currently using microfiche for parts ordering. If car dealers had computers that could tap into inventory programs and order parts electronically from a central point with overnight delivery, many parts problems could be solved and customer service improved.

4. *Priorities–Allocations–Reserves.* Since the resources are usually limited, systems of priorities and allocations are established. The basic point of this logistics consideration is that once a system is established, discipline is necessary to prevent frustration by well-meaning but unaware persons. In like manner, logistics reserves can be used effectively only with proper application of discipline. Logistics reserves are as essential as personnel reserves in a tactical operation.

5. *Coordination and Communication.* The constant exchange of information and coordination is vital for military success.

6. *Flexibility.* Regardless of the level at which planning is conducted, it must provide for the means to be in place at the right time. The plan must also provide for flexibility to allow for the changing military or consumer market situation.

7. *Adequacy–Suitability–Feasibility–Acceptability.* Courses of action must meet the current situation or a situation that might be developed. Courses of action must be considered in terms of (a) adequacy (accomplishment of the objective), (b) suitability (adaptability to various circumstances), (c) feasibility (ability to provide the right means at the right place at the right time, and in usable condition), and (d) acceptability (e.g., affordability).

8. *Command Control.* Logistics considerations 1–7 allude to the key fact that command control must be exercised with sound judgment, understanding, competence, and restraint. Unless positive command control is maintained, the various logistics installations and operations tend to expand to unmanageable size. Figure 1.1 shows the interrelationships between tactics, strategy, and logistics. This same interrelationship is also applicable in the development of new consumer products.

1.6 ILS PROCESS

The fundamental characteristics of the integration process are twofold: establishment and use of appropriate acquisition process interrelationships

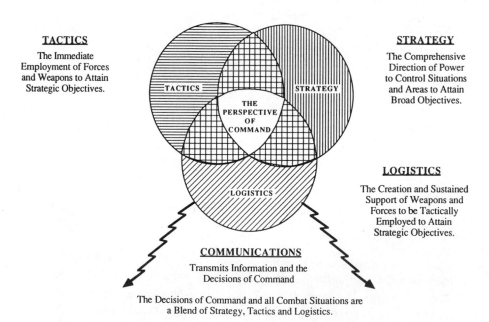

Figure 1.1 The Interrelationships between strategy, tactics, and logistics. (*Source*: FM 701-58.)

among requirements, support, and engineering considerations (Figure 1.2) and iteration to identify and assess the acceptability of the life-cycle commitment of resources. This process is discussed further in Chapter 2.

The objectives of the ILS process are twofold: design for support and design of support. These objectives are illustrated in Figures 1.3 and 1.4, respectively.

Guidelines

An effective logistics program must be guided by basic criteria, including:

1. Planning the logistics support required in the conceptual stage and any special problems identified early in the program.
2. Formalizing the logistics support program by the ILS manager with the project manager prior to the beginning of full-scale development with appropriate performance improvements throughout development, production, and distribution.

1.7   INTEGRATED LOGISTICS SUPPORT ELEMENTS

Design Influence

U.S. Military Service regulations on integrated logistics support explain that design influence is an intangible ILS element but significantly affects overall system readiness, supportability, and affordability. Previously more lip service was given to logistics influence on design than on active, effective weapon system research and development. Design influence includes the relationship of logistics-related design parameters of the material system to its projected or actual operational support resource requirements. These design parameters are expressed in operational terms rather than as inherent values and are specifically related to operational objectives and support costs of the system. In addition, the cost of acquiring the right logistics support at the right time has never been clearly understood. These costs are receiving more scrutiny as the costs of weapons systems increase and may well determine whether a development of system will continue.

Design influence happens very early on in a system development program in the preconcept and concept exploration phases. By the time the system reaches the demonstration–validation phase, the system design, which influences logistics, is nearly fixed. Logistics support analysis (LSA) and logistics support analysis records (LSAR) can effectively influence design only if these records are acquired either by contract or through in-house government services. This has to be done, or the logistician will be caught in an expensive catch-up game. LSA documentation is critical to the logistician's effort to influence design. Where the LSA process analyzes and documents the po-

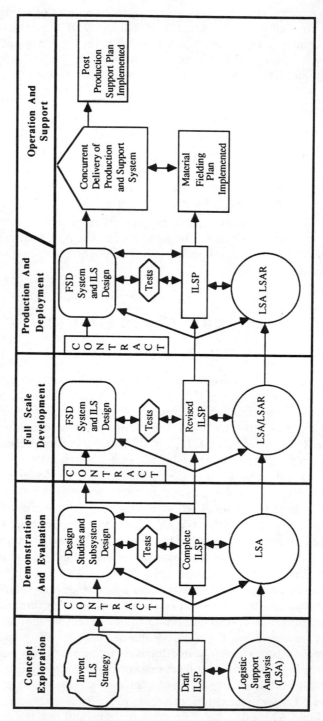

Figure 1.2 The integrated logistic support process.

ILS Objective

Design for Support:

Discard at Failure
Modular Replacement
High Relaibility Parts
Bite/Integrated Diagnostics/Standard TMDE
Standard Test Points
Accessibility
Quick Release Fastners
Standard Parts
Simplicity (Fewer Parts)
Lifting Points for Transportability
Reduced Weight/Cube
Soldier/Machine Interface

Figure 1.3 ILS objective design for support. (*Source: Army Research, Development and Acquisition Bulletin,* January–February 1989.)

Design of Support

Reduce Number of Parts
Reduce Number of Reparables
Reduce Requirement for Common Tools/TMDE
Eliminate Special Tools, TMDE, & Skill Requirements
Reduce Manpower
Reduce Skills Required
Reduce Training Course Lengths
Reduce Training Devices
Increase Modes of Transportation
Reduce Number of Tech Manual Pages

Figure 1.4 ILS objective design of support. (*Source: Army Research, Development and Acquisition Bulletin,* January–February 1989.)

tential influence of logistics considerations on design, the LSAR technically forces and records the results of the logistics analyses. The LSA/LSAR process provides the means for the logistics manager to influence design. However, later in the development cycle, starting at about full-scale development, design decisions will be under engineering change control, reducing design flexibility.

A major breakdown in logistics influencing design evolves around the lack of communication between the design development engineer and the logistician. This is also true in the development of commercial products. Teamwork is essential. The logistician must function as part of the design and development team. Once the logistician is accepted as a team member by the engineering community, effective communication can be established and the logistician will be able to contribute to the design process in time to influence supportability. In order to gain full membership on the team, the logistician must contribute significantly to tradeoff analyses in terms of supportability, not just sit back and criticize without offering solutions. The logistician should share data with the engineers on similar systems. And most important, the logistician must make a serious effort to become completely familiar with the design as it progresses.

Design can be influcnced only through good communication. The development of a supportable item or system requires the skills of the designer and the logistician working together as a team, to structure the equipment in a logical manner for use in a sound maintenance plan and consequently in a practical support plan.

Maintenance Planning

Maintenance planning involves the definition of a maintenance concept, performing LSA, provisioning, assessing and evaluating an overall support capability, and designing a control mechanism for corrective action and modification. The lifeline of any acquisition process is maintenance planning, which continues throughout the life cycle of the equipment. There are three important aspects of maintenance: documentation; basic factors of time, skills, and resources; and analysis. During the early phàses of a program, the maintenance planning activity is crucial. Each logistics element is dependent on the maintenance plan. The emphasis decreases during the middle of the development process, but increases again as the operational phase approaches.

The principal tool for guiding and recording maintenance analysis is contained in MIL-STD-1388-1, which standardizes the approach to maintenance engineering analysis and incorporates operational considerations. However, the depth and detail of any analysis is governed by the design complexity and the resources available. Some government contractors, such as General Electric (GE), Westinghouse, and Raytheon, also use the basic principles if MIL-STD-1388-1 to support their commercial products. All the military services are required to implement the MIL-STD contractually.

Maintenance planning documents the concepts and requirements for each level of equipment maintenance to be performed during the useful life of the system. Factors in maintenance planning include:

1. Definition of actions required to maintain the designed system in a prescribed state of operational readiness. In the commercial arena, is there a local repairman or must you return it to the store to be returned to the manufacturer for repair or replacement.
2. Determination of maintenance functions including checkout, servicing, inspection, fault isolation, replacement, and repair.

Many commercial manufacturers do not give maintenance planning adequate consideration, however, assuming that when a product that is neither new nor expensive breaks or malfunctions the customer will simply purchase a new one rather than return the old one for repair. A little ILS planning would certainly help to increase product sales through better customer support and consumer satisfaction.

The role of LSA in maintenance planning concerns the identification of specific maintenance actions to be performed. This includes the systematic application of analyses to identify and describe tools; testing of equipment, personnel, spares, and repair parts and facilities to support the system; and performance of level of repair analysis.

The maintenance plan should include, as a minimum, the maintenance concept, estimated corrective and preventive maintenance worker–hour requirements for each major assembly, and identification of overhaul considerations. Maintenance planning constitutes a sustaining level of activity beginning with the development of the maintenance concept and continuing through the accomplishment of LSA during design and development, the procurement and acquisition of support items, and the user phase when an ongoing system support capability is required to sustain operations. It covers interim contractor support of the system during the early phases of operation as well as procedures for the upgrading and the installation of modification kits.

Manpower and Personnel

The manpower and personnel element of ILS involves the identification and acquisition of operating and support personnel with the appropriate skills to operate and maintain the system over its lifetime. Although commercial manufacturers know that consumers will use their product, they often fail to plan for the training of repair technicians; thus the lawnmower repair technician is really a jack-of-all-trades for all makes and models. Manpower and personnel guidelines help define requirements for the training program for operations and maintenance personnel and the need for training devices to support training throughout the life cycle. Projections of manpower requirements should reflect the current force structure and forecast attrition rates as well as the capability of the raw personnel inputs. Staffing determinations and their derivation from system design activities must be accomplished to ensure that manpower requirements meet equipment requirements. Manpower requirements must be based on related ILS elements as well as human factors engineering, which will ensure optimum person–machine interface. This area has become known as "MANPRINT" and will be discussed in a later chapter. The initial operational and maintenance analyses in the LSA enable the estimation of manning requirements and also enable the estimation of training requirements. Personnel requirements are derived from the functions or the tasks that must be performed and the time required for their performance.

Supply Support

The supply support element encompasses all management actions, procedures, and techniques used to determine requirements to acquire, catalog, receive, store, transfer, issue, and dispose of inventory. Supply support is essential to the logistics integration effort. Ever wonder why it takes so long to repair your lawnmower? This is the result of poor supply planning and repair training.

Supply planning for spares and repair parts must be based on technical inputs from maintenance planners and engineers for attaining a predetermined state of supply readiness. A number of factors must be considered when planning for supply support; planning inputs (level of repair, repair–discard criteria, and maintenance level facilities); isolation of utilization rates, operating hours, and failure rates; repairable program planning; requirements for provisioning, technical documentation and maintenance plan inputs; inventory management factors; and inventory management controls. Supply support includes all consumables, special supplies, and related inventories needed to support the system. It covers procurement functions, warehousing, distribution of material, and the personnel associated with the acquisition and maintenance of the inventories at all support locations. Thus, because of the vastness of supply support those activities related to it must be designed to meet the readiness requirements of the military systems or commercial products. Few commercial manufacturers adequately consider these activities.

Support and Test Equipment

The purpose of this ILS element is to ensure that the required support and test equipment is available in a timely manner. The ability to perform the required scheduled and unscheduled maintenance depends on the adequacy of the support and test equipment identified or developed with the prime system. Support and test equipment consists of tools, metrology, and calibration equipment; monitoring and checkout equipment; maintenance stands; and handling devices that are categorized into special and common types. It also includes production test or support equipment that is modified and delivered for field use. For (a really simple) example, I recently purchased a ring and selected a stone for mounting. The stone would be in a recessed setting. The jeweler did not tell me he had to order a special tool to set the stone. After 4 weeks I finally got my ring. Had the jeweler had the right tools, I could have had my ring in a couple of days, not four weeks. Generally I would find it difficult to apply ILS to the jewelry industry except for actual production and distribution of the jewelry to the marketplace.

The support and test equipment program encompasses all life-cycle phases. It requires the application of tailored ILS planning techniques. Support and test equipment includes the following factors: automated versus manned; reliability requirements for the tools, monitoring equipment, test, and checkout equipment; special handling devices, and requirements to perform maintenance on the support and test equipment; calibration requirements; and logistic support of the support and test equipment. Support and test equipment may be classified as "peculiar" (newly designed and/or off-the-shelf items peculiar to the system under development) or "common" (existing items already included in the inventory).

Identification of support and test equipment places additional requirements on ILS plans, maintenance plans, and provisioning plans, that is, all the logistic support requirements necessary for the operation and maintenance of the support and test equipment. The main consideration for obtaining support and test equipment must be cost, schedule, performance, and the ability to acquire adequate management capability for follow-on support.

Logistics involvement in support and test equipment includes a determination of what is required, the quantity required, and the schedule of availability (when items will be required). The requirements for support and test equipment are derived through the LSA process published in the maintenance plan and differ for each level of maintenance. Support and test equipment quantities are a function of the quantity of equipment being supported, product reliability and maintainability, the maintenance concept, and the number of maintenance locations.

Support and test equipment is unique in that it does not clearly fall within the category of spares and repair parts or any of the other product support elements. However, support and test equipment provides an inherent ability to monitor, control, test, measure, evaluate, repair, and calibrate the product, thus providing the resources needed for a rapid return to service.

Training and Training Devices

The training and training devices ILS element includes the process, procedures, techniques, and equipment used to train personnel to operate and maintain the system throughout the life cycle of the system. Training requirements must be developed along with training curricula and must reflect the operations and maintenance concepts and the technical data plans. Training devices must be identified, planned for, procured, used for training, and maintained.

The principal factors to be considered for training include schedules for training plans, conferences, and institutions of training; determination of training equipment requirements and their required support; impact of training program leadtimes; and training throughout the life cycle of the system. Initial training estimates are based on experience. The initial operations and maintenance analyses from the LSA enable the estimation of training requirements. Types of training devices must be identified, such as simulators or mock-ups. Classroom facilities and locations must be identified. The types of training must be identified—that is, factory training, instructor and key personnel training, new equipment training, and resident training. Also, courses must be written and schedules developed.

Training differs from education in that it is task-oriented and specific. For example, training places greater emphasis on the acquisition of skills. Training in system operation depends on the planned utilization and operating scenarios, whereas maintenance training relies heavily on reliability, availability, and maintainability analysis (RAM). These data are used by the training staff to evaluate the operations and maintenance tasks and divide them into related task groupings and skill level categories.

Successful implementation of the training course demands proper time phasing with respect to the other elements of the program. Training requirements analysis cannot be completed until maintenance engineering has identified the tasks and times required for task performance. A primary impediment to scheduling training courses is the availability of training devices. The lack of equipment precludes hands-on training. Other factors affecting training include the availability of tools, test equipment, and technical publications.

Technical Data

The technical data element of ILS consists of scientific or technical information necessary to translate system requirements into discrete engineering and logistics support documentation. The purpose of technical data is to provide for the timely development and distribution of technical publications necessary to conduct operations, training, maintenance, supply, modification, repair, and overhaul of the system and equipment. Technical data includes drawings; operational, maintenance, and modification instructions; provisioning and facilities information; specifications; inspection, test, and calibration procedures; instruction cards and computer programs; audiovisual materials; technical manuals; supply bulletins, lubrication instructions, and repair parts special tools lists; components lists; depot maintenance work requirements (DMWRs); and LSAR data. Other documentation such as car owner manuals, operating instructions, and part lists for power tools or appliances such as garage door openers of electric fans are also considered technical data. As can be seen technical data incorporate a vast array of material designed for the expressed purpose of providing information about the system.

Computer Resources Support

The addition of computer resources as a standard element of ILS came was not widely appreciated. As the availability and use of embedded computers become increasingly widespread this area will consume more resources and management time. Computer resources support has the potential to become the most demanding of all the support requirements. Chapter 7 is devoted to this subject.

Most government agencies require that a computer resource plan be developed and maintained during the life cycle of a system. There is no such provision for an automobile manufacturer who uses computerized dashboards. You can only hope that the local dealer has personnel who can repair the problem. Factors to be considered include responsibilities for integration of computer resources into the system; personnel requirements for developing and supporting the computer resources; and computer programs required to support the acquisition, development, and maintenance of computer equipment and other computer programs. The plan should describe the resources necessary to support the software. This plan should be an evolutionary document to be updated during development and distribution of the system.

Computer resource life-cycle management documentation should enable the acquisition manager to understand and be able to meet sound software development practices for a software system acquisition. Software developments have gained great influence owing to characteristics such as cost and schedule overruns, lack of reliability, cost of maintenance, and lack of portability and interoperability. As a result, defense systems are becoming increasingly software-dependent.

A number of factors contribute to software problems, such as widespread lack of properly trained people in key positions, insufficient software management and control, lack of adequate requirements specifications, lack of research into software development techniques to ensure a better more maintainable product, lack of software life-cycle planning, duplication in software effort, lack of standardization, lack of systems engineers, and much more.

The Department of Defense has mandated policies and actions for computer resources. DoD components are required to ensure that requirements analysis and risk analysis are accomplished, computer resource life-cycle planning has been done, configuration management of computer resources is provided for, software language standardization and control is implemented, and delivery of support software is accomplished.

Facilities

The facilities element is composed of a variety of planning activities, all of which are directed toward ensuring that all required permanent of semipermanent operating and support facilities are available concurrently with the system. Planning must include construction funding, real estate requirements, environmental impacts, safety, health standards, and security restrictions. Facilities also includes the need for new construction as well as modifications of existing facilities.

Facility planning activities must take into consideration operations and maintenance requirements identified by the LSA and definition of the types of facilities, locations, and space requirements. Major inputs to any facilities planning activity are based on analysis of design sketches and drawings, specifications, reliability, maintenance constraints, training requirements, test requirements, resources, program schedule inconsistencies, and cost–benefit analysis requirements.

Facility requirements must analyze types and levels of maintenance; schedules pertaining to the physical aspects of the support of an item; and the personnel requirements in terms of skills, skill availability, and training. The availability of a system is greatly affected by location, cost quality, and availability of facilities. Facilities are required to support activities pertaining to the accomplishment of active maintenance tasks, providing warehousing functions for spares and repair parts, and providing housing for related administrative functions.

Transportation

Transportation planning is determining what is to be moved under varying constraints and selecting a mode of transportation to best fulfill a specific requirement. While most of the time-consuming detailed computations can be accomplished in minimal time with the aid of a computer, the planner must interpret and evaluate the computer outputs.

There are five basic modes of transportation: rail, highway, water, pipeline, and air. The relative importance of each mode is measurable in terms of distance, capacity, volume, and type of materials transported.

Packaging, Handling, and Storage

Packaging, handling, and storage include characteristics, actions, and requirements necessary to ensure the capability to preserve, package, and handle all support items.

The following factors must be considered:

- Documentation defining equipment handling and procedures and packaging and preservation needs
- Shelf life
- Constraints of cargo handling equipment
- Cube, weight, and volume
- Special environmental requirements
- Fragility

As packaging, handling, and storage requirements are addressed for analysis, inputs should come from the maintenance and supply analyses that are inputs to the LSA process.

Packaging is one of the most overlooked areas in logistics and can account for approximately 8 percent of the total amount spent on logistics. Packaging has three basic functions: (1) protection of the product, (2) identification, and (3) handling. Equipment packaging significantly influences other factors such as type of transportation and handling equipment, type and quantity of spare parts, extent of coverage in maintenance procedures, and facility space for storage of spare and repair parts.

Handling in the logistical system is concentrated in and around the warehouse facility, except during transportation. In particular, four warehouse handling activities must be performed: (1) receiving, (2) transfer, (3) selection, and (4) shipping. Handling within production and assembly plants is part of basic manufacturing and not an element of logistics movement. Handling can account for as much as 15 percent of logistics costs.

Storage is a labor-intensive operation and is also one of the more costly logistics activities, involving various types of warehousing facilities. In addition, storage requirements exist other that those necessitated by material movement.

Standardization and Interoperability

Standardization affects the type and quantity of test and support equipment; type and quantity of spare and repair parts; personnel quantities and training requirements; the extent of coverage in maintenance procedures; computer software language requirements; calibration and reliability; packaging, handling, and storage requirements; transportation requirements; and testing.

Interoperability affects interchangeability; packaging and handling storage, transportation, testing; maintenance procedures, software requirements, reliability, and calibration requirements.

1.8   WORK-BREAKDOWN STRUCTURE

One of the most useful management tools for the logistician is the work-breakdown structure (WBS). A WBS provides the framework for the required management visibility, cost estimating, and data reporting in a manner directly related to the systems engineering process. As the term implies, a WBS divides a total job or program into its component elements. These elements can then be displayed to show their relationship to each other and to the program as a whole. Each WBS element description should include definition of the item identified by the element, objective and requirement of the element, and a synopsis of the effort required to complete the elements. The WBS provides a schematic portrayal of the products that completely define a program. It provides a means of effective management planning and implementation by providing the various functional managers with a common reference framework for communicating and making decisions. The purposes of the WBS are to provide a product-oriented family tree composed of hardware, services, and data, which result from project engineering work efforts during the development or production of a product or system; completely define the program; and completely define the cost, schedule, and performance reporting criteria (see Figure 1.5).

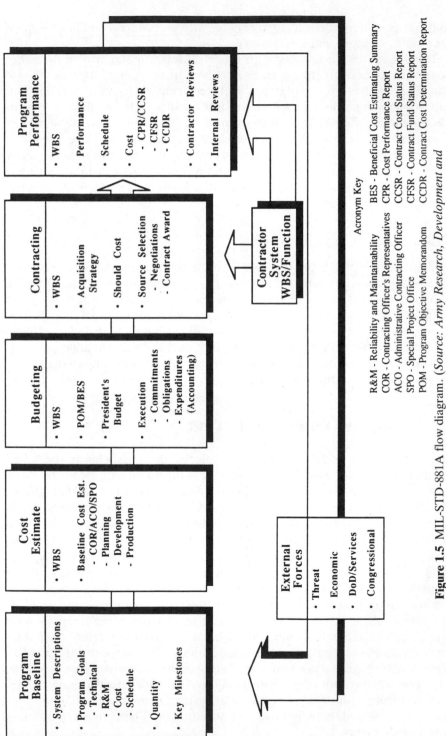

Figure 1.5 MIL-STD-881A flow diagram. (*Source: Army Research, Development and Acquisition Bulletin*, November–December 1987.)

Program Baseline
- System Descriptions
- Program Goals
  - Technical
  - R&M
  - Cost
  - Schedule
- Quantity
- Key Milestones

Cost Estimate
- WBS
- Baseline Cost Est.
  - COR/ACO/SPO
  - Planning
  - Development
  - Production

Budgeting
- WBS
- POM/BES
- President's Budget
- Execution
  - Commitments
  - Obligations
  - Expenditures (Accounting)

Contracting
- WBS
- Acquisition Strategy
- Should Cost
- Source Selection
  - Negotiations
  - Contract Award

Program Performance
- WBS
- Performance
- Schedule
- Cost
  - CPR/CCSR
  - CFSR
  - CCDR
- Contractor Reviews
- Internal Reviews

Contractor System WBS/Function

External Forces
- Threat
- Economic
- DoD/Services
- Congressional

Acronym Key

R&M - Reliability and Maintainability
COR - Contracting Officer's Representatives
ACO - Administrative Contracting Officer
SPO - Special Project Office
POM - Program Objective Memorandom

BES - Beneficial Cost Estimating Summary
CPR - Cost Performance Report
CCSR - Contract Cost Status Report
CFSR - Contract Fund Status Report
CCDR - Contract Cost Determination Report

There are four types of WBS: summary WBS, project summary WBS, contract WBS, and project WBS. The summary WBS (SWBS) consists of the top three levels of WBS elements, their structure, and their definition. This forms only the starting point for building a complete WBS down to the required level of detail. A project summary WBS (PSWBS) is a tailored version of the SWBS to match the objectives of a particular project. Once developed, it should not change during the life cycle of the program unless program objectives change. A contract WBS (CWBS) is an extension of the PSWBS that further defines the scope of the effort to be accomplished. The CWBS provides a direct structured relationship between the contracted effort and the total project, thus enhancing management visibility and control. A project WBS (PWBS) is a composite of all PSWBS and extended CWBS. A program WBS (PWBS) is proposed by a government purchasing office, using as a guide, one or more the appropriate MIL-STD-881A appendixes. The PWBS is used as a framework for reporting and managing the government side of the program and to develop specific performance work statements and reporting requirements for a "contract statement of work." The various types of WBS are developed in agreement with the system acquisition phases.

The WBS serves as a valuable link throughout the acquisition process. It is the one common link in a process that includes the formal baseline, cost estimating, budgeting, contracting and performance discipline, and resulting historical data. Commercial manufacturing could profit from the use of the WBS. The WBS serves as a formalized structure for identifying the required work and the organizational structure for performing the work. The WBS is a logical separation of work-related units. It links objectives and tasks with resources and is an excellent management tool for program planning and budgeting. The WBS is structured and coded in such a way that program costs may initially be targeted and collected against each element in the WBS. Costs may be accumulated both vertically and horizontally to provide a summary of accounts for various categories of work. These cost data are combined with program milestone charts and networks to provide management with the necessary tools for program planning, evaluation, and control. Both the WBS and the CWBS efforts should be coordinated to the maximum extent possible to ensure compatibility in cost estimating. Events and activities can be broken down into work packages, permitting analysis for cost, time, resources, complexity, and schedule.

In establishing the lower-level CWBS, it is essential to accommodate the differences between the organization, and its performance, and the management control of work in the development and production phases. The systems engineering process ensures that as the lower levels of the product elements of the WBS are developed, they continue to satisfy the operational needs specified in the system specification. The systems engineering process also ensures that any alterations of the portions of the WBS under contractor control use the tradeoff process and criteria that maintain system integrity.

The WBS has recently been made more accessible to the ILS community as shown in Figure 1.6. ILS managers will be able to closely monitor specific logistics tasks and subtasks associated with development and production. The redefinition of management, engineering, ILS, equipment, data, and test will provide visibility for all elements of ILS and technical accomplishment.

1.9 SUMMARY

The ILS concept is relatively recent, although logistics has been around for a long time. Logistics costs have always been relatively high because it has been planned for after the fact. However, recent high costs in weapons systems and commercial products have forced the need to get a handle on the logistics costs while improving performance and supportability. ILS strives to do just that through improved management techniques.

ILS is made up of interrelated and interactive components, through which

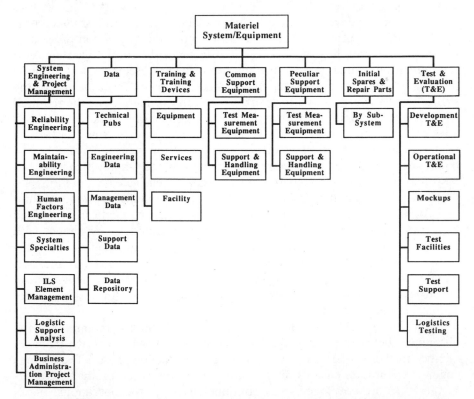

Figure 1.6 Proposed (May 1987) service–industry coordinated common work-breakdown structure for MIL-STD-881B. (*Source: Army Research, Development and Acquisition Bulletin,* November–December 1987.)

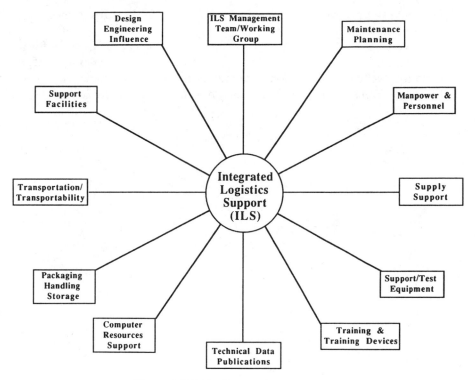

Figure 1.7 ILS elements–management.

integrated logistics improves supportability. Support for any system is viewed as a composite of all elements necessary to assure the effective and economic support of a system throughout its programmed life cycle. These elements of support must be developed on an integrated basis with all other elements of the system. The ILS elements interact with each other, and the effects of these interactions must be reviewed and evaluated continuously. Figure 1.7 illustrates the ILS elements. The proper relationships between system performance characteristics, supportability characteristics, and so on must be established. Attaining that balance requires consideration of logistics support throughout the system–product life cycle, particularly in the early phases of planning and conceptual design when decisions concerning the overall system configuration are made.

Planning is one of the most important functions of a logistician. To be effective, each element plan should reflect exactly what is expected to be accomplished at each echelon of an organization and when and by whom. It is important to remember that planning cannot be accomplished in a vacuum;

plans should be reviewed and updated continuously; as data are refined, they should be included in the plans.

The role of the WBS in logistics has also been covered. The WBS forms the foundation for program and technical planning, cost estimation and budget formulation, schedule definition, statements of work and specifications of contract line items, and progress status reporting and problem areas.

2
THE ACQUISITION PROCESS AND LOGISTICS ORGANIZATION– MANAGEMENT

Major systems acquisition–product development requires large sums of money and represents a substantial share of military and commercial expenditures. As a result, there is an increasing awareness that supportability factors, such as manpower, skills, and maintenance downtime, have become critical limiting factors in weapons system effectiveness. The high costs of systems–products tend to make more difficult the resolution of issues regarding the balance between current and future capabilities, complexity and simplicity of systems, quality and quantity, and operating costs versus research and development costs. Therefore, system–product acquisitions must be directed toward achieving the best balance between cost, schedule, performance, and supportability.

2.1 DEFENSE ACQUISITION

Objectives

The objectives of materiel acquisition process include:

- Maintaining a strong technology base.
- Assuring adequate standardization and interoperability.
- Achieving an appropriate balance between low risk, evolutionary development, and more visionary efforts required to maintain technological superiority.
- Developing an acquisition strategy tailored to meet the needs and con-

ditions of the specific materiel alternative and acquisition streamlining objectives.

- Establishing at the beginning of the program a formal and integrated MANPRINT and logistics support program to ensure that optimally supportable systems meet readiness objectives.
- Developing effective systems that operate in conjunction with other battlefield systems, are transportable, are survivable, and have eliminated or controlled all identified health, safety, and environmental hazards.
- Acquiring materiel systems that meet approved materiel requirements within budget, manpower, personnel, training, logistics, and competition requirements to support acquisition necessary for total unit materiel fielding.
- Emphasizing early life-cycle planning for budgeting, logistics support analysis, operational and human performance, safety, supportability, transportability, procurement, and producibility.

Framework

Because the logistics support and operations channels are not identical there must be close coordination between responsible parties and a system for establishing priorities. Logistics is multifaceted, and its complexity is increased by the requirements to support military operations everywhere. In addition, the logistics support structure must be tailored to meet the requirements of a particular theater of operations; therefore, logistics is designed to meet force needs and may vary from theater to theater. All of these requirements must be considered by the Defense Department in planning its budgetary needs.

Since its inception in 1961, the Planning, Programming, and Budgeting System (PPBS) has been the main management mechanism for developing programs and determining defense budgetary needs. The acquisition of weapons systems, therefore, is closely coordinated with the PPBS process. In addition, the Office of Management and Budget (OMB) provides policy guidance to all executive branch agencies in the acquisition of major systems. Commercial manufacturing has its own budget cycle for new product development but the basic planning is the same.

Department of Defense Directive (DoDD) 5000.1 (*Major and Non-Major Defense Acquisition Programs*) and DoDD 5000.2 (*Defense Acquisition Program Procedures*) provide policies and procedures and management for defense acquisition programs. Defense acquisition is the process of procuring for the military equipment having a stated performance capability, on a timely basis and at minimum cost. Systems are normally called "major" systems when a new start is authorized in the "Program decision memorandum" (PDM). The PDM spells out force levels, system acquisition rates, and levels of support, which the Secretary of Defense approves from the U.S. Military Service "program objective memoranda" (POMs).

Acquisition Strategy

An acquisition strategy (AS) is required for all acquisition programs. It is the heart of program planning and sets the basic course of action to be followed. The AS documents how the particular acquisition program will be tailored. It identifies potential risks and plans to reduce or eliminate risks. It also serves as the foundation for preparation of supporting program–functional plans and documents needed for a sound, coherent program.

A primary goal in developing an acquisition strategy is to minimize the time it takes to satisfy the identified need consistent with common sense, sound business practices, and the basic management policies. During the initial phases of development, studies need to be conducted to identify trade-offs between cost and performance requirements, to assess technological risk, and to identify the cost drivers and producibility factors associated with using new or underdeveloped technologies. Commensurate with risk are such approaches as developing separate alternatives in high-risk areas, using early funding to design-in reliability and support characteristics, reducing leadtime through concurrency, using competitive prototyping of critical components, combining acquisition phases and making use of evolutionary acquisition procedures, and combining developmental and operational tests and evaluations that must be considered and adopted when appropriate.

Provisions for obtaining competition in each phase of the acquisition process must be described in the AS. This includes planning for competition for ideas and technologies in the early phases, and the use of commercial-style competition procedures that emphasize quality and established performance as well as price during the production phase. The strategy should contemplate narrowing the number of competing alternatives to eliminate concepts no longer considered viable as the acquisition process proceeds. This narrowing of competing alternatives must be accomplished without unduly interrupting the remaining contracts, and need not be timed to coincide with milestone decision points. Competitive prototyping of critical components, subsystems, or systems and early operational test and evaluation beginning in the concept demonstration–validation phase are encouraged and emphasized. Whenever possible and appropriate, consideration needs to be given to maximizing the use of "off-the-shelf" commercial products and streamlining the military specifications so that only those military specifications that are directly relevant to the items being procured are applied.

Logistics supportability requirements, in the form of readiness goals and related design requirements and activities, must be established early in the acquisition process and considered in the formulation of the acquisition strategy. These requirements must receive emphasis comparable to that accorded to cost, schedule, and performance objectives and requirements.

The following principles should guide the tailoring of the acquisition strategy:

- User requirements stated in performance–capability-oriented terms.
- Maximize using already developed systems, components, and so forth.
- Identification, selection and maturation of technologies with future growth capability planned.
- Tailored life-cycle phases.
- Early involvement of the logistics community.
- Integrated testing and continuous evaluation.
- System evolution through block improvements.

Acquisition Plan

The content of the acquisition plan (AP) varies based on the nature of the circumstances and phase of the particular acquisition. Generally, however, the AP contains an acquisition background and objectives including a statement of need, applicable conditions, cost considerations, life-cycle costs, delivery requirements, required capabilities, tradeoff considerations and risks, and a plan of action that includes a plan for competition, source selection considerations, special contracting considerations, and budget and funding information.

Acquisition planning must provide adequate leadtime for actions required for a particular acquisition such as the necessity for formal source selection and actions necessary to achieve full and open competition. While activities associated with each circumstance consume large amounts of time, they are integral and concurrent part of the acquisition process and must be considered from the beginning. Leadtimes used in the acquisition planning process must be sufficient for advertisement of the acquisition, preparation of the "request for proposal" (RFP), response by industry, evaluation of the proposals, negotiation with all responsive and responsible sources, business clearances reviews, and source selection.

Acquisition Decision Process

The "materiel acquisition decision process" (MADP) is the review of a program or project at critical points to evaluate status and make recommendations to the decision authority. DoD major programs are designated by the Secretary of Defense (SECDEF), based on development risk, urgency of need, congressional interest, joint service involvement, and resource requirements. A major defense acquisition program is designated as either a Defense Acquisition Board (DAB) or "component" program. Designations are recommended by the Defense Acquisition Executive (DAE) and approved by the SECDEF.

Program initiation is normally accomplished by a "mission need statement" (MNS). A program supported by a MNS possibly may not be selected as a DAB major program; conversely, the Office of the Secretary of Defense (OSD) may designate a program as a DoD major program even though it does not otherwise require the preparation of a MNS.

In the system acquisition process, a DAB program is reviewed by the DAB and requires a SECDEF decision at each milestone review point. A component program is reviewed by an "acquisition review council." Designated "acquisition programs" (DAPs) are determined according to importance, complexity, and resource requirements. A review of operational and organizational (O&O) plans identifies potential DAP. Each MADP review will formally decide whether the next milestone MADP review should be referred to a lower management level. Major management decisions during the acquisition cycle are made at specific "milestones" (Milestones I, II, III) appropriate to the particular program. No single procedure can apply to the acquisition of all materiel systems. Appropriate reviews provide a sound but flexible decisionmaking process. The MADP reviews serve as forums to recommend appropriate action to the decision authority. A MADP review should consider the following:

- Freedom to consider and accept other courses of actions.
- A full interchange of information based on the "test and evaluation master plan" (TEMP), test plans, test reports, evaluations, production readiness assessments, "logistics support analysis" (LSA), "integrated logistics support" (ILS) plans, "design-to-cost" (DTC) status, transportability engineering analyses, "manpower and personnel integration" (MANPRINT), elements of the "concept formulation package" (CFP), environmental, and safety documentation, demonstration proposals, and appropriate economic analysis.

All levels of MADP reviews have essentially the same documentation requirements, although they are tailored to the scope and peculiarity of each program. Once a MADP decision authority has approved an acquisition strategy, that AS will be the official U.S. Army position. The decision is formalized in the "acquisition decision memorandum" (ADM). Authorization to proceed into the next acquisition phase must be accompanied by the decision authority's assurance that sufficient resources are, or can be, programmed to execute the program.

Reasonable stability in acquisition programs is essential to satisfying identified military requirements in the most effective, efficient, and timely manner. Accordingly, program funding and requirements changes should be minimized and not be introduced without assessing and considering their impact on the overall acquisition strategy and the established program baseline. Key factors that enhance program stability include:

1. Conducting meaningful, realistic long-range planning.
2. Considering evolutionary alternatives in parallel with the need for advanced technology insertion so as to strike the most appropriate balance between development and/or production risk and the risk associated with failing to counter the threat.
3. Estimating, programming, budgeting, and funding acquisition programs realistically.
4. Planning for economical rates of prduction, surge and mobilization requirements, and where appropriate, multiyear procurements.
5. Establishing program baselines and assigning to program managers the authority and resources required to achieve these baselines.
6. Developing and executing tailored acquisition strategies consistent with established priorities and affordability constraints.

Review Forums

Defense Acquisition Board (DAB). The DAB is the primary forum used by DoD to resolve issues, provide and obtain guidance, and make recommendations to the defense acquisition executive (DAE) on matters pertaining to the DoD acquisition system; the DAB also makes recommendations on milestone decisions for DAB programs. It is supported by 10 acquisition committees that provide assistance in program review and policy formulation. Each service is authorized one member on each of the committees.

Component Acquisition Review Council (CARC). The CARC is the decision review body for the component acquisition of major systems and "designated acquisition programs" (DAPs). The CARC is conducted at a formal "milestone decision review" (MDR) to provide information and develop recommendations for decisions by the component acquisition executive (CAE).

In-Process Review (IPR). IPRs are conducted at each formal MDR for all programs not designated as DoD major programs or DAP. The IPR provides information and develops recommendations for decision by the program executive officer (PEO).

Materiel Acquisition Review Board (MARB). A MARB is conducted prior to each formal milestone decision review to review and approve key program management documents for execution or release to the appropriate approval level, and to cross-check various documents for consistency with the requirements.

2.2 LIFE-CYCLE PROCESS OF DEFENSE SYSTEMS ACQUISITION

A system or product goes through a life cycle consisting of distinct phases, each beginning at specific milestones. The life-cycle phases are concept exploration and selection, demonstration and validation, full-scale development, production and deployment, operations and support, and retirement as shown in Figure 2.1. ILS is a requirement in all phases. New product development should also go through similar life-cycle planning.

Systems acquisition begins with the identification of a need. Milestone 0 (program initiation–mission need decision) determines mission need and approves initiation and authority to budget for a new major program. Normally a concept exploration–definition phase follows. Preliminary considerations during this milestone include mission area analysis, affordability and life-cycle costs, the ability of a modification to an existing system to provide needed capability, and an operational utility assessment.

The mission needs analyses are performed by the operating commands. The mission analysis defines the need in terms of mission purpose, capability, agency, components involved, schedule and cost objectives, and operating constraints. DoD components conduct continuing analyses of their areas of responsibility to identify deficiencies and to determine a more effective means of performing assigned missions. These analyses may result in recommendations to initiate new acquisition programs to reduce or eliminate operational deficiencies, to establish new capabilities in response to a technologically feasible opportunity, to reduce the DoD cost of ownership significantly, or to respond to a change in national defense policy. The operating commands and agencies perform initial life-cycle cost analysis to ascertain economic feasibility. Affordability, which is a function of cost, priority, and availability, is to be considered at every decision milestone during the PPBS process. A major defense acquisition program–commercial product program is not to be started unless sufficient resources, including manpower, are or can be programmed to support projected development, testing, production, fielding, and support requirements. This is followed by an assessment of the threat and/or the commercial market the proposed system is to meet.

Milestone I (Concept Exploration Phase). The request for a major system start is submitted with the POM for the budget year in which POM funds are requested. When POM funds are available, the major system receives official sanction. In the concept phase, the system is defined and the technical basis for an acquisition program is established through feasibility studies. Also, basic operations and support concepts are identified and early operations and support planning is initiated. Typical areas of this phase include such activities as exploratory development and technology base enhancement. Research in these areas include system design concept studies, exploratory subsystem development, and tests and evaluations. During this phase, the acquisition

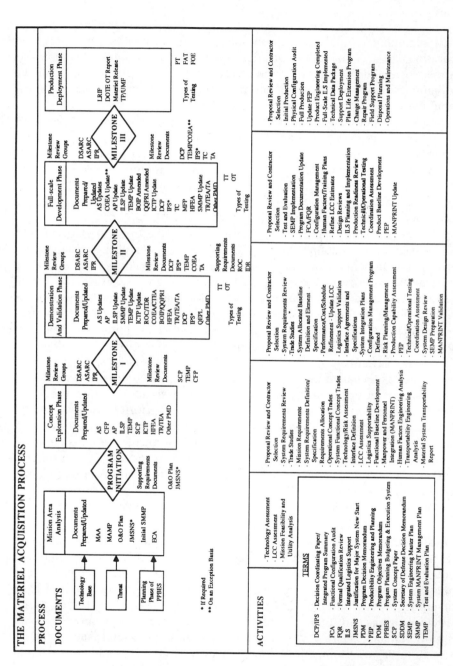

Figure 2.1 The materiel acquisition process.

strategy is prepared, which includes a resource magnitude estimate. Also included is the development of alternative operational and support concepts and evaluation of their implications on support resources. There is an initial assessment of ILS program requirements, resource impacts, and risk reduction measures for alternative acquisition strategy options, including accelerated acquisition strategies. Major effort is also expended to integrate readiness-related requirements into both the ILS section and other appropriate sections of the "statement of work" (SOW).

A number of activities must be accomplished by Milestone I. The baseline operational scenario must be defined for recommended system alternatives with adequate detail for support planning purposes. System readiness objectives and tentative thresholds must be established. A baseline support concept must be developed and integrated with system design criteria, and the consideration given for the use of contractor support. A tentative schedule for phased transition from contractor to organic support must be developed prior to a Milestone I decision. In addition, an ILS plan must have been drafted and milestones developed for each ILS element. Support cost drivers and targets for improvement must have been identified. Initial system transportability requirements must be assessed against the capabilities of existing transportation assets and the impact on strategic deployment. Major items of support-related hardware and software requiring development must be identified. Logistics considerations must be integrated in the SOW, specifications, requests for proposals, source selection evaluation criteria, and contracts. Preliminary facilities requirements must have been identified and properly programmed for construction. For accelerated acquisition strategies, additional resources and management actions must be identified to control logistics risks and properly execute the ILS development program. The Milestone I decision establishes broad program cost, schedule, and operational effectiveness and suitability goals and thresholds, allowing the program manager maximum flexibility to develop innovative and cost-effective solutions.

Milestone II (Demonstration–Validation Phase) This phase starts with a SECDEF milestone decision on concept selection. The requirement to fill a specific need is validated or demonstrated, subject to a preliminary evaluation of the concept, costs, schedule, readiness objectives, and affordability. During this phase, major program characteristics are validated and refined, system configuration is defined, program objectives and performance parameters are considered, major areas of risk are considered, items necessary for support are tested, models and prototypes are fabricated, and a formal logistics support plan is developed. In addition, a determination is made whether all elements, projects, and subsystems will fit together economically and technically. This phase also includes tradeoff studies to ensure that life-cycle DTC requirements are established to maintain the desired relation between cost, performance, and schedule. Hardware to be tested is usually in prototype form. Testing is to check that the design is functional with complete qualification testing later.

The specifications developed during this phase deal with performance. Performance specifications are preferred prior to initiation of full-scale development because detailed design specifications severely limit the latitude of design, result in contract changes, and often require excessive precontract negotiations. Prototypes are geared to design evaluation, performance, and production potential. Tradeoff studies performed should also ensure that the configuration being defined still adequately meets any changes in threat.

Milestone III (Full-Scale Development). This phase starts with a SECDEF major milestone decision for program go-ahead and approval to proceed with full-scale development (FSD). Many of the same activities already discussed must be completed prior to a Milestone III decision. However, all the activities must be updated with considerable depth. The Milestone III decision establishes more specific cost, schedule and operational effectiveness and suitability goals and thresholds, including the approval of the program baseline agreement between the DAE, the service acquisition executive (SAE), the program executive officer (PEO), and the program manager (PM).

Primary considerations in this third phase include affordability in terms of program cost versus military value of the new or improved system and its operational suitability and effectiveness; program risk versus benefit of added military capability; planning for the transition from development to production, which includes independent producibility assessments; realistic industry surge and mobilization capacity; factors that impact program stability; potential common-use solutions; results from prototyping and demonstration–validation; milestone authorization; manpower, personnel, training, and safety assessments; procurement strategy appropriate to program cost and risk assessments; plans for integrated logistics support; and affordability and life-cycle costs.

During this phase a clearly defined systems engineering procedure must be implemented to influence the evolving system design, define automated diagnostic requirements, and further determine ILS element requirements. The final product of this phase is a product baseline configuration design–documentation package reflecting the established cost, schedule, logistics supportability, and performance constraints. Contractual requirements are demonstrated by actual performance testing. In addition, the mission element need is reestablished and the threat updated. Activities during FSD center around specifications, test and evaluation, logistics support, and personnel and training.

During the production phase, the system is produced and delivered as an effective, economical, and supportable system. The decision to start production, Milestone III, is delegated to the DoD components, provided the weapons system program is on target as to schedule and cost. It is also during this phase that personnel are trained and activities that support material flow, product distribution, and warehousing are implemented. Any additional testing that causes changes to the system must be tracked, as configuration control is heavily practiced at this point. Effective control procedures will eliminate the nice but not necessary changes that keep designs in turmoil, lead to litigation, and unnecessarily burden the logistics support system and training program. By this time a preliminary manpower document and supporting analysis has been accomplished, software and related computer support plans have been developed, development status and production leadtimes of ILS elements are commensurate with support capability objectives and deployment needs, the ILS plan has been updated, contract requirements are consistent with ILS plans and support-related objectives and thresholds, facility construction will have been completed in time to support scheduled deployment, transportability approval has been given by the appropriate transportability agent, independent reviews have affirmed the adequacy of training plans, and timely delivery of training equipment is planned to support the schedule deployment. Primary considerations for this phase are similar to those of the previous phase except more detail is provided.

The deployment phase (also part of Milestone III) is one in which the operating system with its associated elements of support is deployed for operational use. This is the moment of truth for new systems. All the acquisition efforts finally culminate in a new mission and support capability. By now plans must be in place for postdeployment reviews, evaluations and analysis of support capability, O&S cost, and manpower in relation to system readiness objectives.

Planning for a smooth phase-in of a new system and the probable phase-out of an old system must be undertaken several years prior to actual deployment. Field support personnel from the contractor is almost always desirable. These technical personnel serve the dual purpose of ensuring proper initial installation, servicing, and use and speeding up feedback from the user to the design agent. In addition, the concept of preplanned product improvements to accommodate projected changes in threat or to reduce risk in initial fielding is essential to seek cost-effectiveness leverage that is sometimes achievable by upgrading existing hardware.

Milestone IV (Operations and Support). Operational support is required throughout the service life of a system. Operational support requirements are initially predicted by the ILS tasks performed during the earlier phases and then constantly corrected by the various sport management effectiveness systems. The support system is set up for long-term continuous operations. It is extremely important to plan the transitional support properly. When a new system is fielded without proper planning or provisioning for initial support, the unsupported equipment suffers abuses that often permanently degrade performance and reliability.

Primary considerations in the O&S phase include logistics readiness and sustainability; weapons support objectives; the implementation of ILS plans;

the capability of logistics activities, facilities, and training and manpower to provide support efficiently and cost-effectively; disposition of displaced equipment; and affordability and life-cycle costs.

The last milestone decision (Milestone V) encompasses a review of a system's or facility's current state or operational effectiveness, suitability, and readiness to determine whether major upgrades are necessary or whether deficiencies warrant consideration of replacement. This milestone decision normally will occur 5 to 10 years after initial deployment. Primary considerations include:

1. Capability of the system or facility to continue to meet its original or evolved mission requirements.
2. The potential necessity of modifications and upgrades to ensure that mission requirements are met and that the useful life is extended.
3. Changes in threat that require increased capability or utility.
4. Changes in technology that present the opportunity for a significant breakthrough in system worth.
5. Disposition of displaced equipment.

A significant question to be decided at this point is whether deficiencies are sufficiently critical to warrant major modification, retirement, and/or new start considerations to start the acquisition process all over again.

The final phase is the retirement phase. Here the system is retired from operational use. The equipment is phased out of the service inventory, recycled, or dispositioned. Retirement is the most overlooked support phase. As the system nears the end of its service life, various support elements become uneconomic and are deleted. Special training courses and maintenance contracts are dropped. Consideration of the phase-out problems that may occur for a system being replaced could make it desirable to alter the phase-in rate and hence the production rate of the new equipment. All the military services use this same basic process.

2.3   ACCELERATED–STREAMLINED ACQUISITIONS

Each materiel acquisition program is unique. As a result of both internal and external influences, adjustments to schedule, requirements, and cost are necessary to conform to constraints. The materiel developer must react to these constraints and still provide the user with the most cost-effective and timely materiel necessary to satisfy the user requirement.

Acquisition streamlining is to be applied to the entire spectrum of acquisition activities. User materiel requirements are to be stated in performance-oriented terms and challenged for realism, value-added, and determined with respect to impact on cost, schedule, and technical feasibility. Contract requirements must be tailored to the unique circumstances of individual acquisition programs and stated in terms of results desired. Test and evaluation requirements must be tailored to the minimum essential to ensure compliance with contractual requirements, DoD policy, and law. Acquisition strategies must be tailored to the shortest possible path commensurate with risk to achieve a quality product.

Two basic approaches that can be used to simplify or eliminate phases in the acquisition process are nondevelopmental item (NDI) acquisition and a streamlined acquisition process (SAP). NDI will be treated later in this chapter and again in more detail in a separate chapter.

The purpose of acquisition streamlining is to promote innovative and cost-effective acquisition requirements and acquisition strategies that will result in the most efficient utilization of resources to produce quality weapons systems and products. Acquisition streamlining is based on the concept that by applying only pertinent contract requirements and allowing early industry involvement in recommending the most cost-effective solutions, the Department of Defense can reduce the cost and/or time of system acquisition and life-cycle cost without degrading system effectiveness. The objective of acquisition streamlining is to provide industry greater flexibility in how best to meet the U. S. Army's objectives for acquisition of cost-effective equipment. This is accomplished by clearly communicating what is required in functional terms at the beginning of development and allowing flexibility in the application of the contractor's experience and judgment. The contractor is encouraged to recommend detailed specifications, standards, and other detailed requirements as the weapon system evolves toward FSD and eventually toward production. In this way, the application and tailoring of specifications and standards become an integral part of the design process.

The Army Streamlined Acquisition Program (ASAP) Example

This program was established to facilitate common-sense streamlining of materiel programs and provide sound guidelines for the application of streamlining principles to requirements, strategies, and business practices. The basic order of precedence for consideration of acquisition alternatives, once the need for a materiel solution has been determined, is the reconfiguration of existing materiel, use of nondevelopmental items, and new development.

The ASAP process should be a primary consideration in the acquisition strategy to ensure that development and production are low risk and that future capability needs can be achieved through preplanned product improvements (P^3Is). Deviations from the traditional process focus on up-front planning and flexibility in the formulation of a development program and result in the overall shortening of the process, without loss of visibility or safeguards important to decisionmaking. As a first priority, DoDD 5000.43 ("Acquisition Streamlining") establishes the policy for streamlining solicitations and contract requirements by:

1. Specifying contract requirements in terms of results desired, rather than "how-to-design" or "how-to-manage."
2. Precluding premature application of design solutions, specifications, and standards.
3. Tailoring contract requirements to unique circumstances of individual acquisition programs.
4. Limiting the contractual applicability of referenced documents to only those that are essential.

The acquisition streamlining approach is intended to utilize contractor ingenuity and experience in arriving at cost-effective designs, while retaining government program manager decisionmaking authority. It is also intended to support the basic requirement to pursue, throughout the system development process, a design that is economically producible as well as operationally suitable and field-supportable. The streamlining approach should also ensure development of complete and definitive production data and specifications, while providing adequate flexibility to the contractor to optimize the system design. Military Handbook (MIL-HDBK) 248 provides detailed guidance for the application of this policy during system acquisition. Key features of this process include:

1. Requirements structured for "near-" and "far-" term capabilities.
2. Early focus of technology on mission area needs and maturation of technology at the component level.
3. Combination of appropriate elements of concept exploration and demonstration–validation phased into a scaled-down proof-of-principle approach to prove out the technical approach and operational concept.
4. Solid prove-out of production along with MANPRINT and ILS prior to entry into the production–deployment phase.
5. Integrated test approach.
6. Minor reorientation of formal milestones.

Phases of ASAP

The major phases of the ASAP are (1) requirements–technology base activity, (2) proof of principle, (3) production and deployment, (4) operations and support, and (5) retirement. Figure 2.2 compares the streamlined process to the regular or current acquisition process. Each ASAP phase is discussed in the following paragraphs.

The requirements–technology base activity focuses technology efforts on mission area differences, with coordinated technology base efforts among laboratories, industry independent research, foreign research, arsenals and depots, and determines suitability for proof of principle. The Technology

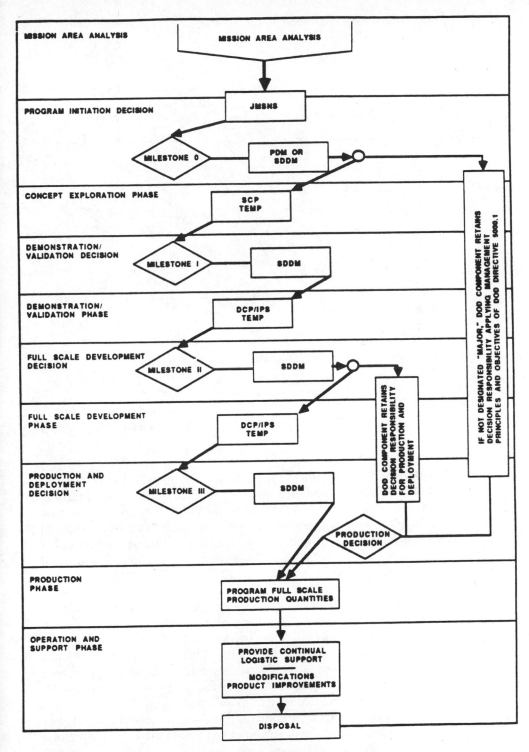

Figure 2.2 The detailed acquisition process.

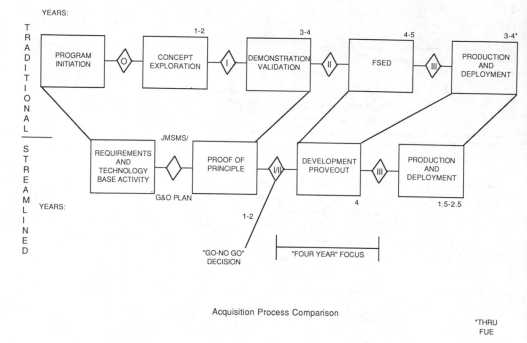

Acquisition Process Comparison

*THRU
FUE

Source: AR 70-1 Systems Acquisition Policy and Procedures, 1986.

Figure 2.2 (Cont.) Traditional vs. streamlined process.

Integration Steering Committee (TISC) evaluates the match of selected technologies with army thrusts and emerging mission needs, and if appropriate, triggers the preparation of an O&O plan, justification for major systems new start (JMSNS), as appropriate, by the Training and Doctrine Command (TRADOC) proponent. If at any time during the development of an O&O plan, estimates indicate that the proposed program will require in excess of $200 million research, development, test, and evaluation (RDTE) monies or $1 billion in procurement (FY80 dollars), the TRADOC proponent must initiate development of the JMSNS. The evaluation of current deficiencies include MANPRINT issues to influence design of a new or modified system early in the acquisition process. A system MANPRINT management plan (SMMP) must be initiated prior to a program initiation when a deficiency requiring a materiel solution is identified. An early comparability analysis (ECA) must be performed during this phase to evaluate key facets of soldier–machine interface of the existing (predecessor) systems.

Technical and resource requirements for proposed systems are established through pertinent studies and the development and evaluation of experimental concepts and early user test and experimentation (EUTE). During this phase, critical technical, training, logistics, operations, MANPRINT, reliability, cost of production, electric power, and environmental control issues are identified for resolution to minimize future development risks. Investigations must also analyze support equipment and training devices of current systems, development requirements for new support equipment and training devices, development of alternative operational and support concepts, and an evaluation of MANPRINT concerns and logistics support resource implications. The consideration of threat is particularly critical during these activities.

Army Materiel Command (AMC) market surveillance activities are conducted during this phase. Primary consideration is given to currently available technology as developed by army laboratories, other services, allies, universities, and United States industry. This surveillance initially identifies the potential for a nondevelopmental item (NDI) solution.

During this period, a number of documents are initiated by AMC and TRADOC or other participating agencies to record the effort and form the basis for entry into the proof-of-principle phase. These documents include an acquisition strategy (AS), an integrated logistics support plan (ILSP), a test and evaluation master plan (TEMP), an international armament cooperative opportunities plan, an individual and collective training plan (ICTP), transportability engineering analysis (TEA), human-factors engineering analysis (HFEA), and a system safety assessment (SSA).

All O&O plans are approved by TRADOC. Approval constitutes program initiation for nonmajor programs. These plans are circulated by the Office of the Deputy Chief of Staff for Operations and Plans (DSCOPS) to allow the AAE to designate programs as DAPs. Unless so designated, a program is managed by IPRs. If a JMSNS is required, it must be submitted to the SECDEF as part of the army's POM cycle. The SECDEF provides appropriate program guidance in the PDM. This action provides official sanction for a new program start and authorizes the army, when funds are available, to initiate the next acquisition phase. A MARB will be convened to review and approve the decision review documents and provide a recommendation to AMC and TRADOC decision authorities regarding entry into the proof-of-principle phase.

The proof-of-principle phase employs prototyping, components, and surrogates in demonstration and experimentation to confirm system operational concepts, acquisition strategy, test planning, and contracting strategy. The basic assumption during this phase is that the acquisition intended to satisfy a requirement will be realized in ways other than full new development. The burden of proof to the contrary rests with the materiel developer (MATDEV) in preparing the AS. The Army Materiel Command (AMC) in coordination with TRADOC conducts the proof-of-principle phase. This phase consists of the steps necessary to verify preliminary design and engineering; accomplish necessary planning; analyze tradeoff proposals; resolve or minimize logistics impacts, MANPRINT, and reliability problems identified during the requirements–technology base activities phase; prepare a detailed requirements document; and prove out the technology and components prior to the formal commitment to prove out the concept.

The joint TRADOC/AMC TISC will direct required technology maturation actions. The TISC, in this instance, determines when the selected technologies are sufficiently mature to pursue a MARB decision to enter the proof-of-principle phase.

In exploring a materiel solution, existing fielded equipment is reviewed for potential enhancements through product improvements. Materiel can be improved by reconfiguring a type classified item that is in production via a Class I engineering change proposal (ECP), reconfiguring a type-classified fielded item via a product improvement proposal (PIP), pursuing a nondevelopmental item (NDI) procurement strategy, or any combination of these. During this phase, AMC conducts a formal detailed market investigation to determine whether these types of acquisition alternatives are viable or if a developmental approach must be pursued. If it becomes evident that an NDI solution will meet the materiel need, then the decision review, Milestone I/II, will become the production decision in that development proveout activities will not be conducted or will be minimized. AMC will develop the basis of issue plan feeder data (BIOPFD) and qualitative and quantitative personnel requirements identification (QQPRI) early in this phase. These data are required prior to Milestone I/II.

During the proof-of-principle phase concepts and components are demonstrated employing brassboard prototypes or surrogate components or systems. Evaluations are performed continuously employing a common and shared database. AMC and TRADOC evaluate the technology by placing a brassboard prototype system in the hands of user troops. The system is employed in accordance with the O&O plan developed by TRADOC to prove out the maturity of the technology, the operational concept, and soldier acceptability. This demonstration needs to be conducted in as realistic a threat environment as practicable. When the proof-of-principle efforts have progressed sufficiently, the TRADOC proponent and the AMC lead Major Subordinate Command (MSC) may agree that the materiel system under development should enter development proveout. At this time, the TRADOC proponent, in coordination with the MSC, prepares the required operational capabilities (ROCs). After approval of the ROC and BIOP/QQPRI package, AMC in coordination with TRADOC, prepares the decision coordinating paper (DCP) for milestone decision review (MDR) I/II. At MDR I/II a firm "go–no-go" decision is made to proceed to the next phase, the development proveout phase (if a development option is the selected acquisition strategy). Required documentation for this review consists of, as a minimum, the updated AS as part of the DCP, the integrated program summary (IPS), ROC, test reports and independent evaluation reports (IERs), TEMP, transportation approval, HFEA, and an updated ILSP. These documents are used in the MADP reviews. For IPR programs, the IPR decision authority is delegated to the AMC MSC and TRADOC proponent. The IPR reviews the program. If all parties concur, the IPR recommendation is forwarded to the decision authorities who document their decision in a Secretary of the Army decision memorandum (SADM). If the IPR cannot reach agreement and intervening command levels cannot resolve the issue, the minutes are forwarded to Department of the Army for resolution and issuance of the SADM. For DAPs, decision review is accomplished by the ASARC. Based on the ASARC's recommendations, the AAE issues a SADM. For DoD major programs, documentation if first submitted to the ASARC and then to the JRMB. JRMB recommendations are submitted to the SECDEF, who issues a Secretary of Defense decision memorandum (SDDM). Once the SADM or SDDM has been issued, constituting approval to enter the next phase, the decision is distributed.

The development proveout phase encompasses full-scale development (primarily system integration) to include MANPRINT, RAM, and ILS; producibility engineering and planning to include hard-tooled prototypes and initial production facilitization whenever possible; continued developmental testing and required operational testing; and comprehensive evaluation. During this phase, the system, with all items necessary for its support including training devices and computer resources, is fully developed, engineered, fabricated, and tested and evaluated. Nonmateriel aspects required to field an integrated system are developed, refined, and finalized. An IPR may be employed to approve initiation of procurement-funded portions of this phase. These activities include initial production facilities (IPF), procurement of long-lead-time (LLT) items, production of hard-tooled prototypes, preproduction testing (PPT), and a production readiness review (PRR).

Documentation during this phase is similar to that prepared during the proof-of-principle phase. The AS, AP, ILSP, TEMP, ICTP, and SMMP are updated as shown in Figure 2.3. In addition, the BOIP and QQPRI are finalized. During this phase, AMC prepares the type classification (TC) documentation. TRADOC prepares the initial materiel fielding plan (MFP). The HFEA is updated and transportability approval is obtained from the Military Traffic Management Command (MTMC). AMC also prepares the Milestone III review documentation. Whenever possible, within risk and affordability constraints, this phase will include preproduction testing (PPT) on limited production prototypes to provide operational test results prior to Milestone III and approval of TC standard.

The second formal milestone decision point is milestone decision review III (MDR III), the production decision. MDR III processes and procedures are basically the same as those for MDR I/II. This includes approval for subsequent deployment. Normally the decision authority delegates the Milestone III decision to the lowest level in the organization where there is a comprehensive view of the program and if there are no breaches of funding thresholds that would change the program category.

The production–deployment phase includes low-rate initial production and

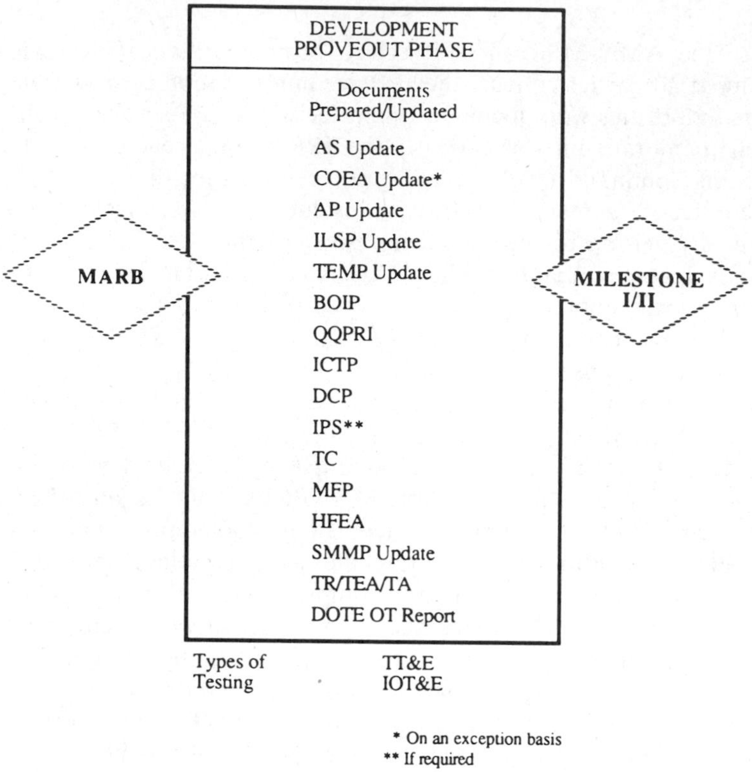

Figure 2.3 The development prove-out phase. (*Note*: For acronyms, see list in Appendix 2.)

first article test, full-rate production, and initial fielding (see Figure 2.4). Milestone IIIA, low-rate initial production, may be conducted to verify production capability and provide the resources necessary to conduct interoperability live fire and initial operational testing. The low-rate initial production decision may be made in conjunction with Milestone II; however, adequate safeguards must be applied, and will normally entail a subsequent checkpoint with specific approval to activate the low-rate initial production decision. Milestone IIIB (full-rate production–initial deployment) covers the sustaining rate production and initial fielding of the materiel system together with its full complement of support equipment, publications, and services.

During this phase, operational units are trained, equipment is procured and distributed, and logistics support is provided. Also, production testing and evaluation are accomplished. Product improvements that have been preplanned are applied to the equipment as required. This phase is supported primarily by procurement funds. The army has a materiel release process to ensure that materiel released to the field is suitable in terms of safety and health, human factors engineering, performance, reliability, quality, envi-

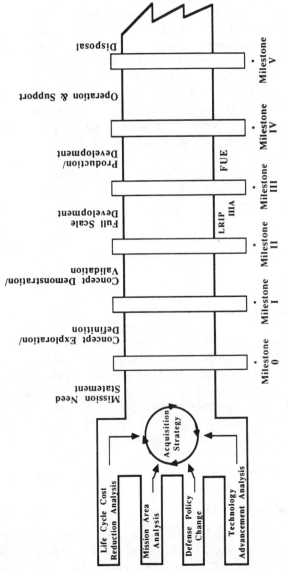

Figure 2.4 Defense systems acquisition life-cycle model.

ronmental factors, and availability and adequate of logistics support, including test measurement and diagnostic equipment and qualified operator and maintenance personnel. AMC has implemented total package–unit materiel fielding for selected systems. This concept minimizes the workload of the gaining unit by gathering the end item and all required support into a single package that is identified, assembled, funded, and deprocessed by the materiel developer. Also during this phase, the materiel system is operated, supported, and maintained. The system is sustained in the active inventory until a decision is made for upgrade, replacement, or disposal.

Within 2 years after initial deployment, a logistics readiness and support review is conducted to assess how well operational readiness and/or support objectives are being achieved and maintained. This is Milestone IV and is usually forgotten in discussions. This is followed by Milestone V. Within 5 to 10 years after initial deployment, consideration is given to conducting a major upgrade–replacement review to assess the current state of the operational effectiveness, suitability, and readiness to determine whether a major upgrade and/or P^3I acceleration is necessary or deficiencies warrant consideration of replacement.

2.4   ACQUISITION OF NONMAJOR SYSTEMS

Background

DoDD 5000.1 establishes acquisition management principles and objectives applicable to major and nonmajor systems. DoDD 5000.39 sets the general policy for the acquisition and management of ILS for all systems, while delegating responsibility for application of ILS policies for nonmajor systems to the U.S. Military Services. Guidelines in DoDD 5000.1 for designation as a major system include program cost thresholds, risk, urgency of need, joint acquisition, and Congressional interest. The ultimate criterion is selection by the Secretary of Defense. Systems not designated as major systems are generally single-service systems that are less costly and by themselves less critical to national defense. However, nonmajor systems may have a large aggregate impact on the capabilities of combat units and their logistics burdens.

Service Management Procedures

The Military Services have delegated management responsibility for nonmajor systems "to the lowest levels of the component at which a comprehensive view of the program exists." Materiel systems are assigned to program categories based on such criteria as combat role and program cost. The decision authority, funding criteria, and examples of programs in each category are shown in Figure 2.5. Nonmajor systems many also be categorized as either developmental or nondevelopmental.

SERVICE	PROGRAM CATEGORY	FUNDING CRITERIA (RDT&E/PROCUREMENT)	DECISION AUTHORITY	EXAMPLE
Army	Major System Program	$200M/$1 Billion	Secretary of Defense	M-1 ABRAMS Tank
	Designated Acquisition Program	*	Asst. Secretary of the Army (Research, Development and Acquisition)	STINGER Missile
	DA In-Process Review Program	*	Deputy Chief of Staff (Research, Development and Acquisition)	VRC-47 Radio
	In-Process Review Program	*	Materiel Developer (Usually AMC)	PRC-77 Radio
Navy/Marine Corps	ACAT I (Major System)	$200M/$1 Billion	Secretary of Defense	F/A-18 Aircraft
	ACAT IIS	$100M/$500M	Secretary of the Navy	AV-88 Harrier
	ACAT IIC	$100M/$500M	CNO/CMC	AEGIS Area Air Defense
	ACAT III	$100M/$500M	DCNO/Asst. CMC	Acoustic Sensor System
	ACAT IVT	*	Syscom Commander/Asst. CMC	Life Support Equipment Engineering Development
	ACAT IVM	*	Syscom Commander/Asst. CMC	Laser Eye Protection System
Air Force	Major System	$200M/$1 Billion	Secretary of Defense	F-16 Aircraft
	AF Designated Acquisition	*	Secretary of the Air Force	BMEWS Modernization
	Delegated Program	*	Implementing Command (Usually AFSC)	"Have Quick" Anti-Jam Communications Upgrade

Figure 2.5 Acquisition program categories.

* None. Case by Case Decision Rendered Upon Factors Such as Mission Criticality and Congressional Interest.

Developmental Systems. Developmental programs for nonmajor systems range from full development to ruggedization of commercial items prior to deployment, as depicted in Figure 2.6. Specific ILS procedures for influencing the design and defining and acquiring the support parallel those for major systems but are generally characterized by a reduced scope, fewer iterations, fewer personnel, and smaller budgets. The ILSP, for example, may be part of the program management plan rather than a separate document. Logistics support analysis (LSA) requirements for nonmajor systems, particularly those requiring only minor development, are often significantly reduced by tailoring.

Nonmajor systems do not have the intense management and detailed reviews enjoyed by major systems. Managers and their staffs may be assigned several nonmajor systems and handle a variety of actions covering a wide spectrum of acquisition functions. Less supervision and the requirement for dealing with many areas can result in some actions being overlooked. Logistics personnel generally have to assert themselves to ensure that ILS receives the resources and attention required. In fact, the impetus is on the staff to ensure that the required planning, coordination, and programming are accomplished.

Small programs have a small logistics burden; however, as was pointed out above, they have a large aggregate impact. The army, for instance, has approximately 30 major systems and in excess of 300 nonmajor systems currently under development. It is important that ILS is applied as necessary to each nonmajor system development.

Nondevelopmental Systems. Nondevelopmental systems (Figure 2.6) include commercial items and materiel developed by another U.S. Military Service or government agency or country. The purchase of nondevelopmental items offers the benefits of shortened acquisition time and reduced cost. The logistics support challenges of purchasing nondevelopmental items include:

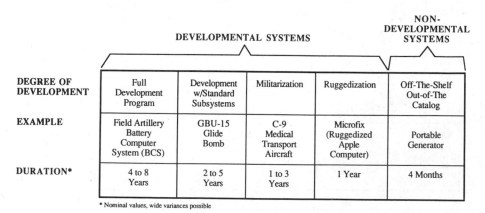

	DEVELOPMENTAL SYSTEMS				NON-DEVELOPMENTAL SYSTEMS
DEGREE OF DEVELOPMENT	Full Development Program	Development w/Standard Subsystems	Militarization	Ruggedization	Off-The-Shelf Out-of-The Catalog
EXAMPLE	Field Artillery Battery Computer System (BCS)	GBU-15 Glide Bomb	C-9 Medical Transport Aircraft	Microfix (Ruggedized Apple Computer)	Portable Generator
DURATION*	4 to 8 Years	2 to 5 Years	1 to 3 Years	1 Year	4 Months

* Nominal values, wide variances possible

Figure 2.6 Acquisition spectrum.

1. *Design Influence.* Design influence is generally limited to the selection process. Source selection criteria should, therefore, include utility of available operation and support manuals; similarity of current and intended use, support environment, and duty cycles; supportability-related design factors; compatibility with current support equipment; compatibility of design with existing manpower skill categories and training programs; and availability of supportability data and experience.

2. *ILS Resources.* Funds must be programmed and budgeted for the performance of ILS tests and analyses normally conducted during development and for acquiring the ILS elements.

3. *ILS Planning.* The planning requirements are also applicable to nondevelopmental systems. ILS plans may be prepared to cover individual items or categories of items. In either case, the contractor's data and field experience will be helpful in structuring the plans.

4. *Maintenance Planning.* The choice between contractor and organic support is based on operational constraints, schedules, resources, and the mission of the user. When the nondevelopmental system is "off-the-shelf" and commercial–contractor support is chosen, minimal LSA and documentation is required. In fact, the use of the contractor's support philosophy and support structure may be a feasible alternative. If not, the support should be tailored to the user's requirements. When organic support is preferred, but leadtimes are insufficient, interim contractor support may be necessary during the period required to establish an organic support capability.

5. *Supply Support.* Nondevelopmental items pose the problem of securing a long-term source of spares and repair parts. Several alternatives are available.

6. *Test and Evaluation.* An evaluation of the military suitability and supportability of nondevelopmental items is required if marketplace testing or other developmental data are inadequate or fail to address the intended military environment.

7. *Technical Manuals.* Commercial manuals should be used if feasible and if they satisfy the requirements of the intended user. The alternative is the commitment of considerable time and money to convert the manuals to military specifications. If commercial manuals are used, a management surveillance system is required to make sure that the contractor updates the manuals when the equipment is changed. The decision to use contractor support facilitates the use of commercial manuals.

Regardless of the type of acquisition, the acquisition strategy must consider economic and time constraints and realities when determining needs and tradeoffs. No acquisition, including NDI, is exempt from minimal essential test and evaluation necessary to verify MANPRINT, quality, safety, reliability, performance, supportability, transportability, and availability characteristics of a system to include life-cycle cost unless previous test and performance data are adequate for verifying operational effectiveness and suitability of the system.

2.5   LOGISTICS SUPPORT CONSIDERATIONS IN ACQUISITIONS

Since system readiness is a primary objective of the acquisition process, it is DoD policy to ensure that resources to achieve readiness receive the same emphasis as those required to achieve schedule and performance objectives. These resources include those necessary to design desirable support characteristics into systems and equipment as well as those to plan, develop, acquire, and evaluate the support. As a result, all acquisition programs are to include an ILS program that begins at program initiation and continues for the life of the system. The primary objective of the ILS program is to achieve system readiness objectives at an affordable life-cycle cost.

Early ILS program activity must focus on designing desirable support characteristics into systems and on determining support requirements. Subsequent activity would focus on acquisition, evaluation, and deployment of support resources. The scope and level of detail must be tailored to meet specific program needs at each phase of the system life cycle. Manpower, personnel, and training (MPT), as essential elements of ILS, are to be given explicit attention early in the acquisition process. Principal activities required include determining and specifying requirements based on previous experience with similar systems and demographic expectations, conducting design tradeoffs, and establishing contractor incentives to meet MPT objectives. In addition, peacetime and wartime readiness objectives and proposed thresholds must be established by Milestone I; firm thresholds must be established by Milestone II. The approach to achieving readiness objectives and thresholds, including explicit planning and resources, must be part of the system acquisition strategy.

In the development of acquisition strategies, programs must emphasize the early identification of reliability and maintainability (R&M) and supportability requirements; evaluation of alternative support concepts and techniques to minimize cost and support risks; test articles to support reliability and maintainability; and ILS development, test, and evaluation (DT&E). Development strategies should also consider early establishment of system readiness and supportability thresholds for verification or assessment during test and evaluation before decision milestones, a funded R&M improvement program to achieve mature system R&M thresholds, contractor incentives for timely attainment of support-related design objectives, and realistic budgets for acquisition of support resources.

In reviewing alternative acquisition strategies, decision authorities should consider ILS risks. Accelerated strategies must place additional emphasis on supportability design requirements and provide additional front-end funding to achieve readiness objectives within the shortened development cycle. When determined necessary, interim contractor support should be planned to avoid compressing support delivery schedules. Cost, schedule, deployment needs, and design stability factors must be assessed for each program and a schedule established for support element delivery to attain the best balance. Transition to government support normally would be scheduled to occur after the system design is stable, the capability to support the system has been demonstrated, and the planned ILS resources for the mature system can be delivered.

Starting with concept exploration, program contracts must include design requirements and engineering efforts for research and development and demonstration leading to system characteristics that best meet readiness and support cost objectives in fielded systems. The continued attainment of readiness objectives following the end of production requires specific attention to postproduction support (PPS) in the acquisition process. Programs must identify PPS requirements and provide the necessary planning and resources to meet them.

2.6 SUMMARY

Defense system acquisition must achieve the best balance between cost, schedule, performance, and supportability. A system or product undergoes a life cycle that contains distinct phases, each beginning with specific milestones. Acquisition streamlining can be used to provide industry greater flexibility in how best to meet U.S. Army objectives for the acquisition of cost-effective equipment and to take advantage of parallel initiatives. ILS considerations apply to all acquisition programs regardless of whether the regular or the streamlined acquisition process is used.

Procurement of nondevelopmental items may offer substantial reductions in total program cost and acquisition time; however, the reduction in time requires that logistics planning be performed concurrent with development of the acquisition strategy. Acquisition of nonmajor developmental items poses special considerations because of the more general management and the review procedures used. Logistics personnel have less program supervision and broader responsibilities for systems.

As can be seen, logistics planning must start early in the program life cycle and be carried on throughout the life cycle of the system. Logistics represents a broad spectrum of activity cutting across many disciplines and organizations, all of which must be integrated to achieve system supportability.

PART 2
DEVELOPING THE ILS

Reliability and maintainability are a means to an end, and that end is increased system capability. Reliable systems reduce ownership costs, reduce spare parts requirements, and require fewer support personnel. The reliability, availability, and maintainability (RAM) process, a functional element essential to system–product acquisition, spans numerous activities from initial requirements definition to assessment of the fielded item. If performed correctly, the RAM process will optimize system readiness and sustainability. In addition, the RAM characteristics of a weapon system are also key leverage points in determining the system's total cycle costs and operational effectiveness. It is estimated that approximately 30 percent of life-cycle costs can be traced directly to RAM characteristics of the weapon system's design.

RAM parameters are the most effective logistics engineering tools for influencing and interacting with the systems engineering process. The ILS program must ensure that the RAM parameters are consistent with planned implementation and the support that will be provided with that implementation.

Systems engineering is concerned with the process of translating mission, test, production, deployment, support, and operational needs into the most cost-effective mix of design requirements. The systems engineering process is iterative and involves functional analysis, synthesis, and optimization with the end goal of achieving a proper balance between economic, operational, and logistics factors. Systems engineering provides the tools that may be applied to integrated logistics support. Systems engineering management is the process of coordinating the engineering and technical effort within a program. It provides a framework for a materiel system to acquire the desired supportability characteristics. The integration of ILS concepts and planning considerations into the systems engineering process is a continual, iterative activity designed to effect optimal balance between performance and support considerations and optimal tradeoffs among costs of ownership, schedule, and system effectiveness. The ILS concept also embodies an analysis of equipment design just as systems engineering does.

Logistics support analysis (LSA) is a set of systematic and comprehensive analyses performed to identify support criteria and operational system support resources. These analyses serve as the interface between design and support planning by establishing the baseline requirements for the incorporation of support criteria into the acquisition process. It is performed on functions, systems, end items, components, and assemblies that require documentation of operational and logistics support requirements for which the government requires an organizational maintenance capability. The application of LSA is mandatory for all DoD materiel systems. The applications must be tailored to the requirements of each acquisition to ensure cost-effective implementation.

3
RELIABILITY, AVAILABILITY, AND MAINTAINABILITY (RAM)

Reliability and maintainability are a means to an end, and that end is increased system capability. Reliable systems reduce ownership costs, reduce the system's dependence on spare parts, and require fewer support people. If systems are more maintainable, maintenance times to repair damaged systems are reduced and fewer people with highly specialized diagnostic skills are needed.

The RAM process is an essential functional element to system acquisition. The process, which spans numerous activities from initial requirements definition to assessment of the item in the field, encompasses the front end activities that principally support research and development (R&D). If done correctly and timely, the RAM process will optimize system–product readiness and sustainability. However, there exists a fundamental funding policy roadblock that threatens the RAM process in support of R&D. The RAM community is not directly funded, and those funds for support of technology base component development during research and development are limited or sometimes even unavailable.

In addition, if a system–product is to be cost-effective over its designated life, one must look beyond the ability of the system to meet its performance requirements. The system's ability to perform when needed and for the duration of its assigned mission directly involves the system's operational readiness and mission reliability. Thus, a proper balance of system RAM is required. Not only is such a balance necessary; to achieve this balance, RAM considerations must begin early in the conceptual and demonstration and validation phases of system acquisition, as part of the overall systems engineering effort. RAM characteristics must be evaluated and assessed throughout all the life-cycle phases.

3.1 RAM PROGRAM OBJECTIVES

RAM programs must be established to achieve a number of major objectives. These RAM programs should include a balanced mix of RAM engineering and accounting tasks. In addition, RAM programs must be tailored to achieve the most cost-effective balance between life-cycle costs and system effectiveness and readiness. RAM engineering should focus on design, manufacture, test, and management practices that will result in delivery of reliable and maintainable items. Acquisition and program plans must stress early investment in RAM engineering tasks to avoid subsequent cost and schedule delays. RAM accounting must provide information essential to acquisition, operation, and support management. Accounting must include properly defined inputs for estimates of operational effectiveness and operational and support costs.

3.2 RELIABILITY GROWTH

Reliability growth is required to achieve reliability requirements. Reliability growth management should (1) aid in allocating resources to achieve reliability requirements on schedule and within cost constraints, (2) focus attention on failure to meet intermediate criteria (corrective actions can then be taken to meet reliability requirements), (3) place into perspective the relationship of achievement of required reliability to the rest of the development and testing program.

3.3 RAM REQUIREMENTS

RAM requirements are composed of three separate and essential elements: (1) the operational-mode summary (OMS) based on the specific mission profiles (MP) the system is required to complete, (2) the failure definition and scoring criteria (FD/SC) consistent with the OMS/MP, and (3) the numerical values of the RAM parameters. RAM requirements or goals should be expressed to encourage innovations and competition in creating, exploring, and developing alternative system design concepts. Variations of any of the elements is recognized as a change in the basic requirements. Test results and evaluations based on alternatives for any of the three elements need to be clearly identified as not directly relatable to the requirements.

RAM requirements (or other suitable measures of effectiveness) that relate directly to operational effectiveness and operating and support costs need to be stated in requirements documents. The parameters selected and the depth of coverage will depend on the peculiarities of the system and the progress of development. Requirements documents need state only those principal performance and RAM characteristics essential to describing the operational and logistics features of the system. Those requirements will represent the developers' best estimates of what is needed and attainable.

3.4   RAM PARAMETERS

RAM parameters should be developed that relate directly to operational effectiveness and operating and support cases. RAM parameters should be measured in at least four separate ways, using units of measurement related to the following: (1) operational readiness, (2) mission success, (3) maintenance manpower costs, and (4) logistics support costs.

RAM parameters should apply to all elements of the system. RAM parameters include the effects of built-in test equipment (BITE) to include malfunctions of the BITE and related maintenance time and operating system downtime. Separate RAM requirements should be established for system peculiar on-site training equipment, support equipment, and TMDE being developed under the system requirements document. However, the effect of that training equipment, support equipment, and TMDE on the RAM of the operating system should not be discounted.

3.5   RAM TESTING

Development Testing. Development Testing (DT) is conducted under controlled conditions utilizing procedures and resources contained or prescribed in the systems support package. The DT RAM emphasis must (1) identify design deficiencies and implement corrections, (2) achieve reliability growth, (3) evaluate adequacy of design consideration for logistics support, (4) estimate the effect of anticipated field operations and environmental conditions, (5) resolve contractual issues, (6) provide a basis for a clear understanding of reliability and maintainability design deficiencies, (7) contribute to the DT/OT RAM database, and (8) provide estimates of RAM characteristics.

Testing for RAM at the system level is designed and conducted to duplicate as closely as possible the operational-mode summary–mission profile. Tailored environmental profiles should be developed and utilized for test of components and subsystems. Environmental profiles should be used only when sufficient environmental information is not available and cannot be generated.

Operational Testing. Operational testing (OT) is conducted to provide estimates of RAM parameters against RAM requirements in a variety of expected operational conditions. OT concentrates on RAM performance of the system when in the hands of typical user troops in an operational environment. OT normally will be conducted in a fixed configuration. Modifications to the equipment should be allowed only if a problem is of such a nature that further testing is precluded. When reliability growth is approved for OT, modifications should be planned as block changes in the test design plan.

OT RAM emphasis must provide operational estimates of RAM and identify operational RAM deficiencies. OT should also evaluate the RAM aspects of operational suitability and represent, to the maximum degree possible, realistic operational conditions based on the OMS/MP. In addition, OT must verify correction of RAM deficiencies. All data collected should contribute to the DT/OT RAM database.

Throughout all phases of the materiel life cycle, the developer should conduct repeated assessments of (1) the effects of the design on RAM, (2) the measured and predicted RAM characteristics of the system, and (3) the impacts of (1) and (2) on mission effectiveness and operating and support costs.

3.6 RAM CONSIDERATIONS IN THE DoD SYSTEM ACQUISITION PROCESS

Concept Exploration Phase. During this phase RAM parameters are selected. These tentative parameters should respond to the projected mission needs of the mission area. Plans and resources need to be identified for RAM engineering tasks required to achieve reliable and maintainable systems. RAM parameters are allocated to the proposed system and affected government furnished equipment or commercial items. Components that require RAM improvement are also identified. This phase ends with a Milestone I decision.

Demonstration–Validation Phase. Operational RAM problems are addressed during design, by proper selection, and tailoring of operational and support concepts. Requirements are established for each applicable RAM parameter. Major elements of the new system, with acceptable RAM performance in similar applications, are identified. Reliability growth plans are developed. RAM engineering, test, and quality-assurance tasks are identified for the design and manufacture of reliable and maintainable systems. DT I is accomplished and RAM data is collected and evaluated to assist in identifying a preferred technical approach. OT I is accomplished to examine the operational RAM aspects of the selected alternative technical approaches and to estimate the potential operational effectiveness and operational suitability of candidate systems. This phase ends with a Milestone II decision.

Full-Scale Development. RAM engineering tasks are accomplished to ensure that RAM thresholds are met at high confidence levels before the production decision. Specific quality-assurance tasks and controls are developed to verify the manufacture and delivery of reliable and maintainable items. DT II is accomplished to ensure that all significant design problems affecting RAM have been identified and solutions to these problems are developed. OT II is accomplished to provide valid RAM estimates of the system's operational effectiveness and the system's operating and support costs. This phase ends with a Milestone III decision.

Production and Deployment. RAM demonstration is included in initial production tests. System assessment and correction of any remaining operational RAM deficiencies is accomplished at this time.

3.7   RAM POLICY

DoDD 5000.40 *(Reliability and Maintainability)* identifies RAM objectives for the Department of Defense. These objectives are to increase the operational readiness and mission success of fielded systems; reduce ownership costs and demand for maintenance–logistics support; limit manpower needs by ensuring that items are designed to be operated and maintained with the skills and training expected to be available in the operational environments; and provide RAM data to acquisition, operation, and support management teams. These objectives can be accomplished through individual policies of the various U.S. Military Services.

In addition to the general policies, there are also specific policies that address documentation and RAM quantification, testing, accounting, non-developmental items, and overhauled materials. Documentation policies include:

1. There must be RAM rationale for each requirements document.
2. RAM sensitivity on life-cycle cost and operational effectiveness on all cost and operational effectiveness analyses (COEA) are necessary.
3. Management documents such as the decision coordinating paper–system concept paper (DCP/SCP), test and evaluation master plan (TEMP), and program management documents (PMD) are necessary.
4. Solicitations and contracts must contain specified values of RAM that will be greater than the technical threshold at maturity.
5. The system technical data package must contain integral RAM characteristics.
6. RAM values must be documented in test documentation such as the TEMP, independent evaluation plan, and independent evaluation reports (IEP/IER), and in essentially all test reports.

RAM quantification policies include individual policies addressing:

1. The operational-mode summary–mission profile (OMS/MP), failure definition–scoping criteria, and numerical values of RAM, as well as

factors that relate to operational effectiveness and operating and support cost in RAM requirements documents.

2. RAM values that will be determined using OMS.

3. Only RAM operational requirements at maturity to be stated in requirements documents.

Specific policies on testing and RAM accountability include:

1. RAM emphasis to identify design deficiencies and corrective actions and the assessment and promotion of reliability growth are to be considered during DT II and OT II.

2. Other areas addressed include first article–initial production testing (FA/IPT), test planning and design, and test criteria.

RAM accounting policies include policies addressing scoping conferences, RAM assessment, data collection, RAM record and audit trail, and RAM reporting.

Basic RAM activities include (1) RAM planning under the O&O plan; (2) RAM requirements and failure definition under required operational capability (ROC) support, (3) RAM issues under TEMP planning, (4) RAM requirements under performance specifications, (5) scoring and assessment under test support, (6) RAM under development specifications; and (7) RAM verification under test.

3.8   METHODOLOGY FOR RAM REQUIREMENTS DEFINITION

The definition of RAM requirements begins with the OMS/MP from the approved O&O plan in the "technology base activities" phase of the system acquisition as shown in Figure 3.1 The process continues until completion and approval of a RAM rationale report and RAM executive summary to support the system ROC document prior to FSD.

3.9   IMPACTS OF RAM

RAM Influence on Effectiveness. RAM characteristics influence operational effectiveness by driving readiness, sustainability, and utilization of personnel and materiel during training. It is recognized that good RAM are force effectiveness multipliers, which offer the means to defeat a numerically superior force by engaging it repeatedly. Reliable systems result in increased capability while employing fewer fielded spare parts and less manpower. Similarly, maintainable systems require that fewer people and specialized skill levels be fielded while achieving reduced maintenance times.

Methodology for RAM Requirements Definition

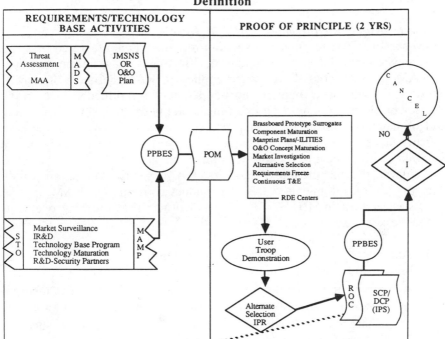

Figure 3.1 Methodology for RAM requirements definition.

RAM Influence on Life-Cycle Cost. The RAM characteristics of a system are also key leverage points in determining the system's total life-cycle costs and operational effectiveness. An estimated 30 percent of life-cycle costs can be traced directly to RAM characteristics of the system's design. These costs occur not only as budgeted line items in the procurement and operations and maintenance appropriations of the particular system but also as indirect costs of the supporting logistics facilities and activities, manpower, attrition replacements, and replenishment spares.

RAM in the Design Process. While conventional stand-alone postdesign RAM "engineering" tasks, such as test, analyze, and fix, have been moderately successful in achieving improved RAM, these approaches are fundamentally limited by their inability to influence the design process itself. The RAM characteristics of a system are, to a large extent, attributes of its design, or more precisely, a direct function of the attention given to them in the design process. They are analyzed into the design after it has been completed only with great difficulty and cost. Additionally, the RAM improvement effort must compete with integration and operational testing for test resources and schedule.

Computer-Aided Engineering (CAE) in Development. The application of RAM-specific CAE resources to system programs in an integrated development environment have the potential for effecting a quantum improvement in RAM characteristics. CAE, when applied to RAM design, will provide the designer with close-coupled, short-cycle analysis and feedback about the efficacy of the design approach in a time frame permitting corrective action and optimization during the design process rather than later. In addition, concurrent design synthesis techniques provide a superior inherent design capability with respect to reliability and maintainability.

Achieving the full potential of integrated RAM-specific CAE requires more than the use of automated "tools" to conduct conventional RAM analysis tasks such as those in MIL-HDBK-217 predictions. It requires improvements to five basic areas:

1. Automated RAM analysis procedures tightly coupled to the parts libraries and materials characteristics databases.
2. Automated RAM synthesis processes based on design rules incorporating lessons learned from prior design experience and field use.
3. Fully characterized (tested and validated) component performance and RAM characteristics databases.
4. Configuration management procedures that link major design decisions affecting the RAM characteristics of the end item to the CAE software and databases used to develop decision criteria and otherwise support the decisions.
5. Supporting structure of hardware, software, and computer networks adequate to support the procedures processes above and to closely couple the RAM specific resources with the rest of the design team.

The ultimate goal of the integration of RAM into CAE is for all major design decisions affecting the RAM characteristics of the end item to be fully supported by automated procedures appropriate to the nature and level of the decision in a concurrent or near-concurrent fashion.

Impact of RAM on ILS. RAM parameters are the most effective logistics engineering tools for influencing and interacting with the systems engineering process. Establishment of effective RAM requirements for the total system and their allocation to lower-level components are a vital influence on mission success and operation and support costs.

Mission success is greatly influenced by mission reliability (mean time between critical failures that impact the mission) and mission maintainability (mean time to restore functions during the mission). Readiness is partially determined by the mean time between downing events and mean time to restore the system. Maintenance manpower requirements and costs are affected by the interval between the worker–hours required to perform maintenance actions. Logistics support costs related to parts are determined by the mean time between removal of repairables and consumables and the total of all costs to remove, replace, transport, and repair components at all levels of maintenance.

The ILS program must ensure that the RAM parameters are consistent with planned operational environments and the support that will be provided. Failure to fully account for the effects of item design, quality, operation, maintenance, and repair can lead to a substantial shortfall in operational performance and an unprogrammed overrun of logistics support costs.

3.10 RELIABILITY

Reliability is a design characteristic that can be defined briefly as the probability that equipment will perform without failure for a specified period time under stated conditions. An analogous definition for maintainability is the probability that an item can be repaired in a specified period of time under stated conditions. These two design characteristics are very important. They combine to produce availability, which is the probability that materiel will be available for use, when required, under stated conditions. They are also the largest generator of support resource requirements, since failures resulting from unreliability generate corrective maintenance workloads, and the level of maintainability determines how economically the maintenance can be accomplished.

Reliability analyses continually provide maintenance engineering with predicted reliability or observed reliability, depending on the materiel program's phase. Predicted reliability data tend to be optimistic when compared to failures that actually occur when materiel is in the hands of the user because reliability engineers normally deal with inherent reliability—the reliability of the paper design—rather than with reliability of the fielded materiel. Inherent reliability does not account for failures that might result from activities such as manufacturing, acceptance tests, user maintenance activities, and operator error.

Reliability planning must be accomplished as an integral part of the overall systems planning effort. The objective is to plan a program that will ensure reliability involvement throughout all aspects of system design and development, production or construction, and system utilization.

Reliability Requirements

Reliability requirements are defined during conceptual design. Reliability analyses and predictions are then accomplished throughout preliminary and detail system–product design. This is followed by reliability testing as part of system test and evaluation. Thus reliability is considered throughout the system life cycle and is particularly relevant during the early phases of system design and development. Figure 3.2 illustrates specific reliability functions related to system life cycle.

The evaluation of any system or product in terms of reliability is based on precisely defined reliability concepts and measures. Reliability requirements are allocated and design criteria are established, reliability analyses and trade-off studies are accomplished to support major design decisions, predictions are made to assess design configurations at various stages in the development process, and test–evaluation is conducted to measure the results of the design effort.

Reliability in System Design

Reliability in system design commences with the establishment of requirements as part of the conceptual design phase. Every system and/or product is developed in response to a need to fulfill some anticipated function, and reliability is a major factor in determining the usefulness of a system.

After an acceptable reliability figure of merit has been established for the system, the requirement must then be allocated among the various components. Constructing a functional system breakdown facilitates allocating a system level requirement. The reliability of a system depends on the reliability of its component parts, and the selection of parts must be compatible with the requirement of the particular application of those parts. To accomplish reliability allocation, the following steps should be considered:

1. Evaluate system functional flow diagrams and identify areas where design is known and failure rate information is available or can be readily assessed. Assign reliability factors and determine their contribution to the whole reliability requirement. The difference represents the portion of the requirement that can be allocated to other areas.

2. Identify the areas that are new and where design information is not available. Assign complexity weighting factors to each functional block. The complexity factor can be based on an estimate of the number and relationship of parts, the equipment duty cycle, whether an item will be subjected to temperature extremes, and so forth. The end result should constitute a series of lower-level values that can be combined to represent the system reliability requirement initially specified.

Major emphasis in design for reliability should consider the selection of standardized components and materials to the maximum extent feasible, the evaluation of all components and materials prior to design acceptance, and the utilization of only those component parts capable of meeting the reliability

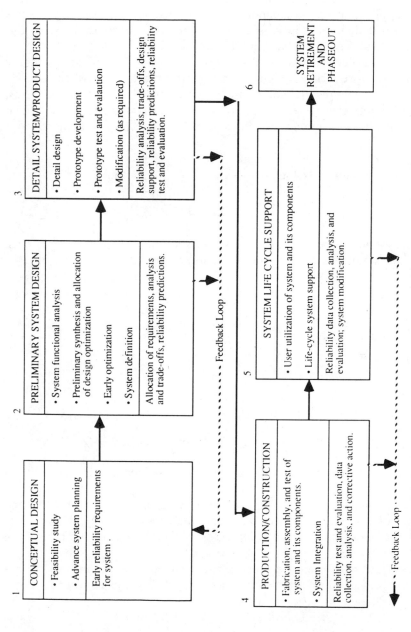

Figure 3.2 Reliability requirements in the system life cycle. (*Source:* MIL.-Handbook-217A.)

objectives. Under certain conditions in system design it may be necessary to consider the use of redundancy to enhance system reliability, providing two or more functional paths in areas that are critical to accomplish a mission. But the application of redundancy does not necessarily solve all problems because it usually implies increased weight and space, increased power consumption, greater complexity, and higher cost.

As stated earlier reliability is an inherent characteristic of design that is concerned with extending system operation and the time between failures. Reliability basically dictates the frequency of corrective maintenance. Thus, system reliability can be seen as the reliability of its component parts. That is, for the system to operate successfully, both elements must work as follows:

Series system

$$R = (R_a)(R_b) \tag{3.1}$$

Or for a system to operate successfully, at least one of its components must work. This is equivalent to saying that neither must fail:

Parallel system $\tag{3.2}$

$$R = 1 - (1 - R_a) \cdot (1 - R_b) = 1 - (1 - R_a - R_b + R_aR_b)$$
$$= R_a + R_b - R_aR_b$$

Parallel or redundant components can dramatically increase the reliability of a system. Given a series system, the sum of the failure rates of the separate components will equal the system failure rate. It should be noted that the reliability of a series system is always less than the reliability of any single component. Also, the reliability decreases as the number of components in the series decreases.

The frequency or rate of corrective maintenance and the mean time between corrective actions are also functions of reliability as shown here:

$$\lambda = \text{failure rate per operating hour} \tag{3.3}$$
$$= \frac{\text{number of failures}}{\text{total operating hours}}$$

Reliability Related to Repair Actions

Reliability is related to the need for repair actions. The product will fail as infinite reliability exceeds practicality. The design of an optimum support package is partially based on the anticipated frequency of failures, and reliability may be considered the starting point for integrated logistics support. Knowing that things will fail, the logistics supportability problem may be simplified by designing a product that is easy to repair when it does fail. Repair level analysis provides a rational approach to the problem of what should be repaired and where the most efficient location for repair is. Life-cycle costing reduces the entire life span of the product to the common denominator of dollars and cents. The activities of reliability, maintainability, repair level analysis, and life-cycle costing provide the connecting link between design of the product and design of the logistics support package.

To reiterate, reliability must be built into the product by reliability engineering during the design–development process. Reliability is nothing more than the probability that a product will perform as expected over a defined period of time under specified conditions. Four elements determine the reliability of a product: probability, performance as expected, time, and specified conditions. Probability is stated as a quantitative expression representing the percentage expectation of satisfactory performance. Performance, as expected, is the second element of reliability. This implies that specific criteria must be established to define the expected performance as accurately as possible. Expectations regarding performance are normally based on product specifications. The third element is time, as it provides a finite factor against which product performance can be measured. The term "time" is somewhat of a misnomer, however, as operating hours represent only one method of measurement. Product performance may also be measured in miles of operation, number of events, number of cycles, or any other convenient standard of repetitive units. Time, in conjunction with other reliability elements, provides all data necessary for predicting failures within a sustained period of operation. Specified conditions make up the fourth element that determines product reliability. The specified conditions define the operational profile and include both environmental factors and the physical attributes of the operating location.

The reliability of a product is also measured by the failure rate. However, it must be understood that the failure rate is not uniform over the life of a product.

Causes of Failure

Reliability measurements indicate how often a product will fail. A failure can result from a variety of causes. These causes have been generally grouped into the following categories: (1) early or burn-in failures, (2) change failures, and (3) wear-out failures. Some of the consequences of failure include loss of life or system, repair costs, and hidden or secondary failures.

Safety-related failures are failures that cause a loss of function or secondary damage with adverse effects on personnel and operating safety. These types of failure are detected when hardware operation stresses exceed design limits in a specific failure mode or a specific inspection standard is exceeded.

The Reliability Problem

When it is proposed to design a system–product to perform a complex and demanding job, it is assumed that the required investment will be justified according to the perfection by which the job is performed or by the large number of times the system and/or product can do the job. This assumption cannot be justified when a system–product fails to perform on demand or fails to perform repeatedly. Thus, it is not enough simply to show that a chasm can be spanned by a bridge; the bridge must continue to span the chasm for a long time to come while carrying useful loads.

In the design of complex systems, the assumption mentioned above is, in fact, not accepted. Instead, considerable effort is made to obtain reliable system performance.

Reliability is a consideration at all levels, from materials to operating systems, because materials to make up parts, parts compose assemblies, and assemblies are combined in systems of ever-increasing complexity and sophistication. Therefore, at any level of development and design, it is natural to find the influence of reliability engineering acting as a discipline founded to devote special engineering attention to the unreliability problem. Reliability engineering is concerned with the time degradation of materials, physical and electronic measurements, equipment design, processes and system analysis, and synthesis.

The Role of Reliability Prediction in Engineering

To be of value, a prediction must be timely. However, the earlier it is needed, the more difficulties will be encountered. It is certainly true that the earlier the unknown nature of a future event must be predicted, the more difficult it is to make a meaningful prediction. An early prediction is made on the basis of very little knowledge in order to form a rational basis for doing something about changing the basis of the prediction.

The art of predicting the reliability of equipment does have practical limitations such as those depending on data gathering and technique complexity. Considerable effort is required to generate sufficient data to report a statistically valid reliability figure. Casual data gathering occasionally accumulates data more slowly than the advance of technology; consequently, a valid level of data is never attained. Thus, it can be seen that derivation of failure rates is empirically difficult and obtaining valid confidence values is practically precluded because of lack of correlation.

The use of failure-rate data, obtained from field use of past systems–products, is applicable on future concepts depending on the degree of similarity existing in both the hardware design and the anticipated environments. Data obtained on a system–product used in one environment may not be applicable to use in a different environment, especially if the new environment substantially exceeds the design capabilities. Other variants that can affect the stated failure rate of a given system are different uses, different operators, different maintenance practices, and different measurement techniques or definitions of failure. When considering the comparison between similar but unlike systems, the possible variations are obviously even greater.

Thus, a fundamental limitation of reliability prediction is the ability to accumulate data of known validity for the new application. Another fundamental limitation is the complexity of prediction techniques. Very simple techniques omit a great deal of distinguishing detail, and the prediction suffers inaccuracy. More detailed techniques can become so bogged down in detail that the prediction becomes costly and may actually lag behind the principal hardware development effort.

Methods of Reliability Prediction

Reliability prediction is based on either (1) the analysis of similar equipment, (2) an estimate of active element groups, (3) an equipment parts count, or (4) stress analysis. The figures derived from reliability prediction are direct inputs to maintainability prediction data, logistics support analysis, and the determination of specific support requirements.

3.11 RELIABILITY-CENTERED MAINTENANCE (RCM)

RCM is a systematic analysis approach whereby the system–product design is evaluated in terms of possible failures, the consequences of these failures, and the recommended maintenance procedures that should be implemented. The RCM analysis is very similar to failure-mode effects and criticality analysis (FMECA) in many respects and should be accomplished in conjunction with FMECA to constitute a major data input for the logistics support analysis.

Purpose

The purpose of an RCM analysis is to identify the essential preventive maintenance tasks required to retain the safety and reliability inherent in system design. The application of RCM results in the identification of failure modes requiring additional design evaluation, establishment of scheduled preventive maintenance tasks for inclusion in technical manuals and/or orders, and the establishment of overhaul selection procedures for end items and components. RCM is a preventive maintenance philosophy encompassing four basic objectives: the realization of inherent safety and reliability, the restoration of safety and reliability after deterioration, the development of design improvements when needed, and the accomplishment of these objectives with minimum impact on the life-cycle cost.

RCM Decision Logic

RCM eliminates some time-honored maintenance practices that do nothing more than increase costs, without adding to the safety or reliability of an item. RCM is a valid response to the unnecessary and wasted maintenance motions that contribute to rising maintenance costs.

RCM decision logic transposes failure data, effects data, criticality analysis data, and field maintenance data into specific maintenance tasks that retain inherent equipment safety and reliability levels at the lowest cost through the life cycle. RCM decision logic provides the detailed logic process to segregate maintenance requirements. By segregating the maintenance requirements, decision logic (1) identifies the system–equipment component that is critical to mission or operating safety, (2) provides a logistical analysis process to determine the feasibility of scheduled maintenance task requirements, (3) highlights maintenance problem areas for design review consideration, and (4) provides the supporting justification for scheduled maintenance task requirements.

The logic process is based on two major principles. Scheduled maintenance tasks should be performed on noncritical components only when performance of the scheduled task will reduce the life-cycle cost of ownership of the system and on critical components only when such tasks will prevent a decrease in reliability and deterioration of safety to unacceptable levels, or when the tasks will reduce the life cycle cost of ownership of the system.

The RCM Program

The RCM program comprises three major elements: equipment design guidelines, preventive maintenance program development, and continuing review and update of preventive maintenance requirements. Thus, any RCM program should have the following objectives: (1) to establish design priorities or guidelines that ease preventive maintenance, (2) to ensure realization of the inherent equipment safety and reliability levels, (3) to restore equipment safety and reliability to their inherent levels when deterioration has occurred, and (4) to obtain the information necessary for design improvement of those items whose inherent reliability proves inadequate.

The RCM program is concerned with identifying those practices that enhance preventive maintenance and those that hinder it. The program seeks to develop design guidelines for use during initial equipment design or modification to ensure that a design is compatible with preventive maintenance.

The RCM program includes a systematic approach for identifying and developing preventive maintenance tasks for specific end-item equipment from a disciplined application of the decision logic process. RCM requires a critical examination of the equipment's design to identify potentially significant items as related to safety, readiness, and economics. A failure-mode–effects analysis, a decision logic process, and operational data and experience are applied to evaluate and classify the consequences of failure of each significant item according to failure severity and to identify applicable and effective maintenance tasks for preventing item failures or for identifying failed hidden-function items. The objective is to develop a minimum preventive maintenance program that includes only those tasks necessary to preserve safety and operating reliability economically and to ensure acceptable economy of operations, maintenance, and logistics support when the maintenance action may not have an impact on safety or reliability.

RCM is applicable during all phases of equipment acquisition and life-cycle support and to all levels of maintenance and serves as the means for justifying new or modified preventive maintenance tasks and for the continuing evaluation of existing tasks.

3.12 AVAILABILITY

Equipment should preferably be in a ready condition when an operational demand occurs. If equipment is to be operationally ready, then availability must be a feature designed into the system. Availability is a measure of the degree to which an item is in an operable state at the start of the mission, when the mission is called for at an unknown (random) time. This is measured by the ratio of system–product uptime to system–product uptime plus downtime as

$$A = \frac{\text{uptime}}{\text{uptime + downtime}} = \frac{\text{uptime}}{\text{total time}} \qquad (3.4)$$

Realizing that this uptime to total time ratio is dependent on the frequency of events causing downtime and the duration of the downtime events, availability can then be seen as a function of both reliability and maintainability.

Inherent availability (A_i) is the probability that a system–product, when used under stated conditions in an ideal support environment, will perform as intended at a given point of time. An ideal support environment includes all the necessary test and support equipment, spares and repair parts, properly trained personnel, and so forth. Scheduled (preventive) maintenance actions, logistics supply time, and administrative downtime are excluded from the calculations as follows:

$$A_i = \frac{MTBF}{MTBF + \overline{Mct}} \tag{3.5}$$

where $MTBF$ represents the mean time between failure and equals

$$MTBF = \frac{1}{\lambda} \tag{3.6}$$

$\overline{Mct}$ is the mean active unscheduled (corrective) maintenance time or the statistical mean for all corrective maintenance tasks. It is also referred to as the mean time to repair $(MTTR)$; $\overline{Mct}$ represents the expected number of failures for each hour of operation.

$$\overline{Mct} = \frac{\sum_{l}^{N} \overline{Mct}}{N} \tag{3.7}$$

where $\overline{Mct_i}$ represents individual corrective maintenance tasks and N equals the number of tasks.

   Achieved availability (A_a) is the probability that a system, when used under stated conditions in an ideal environment, will perform as intended at a given point in time. This differs from A_i in that preventive maintenance time is included. Only logistics supply time and administrative downtime are excluded as follows:

$$A_a = \frac{MTBM}{MTBM + \overline{M}} \tag{3.8}$$

where $MTBM$ represents mean time between maintenance, which includes scheduled (preventive) and unscheduled (corrective) maintenance, and equals

$$MTBM = \frac{1}{(MTBM_s)^{-1} + (MTBM_u)^{-1}} \tag{3.9}$$

where $MTBM_u$ is the mean time between unscheduled maintenance actions and $MTBM_s$ is the mean time between scheduled maintenance actions and M is the mean maintenance time, equal to

$$\overline{M} = \frac{(\overline{Mct}) (MTBM_u)^{-1} + (Mpt) (MTBM_s)^{-1}}{(MTBM_u)^{-1} + (MTBM_s)} \tag{3.10}$$

where $\overline{Mct}$ is the mean active unscheduled (corrective) maintenance time and *Mpt* is the mean active scheduled (preventive) maintenance time.

Operational availability (A_o) is the probability that a system, when used under stated conditions in the actual operating environment, will perform as intended. It is equal to

$$A_o = \frac{MTBM + \text{ready time}}{(MTBM + \text{ready time}) + MDT} \qquad (3.11)$$

where the ready time is when the system is ready for use but not being utilized and *MDT* is the time the system and/or product is not in a condition to perform its intended function. This time includes active repair time, administrative downtime, and logistics supply time.

3.13   MAINTAINABILITY

Maintainability requirements are defined in conceptual design as part of system operational requirements and the maintenance concept. Maintainability predictions are accomplished throughout preliminary and detail system–product design. In addition, the maintainability demonstration is accomplished as part of system test and evaluation. Thus maintainability is considered throughout the system life cycle as shown in Figure 3.3.

Maintainability Program

The purpose of a maintainability program is to improve operational readiness, reduce maintenance manpower needs, reduce life-cycle cost, and provide data essential for management. A major problem confronting all government and industry organizations responsible for a maintainability program is the selection of tasks that can materially aid in attaining program maintainability requirements. Current schedule and funding constraints mandate a cost-effective selection, one that is based on identified program needs. Once appropriate tasks have been selected, the tasks themselves can be tailored. Selection and tailoring of tasks specifying maintainability requirements, addition of supporting details, and establishing contract delivery requirements requires a balanced approach. The tendency is to overtask and to acquire extensive delivered data, which results in unnecessary contractor efforts, increased contract costs, and possible delays in schedule. An extensive maintainability program requirement will not result in equipment meeting their repair time quantitative requirements if the quantitative requirements were unrealistic in the first place.

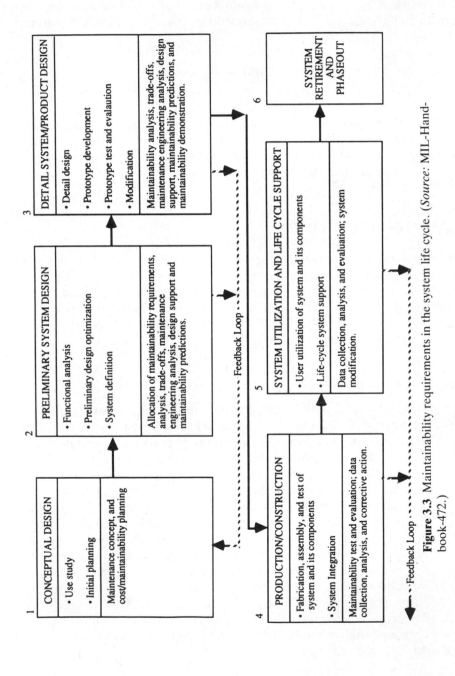

Figure 3.3 Maintainability requirements in the system life cycle. (*Source:* MIL–Handbook-472.)

Maintainability Objectives

The general objective of maintainability is to impact equipment designs to ensure that these designs will satisfy operational availability requirements and can be maintained easily and economically. In relation to logistics, the term "easily" implies low personnel skills, simple diagnostic procedures, and short times to remove and replace and test or verify adequacy of the operation. The term "economically" implies accomplishment of the maintenance at the lowest operational and support cost. Maintainability determines design features such as equipment packaging and diagnostics that economically satisfy both operational requirements and the maintenance concept and incorporates those features into materiel design.

Maintainability became important when the ever-increasing costs of maintenance gained importance. Analysts saw that the techniques used in reliability engineering could be applied to maintainability. The following statistical methods and expressions are used to measure and evaluate maintainability: $MTBM$, mean time between maintenance; $MTBR$, mean time between replacement; $MTTR$, mean time to repair; MDT, mean downtime; M_{max}, maximum corrective maintenance time; MMH/OH, maintenance man–hours per operating hour; TAT, turnaround time; and the most common, M, mean maintenance time.

There is a relationship between maintainability and the LSA process that can best be shown in Table 3.1, which depicts these functional relationships and interfaces. Maintainability is now considered one of the essential ingredients of the logistics system and a factor in both systems effectiveness and cost-effectiveness.

Maintainability planning is essential and must be accomplished as an integral part of the overall systems–product planning effort. A maintainability plan must be developed that considers (1) maintainability quantitative and qualitative requirements for the system; (2) allocation of maintainability requirements to the subsystem level; (3) design techniques and practices; (4) maintainability analysis; (5) maintainability predictions; (6) maintenance engineering analysis; (7) formal design reviews; (8) maintainability demonstration; and (8) data analysis, collection, and corrective action.

Maintainability Design Criteria

In order to translate maintainability requirements and anticipated operational constraints into practical and effective hardware designs, a broad spectrum of design criteria, standards, and policies, both general and specific, must be defined and employed. The design criteria must be developed to assist the maintainability analyst in the selection of maintainability quantitative design features to enhance the incorporation of optimum maintainability into the design of systems and equipment. System design criteria must be structured so that features that enable cost-effective maintenance support throughout a

TABLE 3.1 Matrix of Functions and ILS Plan Interrelationships

Disciplines Plans	Requirements Analysis	Prediction Analysis	Maintainability Design Analysis	Automatic Test Equipment Analysis	Failure Modes and Effects Analysis	Demonstration/ Test of Design
Maintenance plan	X	X	X	X	X	
Support and test equipment	X	X	X	X	X	
Supply support	X					
Packaging, handling, storage, transportability	X		X			
Technical data						
Facilities	X		X	X	X	
Manpower, personnel, and training	X	X	X	X	X	
Interim support				X	X	
Contract maintenance	X			X	X	
Logistics demonstration						X

Source: MIL-STD-1388-1.

deployed hardware life are considered in the design process. Some examples of maintainability design criteria are (1) all repair part items having the same part numbers must be functionally and physically interchangeable without modification or adjustment of the items or the system or equipment in which they're used; (2) maintenance adjustment or alignment should not be required; (3) preventive (scheduled) maintenance requirements, including calibration, should be eliminated; (4) physical and functional maintenance access must be provided to any active component on opening or removal of access entries, and should not require prior removal or movement of other components; (5) devices securing access entrances and maintenance replaceable items should be the captive, "quick-release" type with positive locking features; and (6) special tools should not be required in the performance of user of intermediate-level maintenance tasks.

General design criteria relate to the achievement of various goals or targets, such as (1) to minimize downtime due to maintenance, by using maintenance-free design, standard and proven designs and components, simple reliable and durable design and components, fail-safe features to reduce failure consequences, "worst-case" design techniques and tolerances that allow for use and wear over item life, and modular design; (2) to minimize maintenance downtime, by designing for rapid and positive prediction or detection of malfunction or degradation, localization to the affected assembly, isolation to a replaceable or repairable module, correction by replacement, adjustment or repair, verification of correction and serviceability, identification of parts, test points, and connections; (3) to minimize maintenance costs by designs that minimize hazards to personnel and equipments, special implements for maintenance, consumption rates and costs of spares and materials, and personnel skills; (4) to minimize the complexity of maintenance by designing for compatibility among system equipment and facilities, standardization of design, parts and nomenclature, interchangeability of like components, material, and spares, minimum maintenance tools, and adequate accessibility; (5) to minimize the maintenance personnel requirements by designing for logical and sequential function and task allocations and simple and valid maintenance procedures and instructions; (6) to minimize maintenance error by designing to reduce the likelihood of undetected failure or degradation; and (7) to minimize the frequency of tool failure.

Need for Maintainability Prediction

The prediction of the expected number of hours that a system or device will be in an inoperative or "down state" while it is undergoing maintenance is of vital importance to the user because of the adverse affect that excessive downtime has on mission. Therefore, once the operational requirements of a system are fixed, it is imperative that a technique be utilized to predict its maintainability in quantitative terms or as early as possible in the design phase.

This prediction should be updated continuously as the design progresses to ensure a high probability of compliance with specified requirements.

A significant advantage of using a maintainability prediction procedure is that it highlights for the designer those areas of poor maintainability that justify product improvement, modification, or a change of design. Another useful feature of maintainability prediction is that it permits the user to make an early assessment of whether the predicted downtime, the quality, quantity of personnel, tools, and test equipment are adequate and consistent with the needs of the system operational requirements.

Definition of Maintainability

MIL-STD-721B defines maintainability as:

> a characteristic of design and installation which is expressed as the probability that an item will conform to specified conditions within a given period of time when maintenance action is performed in accordance with prescribed procedures and resources.

This definition has fostered the development of many maintainability prediction procedures for providing an assessment of system maintainability. Each of these uses various quantitative measures to indicate system maintainability. However, all of these measures have a specific relationship to, or constitute some element of the distribution of, the total system downtime. Hence, if a universal method or technique could be developed to determine the "total system downtime distribution" for any type of system, this would facilitate calculating the measure of maintainability currently in use.

Elements of Maintainability Prediction Techniques

Each maintainability prediction technique utilizes procedures that are specifically designed to satisfy its method of application. However, all maintainability prediction methods are dependent on at least two basic parameters: (1) failure rates of components at the specific assembly level of interest and (2) repair time required at the maintenance level involved.

Many sources record the failure rate of parts as a function of use and environment. This failure rate is expressed at the number of failures per unit of time. A typical measure is "failure per 10^6 hours." The major advantage of using the failure rate in maintainability prediction calculations is that it provides an estimate of the relative frequency of failure of those components that are utilized in the design. Failure rates can also be utilized in applicable regression equations for calculating the maintenance action times. Another use of the failure rate is to weigh the repair times for various categories of repair activity, in order to provide an estimate of its contribution, to the total maintenance time. Repair times are determined from prior experience, simulation of tasks, or previous data secured from similar applications. Most procedures break up the "maintenance action," which is a more general expression than "repair action," into a number of basic maintenance tasks whose time performance is summed to obtain the total time for the maintenance action.

Measures of Maintainability

Considerable information is required in order to predict maintainability. The following types of information are essential: (1) location and failure rate of each component of the system; (2) number and types of spares; (3) number of test points; (4) nature of special test equipment; (5) estimates of durations of average mission; (6) manning schedules for operations and maintenance personnel, including all shifts; and (7) estimates for intervals occupied by unscheduled activities.

Maintainability can be measured in terms of a combination of elapsed times, personnel labor-hour rates, maintenance frequencies, maintenance cost, and related logistics support factors. Maintainability is a measure of the ease and rapidity with which a system can be maintained and is measured in terms of time required to perform maintenance tasks. Maintenance constitutes the art of diagnosis and repairing, or preventing, system failures.

Maintainability deals with the characteristics in system design pertaining to minimizing the corrective maintenance requirements for the system when it assumes operational status. Thus, reliability and maintainability requirements for a given system must be compatible and mutually supportive. Maintainability also deals with the characteristics of design that minimize preventive maintenance requirements for the system. As a result, a major objective of maintainability is to provide the proper balance between corrective maintenance and preventive maintenance at the least overall cost.

A frequently used measure of maintainability is maintenance man-hours divided by operating hours (MMH/OH), which defines the total maintenance expended over a finite interval of time. The maintainability measure, MMH/OH, by defining maintenance expenditures as a function of time, indicates the number of maintenance personnel required. This method of measuring maintainability places the maintainability engineer at a disadvantage, since MMH/OH is a function of maintenance time as well as maintenance frequency. Thus, if the reliability engineer fails to meet the $MTBF$ objectives, the maintainability engineer is almost certain to miss MMH/OH goals.

A second maintainability unit of measure is mean downtime, which is the statistical mean of the time a system is out of service for maintenance. This measure indicates inherent system availability.

Mean time to repair ($MTTR$) is yet another measure of maintainability. The system that is easy to maintain can quickly be restored to service; it exhibits a preferred $MTTR$. But $MTTR$ should not be used by itself. A system having an excellent *MTTR* exhibits a questionable virtue if it also possesses a very low *MTBF*.

Units that measure maintainability rely on the time required for maintenance, that is, net versus gross maintenance. For example, a flat tire can be removed and replaced in about 20 minutes. However, there are many other associated tasks that must also be accounted for in gross maintenance time, that is, assuming that there is a spare in the trunk, opening the trunk, removing the spare tire, putting the jack together, jacking up the car, and so on. The associated tasks were necessitated by the failure; therefore it is correct to allocate this time to the repair activity.

Mean time between failures (*MTBF*) is a fundamental quantitative maintenance parameter. This parameter establishes the frequency at which corrective maintenance is performed. *MTBF* is derived for a particular interval by dividing the total functioning life of a population of an item by the total number of failures within the population during the measurement interval. There is a mathematical relationship between the *MTBF* and failure rate λ for materiel. The failure rate is the number of failures of an item per unit measure of life.

Individual corrective maintenance task time $\overline{Mct_i}$ is the time required to complete an individual maintenance task or an individual maintenance action. When maintenance time estimates are based on an average of several observations, or used in prediction analysis, for example, individual maintenance task or action items are denoted by $\overline{Mct_i}$ to indicate the value is an average value for the individual task or action.

In mean time to repair, $\overline{Mct}$ is the mean time required to complete a maintenance action; that is, total maintenance downtime divided by total maintenance actions over a given period of time. Mean time to repair (*MTTR*) is defined as the summation of all maintenance downtime during a given period divided by the number of maintenance tasks during the same period of time, given as

$$\overline{Mct} = \frac{\sum_{i=1}^{N} \lambda \; \overline{Mct_i}}{\sum_{i=1}^{N} \lambda \; i} \tag{3.12}$$

where λ is the failure rate of the individual element of the item for which maintainability is to be determined, adjusted for duty cycle, catastrophic failures, and so forth, which will result in deterioration of the item performance to the point that a maintenance action will be initiated and $\overline{Mct_i}$ is the average repair time required to correct the ith repairable element in the event of its failure.

In medium time to repair, $\overline{Mct}$ is the downtime with which 50 percent of all maintenance actions can be completed. The median downtime is that value that divides all downtime values so that one-half of the values is equal to or less than the median and one-half is equal to or greater than the median. The median value of the maintainability function is related to individual time to repair.

Maximum time to repair (*MaxTTR*) represented by $M_{\max_{ct}}$ is the maximum time required to complete a specified percentage of all maintenance actions. $M_{\max_{ct}}$ is related to individual repair times comprising the underlying log normal probability density functions.

Mean preventive maintenance time represented by $\overline{Mpt}$ is the mean equipment down time required to perform scheduled preventive maintenance on an item, excluding preventive maintenance time expended on the equipment during operation and excluding administrative and logistic downtime. Mean time for preventive maintenance is represented by

$$\overline{Mpt} \frac{\sum_{i=1}^{N} f_i \, \overline{Mpt_i}}{\sum_{i=1}^{N} f_i} \tag{3.13}$$

where f_i represents frequency of individual (*i*th) preventive maintenance actions in actions per operating hour adjusted for equipment duty cycle and $\overline{Mpt_i}$ is average time required for *i*th preventive maintenance action.

The maintenance downtime rate per operating hour consists of downtime due to corrective maintenance and downtime required for preventive maintenance. The corrective downtime rate $\overline{MDT_{ct}}$ is corrective maintenance downtime per hour of operation. The preventive downtime rate $\overline{MDT_{pt}}$ is preventive maintenance downtime per hour of operation. The total downtime rate $\overline{MDT}$ is total maintenance downtime for corrective and preventive maintenance rates combined.

Maintainability Characteristics

Maintainability characteristics of equipment design are reflected in the cost of equipment ownership by the number of man-hours of technical time required to keep the equipment at the specified level of performance. The computation of maintenance man-hours per operating hour (maintainability index) includes the determination of maintenance man-hours required at each level of maintenance per hour of equipment operation.

The maintainability index for corrective maintenance MI_c is the mean corrective maintenance man-hours per equipment operating hour:

$$MI_c = \sum_{i=1}^{N} \lambda \, i \, \overline{Mc_i} \tag{3.14}$$

where MI_c represents mean corrective maintenance man-hours at the designated level of maintenance required per hour of equipment operation; λ is failure rate of the individual (*i*th) repairable element in failures per 10^6 hours of operation, weighted by duty cycle, tolerance, and iteration malfunction rates; and $\overline{Mc_i}$ is average maintenance man-hours at the designated level of maintenance required to complete the individual (ith) corrective repair action. The maintainability index for preventive maintenance MI_p is the mean preventive maintenance man-hours per equipment operating hour:

$$MI_p = \sum_{i=1}^{N} f_i \, \overline{M}p_i \tag{3.15}$$

where MI_p represents mean preventive maintenance man-hours at the designated level of maintenance required per hour of equipment operation; f_i is frequency of the ith preventive maintenance action, in actions per 10^6 hours of operation, weighted for duty cycle; and $\overline{M}p_i$ represents average maintenance man-hours at the designated level of maintenance required to complete the ith preventive repair action.

The maintainability index MI is a measure of the total maintenance man-hours required to maintain a system in operational status per hour of operation:

$$\begin{aligned} MI &= MI_c + MI_p \\ MI &= \sum_{i=1}^{N} \lambda \, i \, \overline{M}c_i + \sum_{i=1}^{N} f_i \, \overline{M}p_i \end{aligned} \tag{3.16}$$

Maintenance man-hours per task is the relationship between maintenance man-hours per operating hour and maintenance man-hours per maintenance task:

$$\overline{M}_c = \sum_{i=1}^{N} \lambda \, i \, MI_c \tag{3.17}$$

and

$$\overline{M}p = \sum^{N} f_i \, MI_p \tag{3.18}$$

where $\overline{M}_c$ is mean corrective maintenance man-hours per corrective maintenance actions and $\overline{M}_p$ is mean preventive maintenance man-hours per preventive maintenance action.

3.14 SUMMARY

RAM analysis aids in providing information necessary to determine the parts that are likely to fail, the requirements for tools, test and support equipment, and maintenance tasks, and the estimated time required to perform those tasks. The RAM process is an essential functional element to system and/or product acquisition and the LSA process. The front-end efforts of the RAM process in the R&D phase of acquisition have the most impact on system operational readiness and sustainability.

Reliability is an expensive discipline. The number of tasks and resources applied and the amount of management attention required are very costly. However, lack of reliability is even more expensive.

The RCM program encompasses and interfaces with many other disciplines and programs. It is, in fact, nothing more than a logical attempt to ensure that maintenance engineering efforts include scheduled maintenance considerations during the design process. But RCM is not a panacea for all maintenance determinations. It should be viewed as a valid response for the logical elimination of time honored maintenance practices that increase costs without adding to the safety or reliability of a system.

Availability is a function of both reliability and maintainability. There are three categories of availability: inherent availability, achieved availability, and operational availability.

The selection and application of the proper maintainability techniques results in many economies measured in terms of man-hours, material, and money. These savings are attributable to the fact that maintainability prediction is considered to be a tool for design enhancement because it provides for the early recognition and elimination of areas of poor maintainability during the early stages of the design cycle. Otherwise, areas of poor maintainability would become apparent only during demonstration testing or actual use, after which time correction of design deficiencies would be costly and unduly delay schedules and missions. Maintainability prediction, therefore, is a most useful instrument to both manager and engineer because it provides for improved system effectiveness and reduces administrative and maintenance costs.

Maintainability objectives with regard to operational availability (or systems effectiveness) are concerned primarily with the rapidity of diagnostics and return to a specified level of operational capability of the system as a whole. Maintainability prediction is of vital importance to the user because of the adverse affect that excessive downtime has on mission.

4
ILS IN THE SYSTEMS ENGINEERING PROCESS

Systems engineering (SE) is concerned with the process of translating mission, test, production, deployment, support, and operational needs into the most cost-effective mix of design requirements. The systems engineering process is iterative and involves functional analysis, synthesis, and optimization with the end goal of achieving a proper balance between economic, operational, and logistics factors. SE provides tools that may be applied to integrated logistics support. It encompasses the inventions, design, and integration of the entire assembly of equipment as distinct from the invention and design of the parts. This approach is used in solving problems of designing systems of machines in such a way that machines contribute to accomplishment of component objectives, components contribute to subsystem objectives, and so forth. It is essentially a blending of human engineering, industrial engineering, and conventional engineering to design pieces of equipment and a physical setting within which they can be related in a manner that will lead to overall goal achievement without conflict. The *Systems Engineering* volume in this same series by Ben Blanchard covers this aspect in greater detail.

4.1 PURPOSE

When weapons–product development programs were relatively simple to manage, engineering effort could be directed by a few top managers. Communications between participants were uncomplicated; functions and responsibilities were easily stated; and decisionmaking in regard to cost, performance, and schedule goals were fairly straightforward. As the state of the art advanced, the engineering profession expanded along highly specialized functional lines, acquired increased importance and complexity, and required more sophisticated management.

The problems of coordination, communication, direction, and control of these specialties and among geographically separated personnel have become increasingly severe. Some specialties are grouped into "functional" organizations to coordinate the state of the art across more than one program and to time-share between programs. In other instances, specialties are divided into program-oriented organizations, or a compromise bilateral organization may be adopted. With the advanced and increasingly complex new programs, the need for improvement in the following technical and management areas has been apparent:

1. Control of the design interfaces among systems, equipment, personnel, facilities, and computer programs.
2. Use of tradeoff analysis techniques in allocation of functions, selection among design approaches, and resolution of conflicting design objectives and constraints.
3. Assurance that the performance specifications, detail design, and production data packages are consistent with the fundamental mission requirement and with balanced consideration of such factors as producibility, operability, supportability, reliability, safety, and compatibility with interfacing systems, equipment, personnel, facilities, and computer programs.

The development of solutions to the problems of communications, direction, and control also requires methodical, analytical approaches to the development of total systems–products. These approaches are termed "systems engineering." The total management effort is termed "systems engineering management." Systems engineering is the technical process of transforming an operational need into a description of system performance parameters and a system configuration; and systems engineering management is the process of coordinating the engineering and technical effort within a project or program.

The total design process involves systems analysis, definition, and synthesis of requirements, preliminary design, and detail design. Systems analysis is the analysis and transformation of material requirements into a theoretical model with quantitative terms, and the manipulation of the model in simulation of the operational environment. Definition and synthesis of requirements is the translation of performance objectives of a selected system approach into design criteria (design to requirements) for the individual elements that will make up the system. Preliminary design develops the design approach for the system and its elements based on the criteria provided by the definition and synthesis of requirements. Detail design translates the design approach into a manufacturing configuration that can be produced and supported within the state of existing or economically achievable manufacturing technology and support capability. The relation of systems engineering to the total design process and the material life-cycle phases integrates the entire engineering effort.

4.2 OBJECTIVES

The objectives of the systems engineering process include:

1. Ensuring that the engineering effort is fully integrated, to reflect adequate and timely consideration of design, test and demonstration, production, operation, and support of the system–equipment.
2. Ensuring that the definition and design of the system–equipment item is conducted on a total system basis, reflecting equipment, facilities, personnel data, computer programs, and support requirements to achieve the required effectiveness and acceptable risk, cost, and schedule considerations.
3. Ensuring that the design requirements and related efforts of reliability, maintainability, integrated logistics support, human-factors engineering, health, safety, and other specialties with respect to each other are taken into consideration in the engineering effort.
4. Ensuring that compatibility of all interfaces within the system, including the necessary supporting equipment and facilities; and ensuring that the system is compatible and properly interfaces with other systems and equipment that will be present in the operational environment.
5. Providing the means for the evaluation of changes that will reflect consideration of the effect of the change on the overall system performance, effectiveness, schedule, and cost; and ensuring that all affected activities participate in the evaluation of changes.
6. Providing a framework of coherent system requirements to be used as performance, design, and test criteria and to serve as a source of data for development plans, contract work statements, specifications, test plans, design drawings, and other engineering documentation.
7. Measuring and judging technical performance for the timely identification of high-risk areas and other problem areas.
8. Providing a means of documenting major technical decisions made during the course of the program.

4.3 THE SYSTEMS ENGINEERING PROCESS

Systems engineering is essentially a front-end process that must lead the design effort. Each acquisition phase will involve the functional analysis of input requirements, synthesis of requirements, evaluation and decision, and description of the system elements as shown in Figure 4.1. Functional analysis answers the "what" and "why" questions relative to system design. Synthesis supplies the "how" answers to the "what" outputs of the functional analysis. The process provides the logic and timing for a disciplined approach, with certain internal assurances of technical integrity such as traceability. Technical

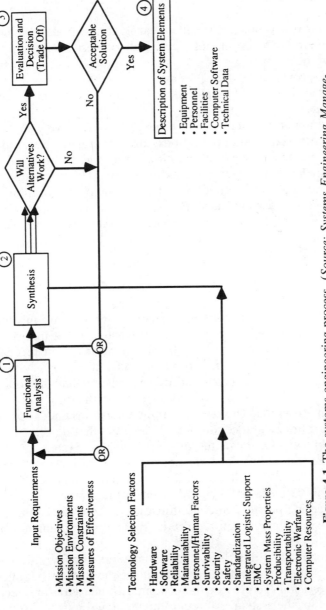

Figure 4.1 The systems engineering process. (*Source: Systems Engineering Management Guide*, DSMC, 1986.)

87

integrity ensures that the design requirements for the system elements reflect the functional performance requirements, that all functional performance requirements are satisfied by the combined system elements, and that such requirements are optimized with respect to system performance requirements and constraints.

Although no two problems are identical in terms of developmental requirements, there is a uniform and identifiable process for logically arriving at engineering decisions regardless of system purpose, size, or complexity. The SE process converts input requirements related to user need into output information that describes the optimal combination of system elements that will satisfy that need.

Figure 4.2 shows the systems engineering process iteratively applied to the interrelated functional areas of operations, logistics support, test, production, and deployment. At each level of definition from the system level down to the component level, the requirements imposed by logistics support, test, production, and deployment are considered in system optimization. At each level, the process is accomplished for each functional cycle to the extent necessary to identify risks, achieve delineation of system elements and product elements of the work-breakdown structure (WBS), and validate the decisions that must be made at that level.

Input Requirements

The four initial inputs to the systems engineering process are mission, environment, constraints, and measures of effectiveness. Input information describing the operational mission must be sufficient to permit recognition of major functions and functional requirements to be met by the system. Input information should be screened rigorously to ensure that the total system objective is adequately defined and is consistent with the system identification and interface information. The system–product performs under both internal and external environmental conditions. Parameters that describe the operational environments must be stated in specific values. System–product constraints originate from policy, experience, budget limitations, and prior analyses. These constraints usually affect the characteristics and composition of the system as its elements are being derived in the synthesis posture of the systems engineering process. The identification of constraints and their potential impact on system design require early input from human-factors engineering.

Each decision made within the systems engineering process must be guided by standards of measurement for evaluation of the various parameters involved in the decision. To provide these standards, measures of effectiveness are established for the system. All requirements stated for the system should be related to some measure of effectiveness. A measure of effectiveness is a particular value of system effectiveness pertinent to one or more mission objectives. A measure of effectiveness is related to the inherent value or

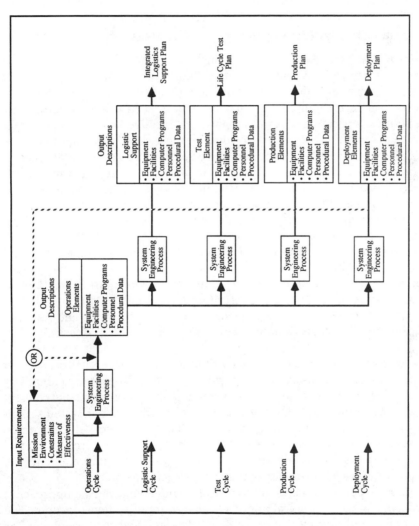

Figure 4.2 Application of systems engineering process to functional areas. (*Source:* TM 38-760-1.)

utility of the system. It may or may not involve cost-effectiveness. In military systems, the measures of effectiveness should be based on mission objectives. They may be related to any or all of the effectiveness factors of availability, dependability, and capability. In addition, measurement of effectiveness will change with the life-cycle phases because the basis for optimizing decisions changes.

Description of the Systems Engineering Process Steps

Functional Analysis. Systems engineering applies the inherent and essential relationships of functions and elements to the development of systems. The objective of functional analysis is to define a baseline of functions and functional performance requirements that must be met in order to adequately accomplish the operation, logistic support, test, production, and deployment requirements of the system, and to identify those functions where system–product life-cycle costs are expected to be sensitive to incremental changes in performance requirements. These functions and their performance requirements provide a common denominator of selection and design criteria for the system elements, and initially identify areas where tradeoffs between input requirements and engineering development require future consideration. Functional analysis supports mission analysis in defining the functional areas, sequences, and interfaces. It is also used by engineering specialists and support organizations to develop requirements for equipment, software, personnel, and operational procedures to complete implementation and deployment of the system. Functional identification is accomplished through the use of functional flow block diagrams depicting the task sequences and relationships, and time-line analysis depicting the time dependence of tasks.

Synthesis. Synthesis is the classic activity known as "conceptual design." It is the point in the systems engineering process at which engineering creativity and technology is brought to bear in the creation of a system or design concept to meet stated objectives. One of the main objectives of the systems engineering process, as stated earlier, is to ensure that design is actively performed with the benefit and full cognizance of functional performance requirements, system constraints, and effectiveness criteria, and that system elements are given proper consideration in arriving at a design concept.

Synthesis is performed initially to postulate possible technical approaches and, supporting each technical approach, one or more system concepts. The configuration and arrangement of system elements and the techniques for their uses are portrayed in any suitable form such as sets of schematic block diagrams.

Synthesis must consider the results of various technical and design studies as well as the requirements delineated by functional analysis. Engineering creativity is a key factor in the accomplishment of effective synthesis. Within each synthesized solution, characteristics of the equipment, facilities, personnel, and procedural data are balanced in accordance with the established measures of effectiveness.

The portrayal of the synthesized system in terms of its elements will provide a source of data for equipment design documentation, interface control documentation, consolidated facility requirements, contents of procedural handbooks, task loading of personnel, consolidated computer programs, the specification tree, and product elements of the WBS.

Evaluation and Decision. Evaluation is continual in the systems engineering process to select the best combination of system elements to meet mission objectives and support requirements. Evaluation and decision are always required to establish that a feasible and adequate design concept has been synthesized. Evaluation leads to a decision that selects a recommended system design concept; determines that additional analysis, synthesis, and/or trade studies are required to make a selection; or establishes that the state of the art in technology does not provide an acceptable solution.

Description. The description step of the systems engineering process produces engineering data that define the configuration, arrangement, and usage of all system elements and their effectiveness in achieving functional performance. These descriptions define the product elements of the WBS.

Documentation

Systems engineering documentation provides analytical tools for each of the process steps as applied to each of the functional areas. The documents encompass the minimum information required at each step of the process to adequately perform that step and to ultimately define the system elements. Figure 4.3 identifies the basic and special purpose documentation required for systems engineering.

One of the mandatory features of the systems engineering process is that it provides traceability. Traceability ensures technical integration in the application of the systems engineering process. Most of the data developed by the systems engineering process is required, in some format, by other activities engaged in the development project. Figure 4.4 shows the multiple application of systems engineering data elements.

4.4 ROLE OF ILS IN THE SYSTEMS ENGINEERING PROCESS

DoD Directive 5000.39 ["Acquisition and Management of Integrated Logistic Support (ILS) for Systems and Equipment"] emphasizes the identification of supportability design requirements through integration with the mainstream engineering effort. One way to achieve this is to establish a rigorous formal relationship between the ILS process and the systems engineering process.

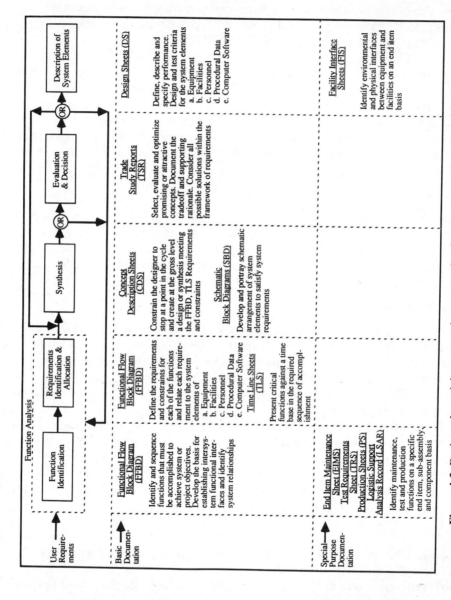

Figure 4.3 Basic and special-purpose documentation for systems engineering. (*Source: Systems Engineering Management Guide, DSMC, 1986.*)

Output Data for Plans and Activities

Program Plans and Activities

Plans and activities shown here are typical examples, and may vary with different projects.

SYSTEM ENGINEERING DATA ELEMENTS	Design	Development	Production	Product Assurance	Facilities Construction	Test & Evaluation	Training	Reliability	Deployment	Operational	Depot Maintenance	Maintenance	Logistic Support	Product Improvement	And Others
Operational Functions/Requirements	X	X		X	X	X		X		X				X	
Logistic Support Functions Requirements	X	X	X	X	X	X	X	X	X	X	X	X	X	X	
Reliability/Maintainability Requirements	X	X		X	X			X	X			X	X	X	
Equipment Design Criteria	X	X	X		X	X	X	X		X	X	X		X	
Equipment Design Solutions	X	X	X	X	X	X	X	X	X	X	X	X	X	X	
Facilities Requirements	X	X		X		X	X		X			X	X	X	
Function Performance Times		X			X	X	X		X	X		X		X	
Task Description/Skill Requirements		X		X	X	X	X	X				X		X	
Operating Procedures		X		X	X	X	X	X	X	X				X	

Figure 4.4 Multiple applications of systems engineering data elements. (*Source:* TM 38-760-1.)

The success of the integrated logistics support program hinges on how the readiness and supportability characteristics are designed into a system–product. These characteristics must be designed-in early, during concept exploration and demonstration–validation (CE and D/V), and continuously through full-scale development (FSD). They definitely must be considered in performing functional analyses and tradeoff analyses. The systems engineering process provides a framework for a DoD materiel system to acquire the desired supportability characteristics. In addition, ILS planning impacts on and in turn is impacted by the engineering activities throughout a system life cycle. Initially, support descriptions in the form of criteria and constraints are furnished with the top-level system operational needs. These descriptions include such items as basing concepts, personnel or training contraints, repair level constraints, and similar support considerations. Systems engineering in its evaluation of functional and detail design requirements, has as its goal the achievement of the proper balance among operational, economic, and logistics factors. This balancing and integrating function is an essential part of the system–cost-effectiveness tradeoffs and studies. The integration of ILS concepts and planning considerations into the systems engineering process is a continual and iterative activity, with the objective of optimal balance between performance and support considerations and optimal tradeoffs among costs of ownership, schedule, and system effectiveness.

Figure 4.5 illustrates the analytic and decisionmaking process involved in the application of systems engineering to acquisition management. To achieve the necessary balance of ILS factors within the systems engineering process, the contractor must define tradeoff and decision criteria that adequately address support requirements. A balanced integration of logistics support requirements into the systems engineering process will achieve the following objectives:

- Accomplish readiness objectives that are challenging but attainable.
- Accomplish realistic RAM requirements to achieve these objectives.
- Identify support and manpower drivers.
- Assign appropriate priority to ILS element requirements in system design tradeoffs.

4.5 THE SE RELATIONSHIP WITH ILS

The ILS concept embodies an analysis of the equipment design with the following objectives:

1. Earlier consideration of support requirements in design and development of new materiel.
2. Improved maintenance support and reduced skills requirements.
3. Optimum balance among support elements achieved by considering possible tradeoffs.
4. All support elements on hand when required.
5. Minimum life-cycle cost for support.

Systems engineering procedures allow for logistics support elements of the system to be determined to the degree of detail appropriate at given points in the life cycle. ILS also uses the system approach but covers the multitude of detailed actions to be accomplished and procedures to be followed to ensure the preparation of detailed management support plans and the development of total support requirements.

The systems engineering process provides the operational data on which the maintenance engineering analysis is based. The maintenance engineering data are used by the systems engineering process of provide or revise requirements for equipment, personnel, facilities, procedural data, and computer programs in appropriate functional areas. Based on analysis of the design characteristics of the proposed system equipment and facility elements and on logistics support concepts, the systems engineering process is employed to define the requirements for and provide descriptions of operations and

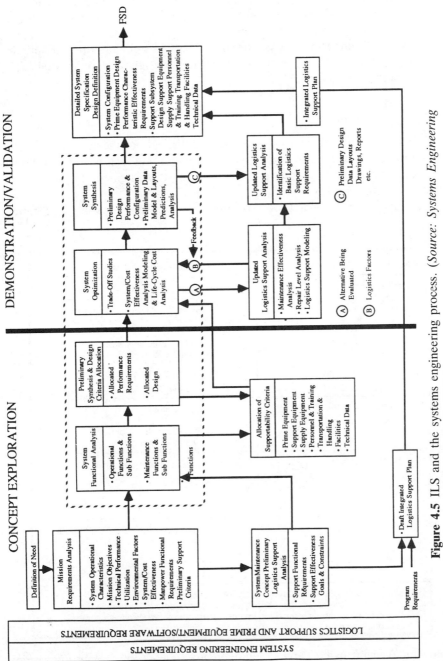

Figure 4.5 ILS and the systems engineering process. (*Source: Systems Engineering Management Guide*, DSMC, 1986.)

maintenance elements. Based on the maintenance engineering analysis data and description of the proposed maintenance elements provided by the systems engineering process, ILS updates the maintenance support plan to provide technical direction and management to the ILS effort. In performing the maintenance engineering analysis portion of systems engineering, the impact of integrated logistics support on equipment design and on system/cost-effectiveness is considered.

Role of Trade Studies

Control of systems engineering trade studies throughout the acquisition cycle is the primary means of executing systems engineering responsibilities. Some applications of the trade study method in the system engineering process are shown in Figure 4.6.

The tradeoff analysis methodology provides a structured, analytical frame-

Acquisition Process Phase	Trade-Off Analysis Function*
Mission Area Analysis	• Prioritize Identified User Needs
Concept Exploration	• Compare New Technology with Proven Concepts • Select Concepts Best Meeting Mission Needs • Select Alternative System Configurations
Demonstration/ Validation	• Select Technology • Reduce Alternative Configurations to a Testable Number
Full Scale Development	• Select Component/Part Designs • Select Test Methods • Select OT&E Quantities
Production	• Examine Effectiveness of all Proposed Design Changes • Perform Make-or-Buy, Process, Rate, and Location Decisions

* In addition, trade studies are used to balance considerations such as producibility, testability, survivability, compatibility, supportability, stability, and reliability during each phase of the acquisition process. Each source selection using trade-off analysis methods.

Figure 4.6 Tradeoff analysis in the acquisition process. (*Source: Systems Engineering Management Guide,* DSMC, 1986.)

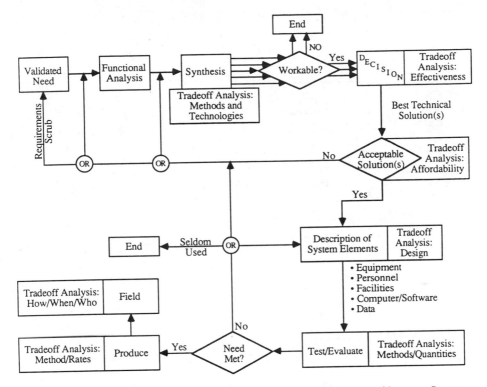

Figure 4.7 Tradeoff analysis in the systems engineering process. (*Source: Systems Engineering Management Guide,* DSMC, 1986.)

work for evaluating a set of alternative concepts or designs. Figure 4.7 shows the basic steps of the analysis process in systems engineering and Figure 4.8, the tradeoff analysis methodology.

ILS-SE Management Interaction

The typical areas of management interaction between systems engineering and ILS program elements are illustrated in Figure 4.9, which shows a broad array of logistics-related functional disciplines in organizational cells on the left side of the linkage diagram and illustrates the complexity of integrating support into the design process of large programs. These linkages must be formally addressed in the contractor's systems engineering management plan (SEMP) and integrated logistics support plan (ILSP). The organization of the ILSP and its subplans is shown in Figures 4.10 and 4.11. A discussion of the SEMP follows.

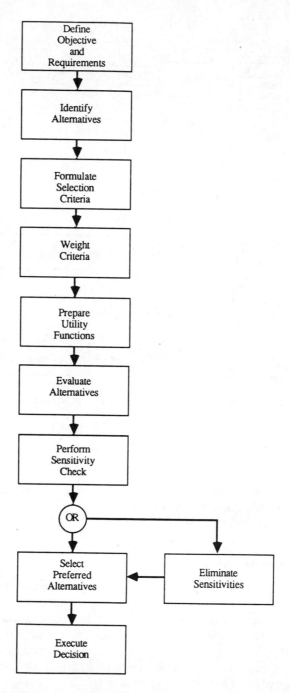

Figure 4.8 Tradeoff analysis methodology. (*Source: Systems Engineering Management Guide,* DSMC, 1986.)

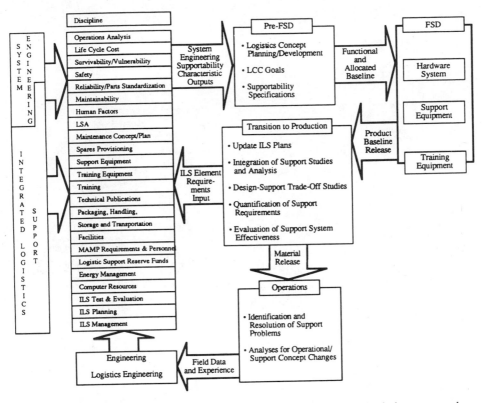

Figure 4.9 Typical contractor systems engineering linkages with logistic support elements. (*Source: Systems Engineering Management Guide,* DSMC, 1986.)

Systems Engineering Management Plan (SEMP)

The basic systems engineering process described earlier in this chapter is applicable to any development process regardless of size or complexity. A SEMP is prepared to accomplish systems engineering as part of development projects. A SEMP is a concise top-level management plan for the integration of all system activities. Its purpose is to make visible the organization, direction and control mechanisms, and personnel for attainment of cost, performance, and schedule objectives. Additionally, the SEMP identifies, defines, and integrates the systems engineering efforts to fully define the system concept; provides a basis for directing and monitoring the systems engineering and integration tasks; develops the planning necessary to monitor and use the results from various experiments and disciplines; and establishes traceability.

The systems engineering decisionmaking process should be fully described in the SEMP. The SEMP contains a detailed description of the process to be used, including the specific tailoring of the process to the requirements of the

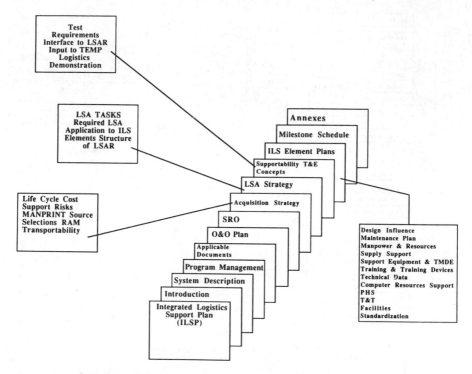

Figure 4.10 Integrated logistics support plan organization.

system and project, the procedures to be utilized in implementing the process, in-house documentation, the tradeoff study methodology, and the types of mathematical and/or simulation models to be used for system and cost-effectiveness evaluations.

Functional and allocated baselines are developed during CE and D/V, respectively. The successful integration of ILS into system design is partially demonstrated by the extent of effective supportability characteristics and requirements in system specifications (Type A) and development specifications (Type B). The format of the Type A specification provides for identification of supportability characteristics (RAM) and logistics concept requirements (maintenance, supply, and facilities). Requirements of the system specification are allocated to Type B development specifications for major configuration items, components, and software. These specifications (Types A and B) are the requirements that control the engineering design activities during FSD. Timely release of major configuration items and their support and training equipment designs is required for scheduling logistics activities such as the preparation of final technical manuals, preparation and processing of provisioning documentation, and development of packaging requirements.

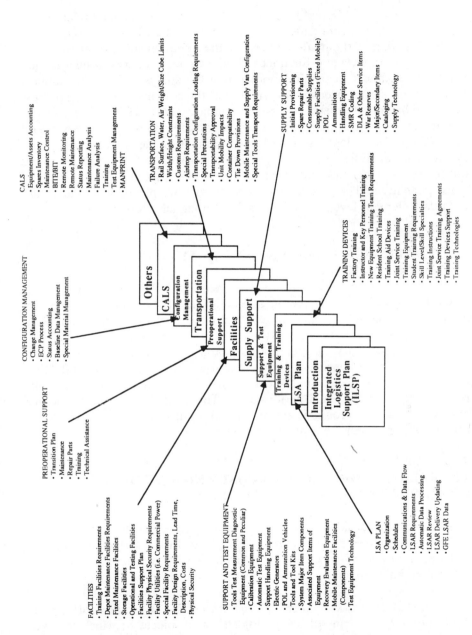

Figure 4.11 Subplans of the integrated logistics support plan.

CALS
· Equipment/Assets Accounting
· Spares Inventory
· Maintenance Control
· BITE/BIT
· Remote Monitoring
· Remote Maintenance
· Status Reporting
· Maintenance Analysis
· Failure Analysis
· Training
· Test Equipment Management
· MANPRINT

CONFIGURATION MANAGEMENT
· Change Management
· ECP Process
· Status Accounting
· Baseline Data Management
· Special Material Management

TRANSPORTATION
· Rail Surface, Water, Air Weight/Size Cube Limits
· Width/Height Constraints
· Customs Requirements
· Airdrop Requirements
· Transportation Configuration Loading Requirements
· Special Precautions
· Transportability Approval
· Unit Mobility Impacts
· Container Compatibility
· Tie Down Provisions
· Mobile Maintenance and Supply Van Configuration
· Special Tools Transport Requirements

SUPPLY SUPPORT
· Initial Provisioning
· Spare Repair Parts
· Consumable Supplies
· Supply Facilities (Fixed Mobile)
· POL
· Ammunition
· Handling Equipment
· SMR Coding
· DLA & Other Service Items
· War Reserves
· Major/Secondary Items
· Cataloging
· Supply Technology

PREOPERATIONAL SUPPORT
· Transition Plan
· Maintenance
· Repair Parts
· Training
· Technical Assistance

TRAINING DEVICES
· Factory Training
· Instructor and Key Personnel Training
· New Equipment Training Team Requirements
· Resident School Training
· Training Aid Devices
· Joint Service Training
· Training Equipment
· Student Training Requirements
· Skill Level/Skill Specialties
· Training Instructions
· Joint Service Training Agreements
· Training Devices Support
· Training Technologies

FACILITIES
· Training Facilities Requirements
· Depot Maintenance Facilities Requirements
· Fixed Maintenance Facilities
· Storage Facilities
· Operational and Testing Facilities
· Facilities Support Plan
· Facility Physical Security Requirements
· Facility Utilities (i.e. Commercial Power)
· Special Facility Requirements
· Facility Design Requirements, Lead Time, Description, Costs
· Physical Security

SUPPORT AND TEST EQUIPMENT
· Tools Test Measurement Diagnostic Equipment (Common and Peculiar)
· Calibration Equipment
· Automatic Test Equipment
· Support Handling Equipment
· Electric Generators
· POL and Ammunition Vehicles
· Tools and Tool Kits
· System Major Item Components
· Associated Support Items of Equipment
· Recovery Evaluation Equipment
· Mobile Maintenance Facilities (Components)
· Test Equipment Technology

LSA PLAN
· Organization
· Schedules
· Communications & Data Flow
· LSAR Requirements
· Automatic Data Processing
· LSAR Review
· LSAR Delivery Updating
· GFE LSAR Data

Others

CALS

Configuration Management

Transportation

Preoperational Support

Facilities

Supply Support

Support & Test Equipment

Training & Training Devices

LSA Plan

Introduction

Integrated Logistics Support Plan (ILSP)

4.6 SE AND LOGISTICS SUPPORT ANALYSIS (LSA) LINKAGES

The contractor's SEMP and LSA plans must define the detailed relationship of both the analysis and data developed under MIL-STD-1388-1 and MIL-STD-499A. A general area of concern is the nonduplication of analysis and data and the requirement for traceability between systems engineering and LSA data elements. It should be noted that the analysis of logistics support functions has always been an integral part of the systems engineering process described by military specifications and standards. MIL-STD-1388-1 provides a contemporary focus on specific requirements. In addition, the LSA record (LSAR) provides a depository for data resulting from the systems engineering maintenance analysis. The requirement for a LSAR would normally be called out for detailed logistics analysis in the SEMP and LSA plan for any detailed logistics analysis. The contractor's SEMP and LSA planning should detail, on a data element level, the relationships among all interfacing data.

4.7 SYSTEMS ENGINEERING–SPECIALTY INTERRELATIONSHIPS

The integration of the engineering specialties into the total engineering program requires close coordination and is a major objective of systems engineering management. In the early stages of a program, systems engineers work in conjunction with operations research and operations analysis specialists to establish appropriate measures of effectiveness. System effectiveness is normally influenced by factors of reliability, maintainability, and other parameters of total system performance. Thus, reliability, maintainability, and other specialty programs are incorporated into the systems engineering process at logical pertinent points in time.

4.8 RELATIONSHIP OF SE TO COST AND SCHEDULE CONTROL MECHANISMS

Cost is a major consideration every time an engineer or designer conceives possible alternatives to an operational problem. In the systems engineering process, life-cycle costs are considered along with the design constraints, reliability, maintainability, safety, and other parameters and engineering specialties.

The use of the work-breakdown structure (WBS) as a framework for visibility provides a means of identifying the small pieces to which life-cycle costs are assigned. Cost control and prevention of cost escalation are directly tied to constraining the design by identifying the risks and life-cycle cost implications every time synthesis of alternatives is accomplished.

The systems engineering process provides the details to which are applied various management and control systems and criteria in order to achieve firm, fixed, and credible estimates of costs. Systems engineering provides an achievable technical foundation on which costs can be based.

4.9   RELATIONSHIP OF SE TO COST AND TECHNICAL PERFORMANCE MEASUREMENT

Systems engineering employs engineering analysis, test, and evaluation to make periodic assessments of the status of the technical program in achieving the performance parameters it has established for the system. Systems engineering provides the technical basis for allocating funds to program tasks against time and for relating the earned value of cost schedule control systems to demonstrated values of performance parameters. The information generated by this relationship enables the manager to efficiently plan and control the technical program for the design, development, test, and evaluation of the system.

4.10   SUMMARY

Thus we have seen that systems engineering is concerned with the process of translating mission, test, production, deployment, support, and operational needs into the most cost-effective mix of design requirements. More specifically, mission needs are translated into engineering functional requirements and those functional requirements are expanded into detailed design requirements. The systems engineering process is iterative and involves functional analysis, synthesis, and optimization with the end goal of achieving a proper balance between economic, operational, and logistics factors. The aspect of logistics support is a significant factor in this process. Systems engineering provides tools that may be applied to integrated logistics support design and recognizes the interplay between design engineering elements such as reliability and maintainability, and the support design elements.

Systems engineering has many interrelationships, such as cost and schedule control mechanisms, other engineering disciplines, cost and technical performance measurement, and LSA. For more information on systems engineering see Benjamin Blarchard's book in this same series on systems engineering.

5

LOGISTICS SUPPORT ANALYSIS

Logistics support analysis (LSA) is a set of systematic and comprehensive analyses performed during the conceptual design and development phases of a program to identify support criteria and operational system support resources. These analyses serve as the interface between design and support planning by establishing the baseline requirements for the incorporation of support criteria into the acquisition process. LSA identifies qualitative and quantitative support requirements; influences system equipment design to ensure supportability; optimizes system support requirements and resources; and provides justification and source data for allocation and/or acquisition of support equipment, spares, repair parts, consumables, technical data, support facilities, personnel, and training. LSA is performed in conjunction with design and interacts with and supports other functional areas to ensure commonality of analysis and nonduplication of efforts. The LSA consists of five basic types of activities: functional analyses, studies–analyses, design analyses, engineering interfaces, and LSAR (documentation) and data verification.

LSA is performed on functions, systems, end items, components, and assemblies that require documentation of operational and logistics support requirements for which the government requires an organic maintenance capability. Many commercial manufacturers that have established repair centers also have product documentation and warranty service. Interface with design engineering provides LSA inputs that integrate ILS into the engineering process.

5.1 OBJECTIVE

LSA is an analytical effort for influencing the design of a system and defining support system requirements and criteria. It is a requirement on all government contracts for hardware. Some commercial manufacturers have implemented an informal LSA to provide better customer support. The objective of LSA is to ensure that a systematic and comprehensive analysis is conducted on a repetitive basis through all phases of the system life cycle in order to satisfy readiness and supportability objectives as shown in Figure 5.1. The selection, level of detail, and timing of the analyses are to be structured and tailored to each system and program phase. The LSA record (LSAR) is designed to be a standardized medium for systematically recording, processing, storing, and reporting data. LSA data are the basis for determining and budgeting for the logistic support resources (maintenance manpower, training requirements, supply support, etc.) required to attain system–product readiness objectives.

5.2 LSA GUIDANCE

What is the LSA process? The LSA process is a planned series of tasks performed to examine all elements of a proposed system to determine the logistic support required to keep that system usable for its intended purpose and to influence the design so that both the system and support can be provided at an affordable cost. Guidelines and requirements for LSA are established by DoDD 5000.39. The guidance for LSA is in MIL-STD-1388-1A, *Logistic Support Analysis,* and the guidance for LSAR is in MIL-STD-1388-2A, *DoD Requirements for a Logistic Support Analysis Record.* The goal of MIL-STD-1388-1A is to provide a single, uniform approach for the military services to conduct those activities necessary to (1) cause supportability requirements to be an integral part of system requirements and design, (2) define support requirements that are optimally related to the design and to each other, (3) define the required support during the operational phase, and (4) prepare attendant data products.

These two military standards have expanded significantly on information previously provided in the earlier version of the 1388 series. Specifically, MIL-STD-1388-1A provides for definitive analysis requirements; front-end analysis requirements are clearly defined, LSA task inputs are identified to include what the government must provide to the contractor, the expected outputs from each LSA task are specified, data item descriptions (DIDs) are referenced, and instructions for tailoring analysis requirements are provided. These significant requirements must be understood by the ILS managers and utilized in the planning and execution of the LSA process. MIL-STD-1388-2A contains added LSAR input data records and associated automated data processing (ADP) routings that provide ILS managers throughout DoD and the defense industry with a standardized means of handling the logistics data.

5.3 LSA REQUIREMENTS

The LSA process is structured to provide early ILS design influence to obtain a ready and supportable system at an affordable life-cycle cost. The LSA

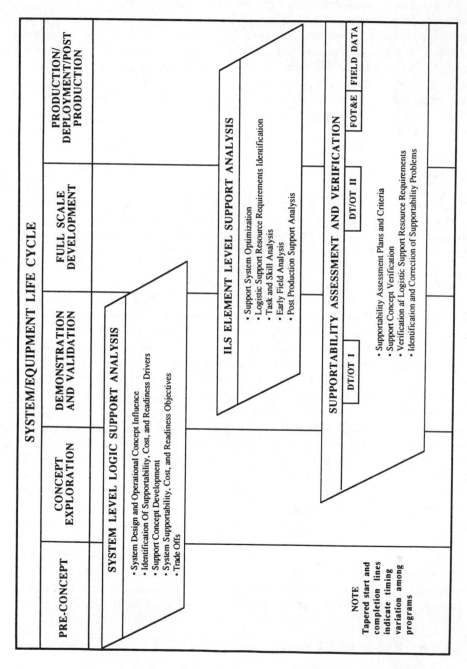

Figure 5.1 Logistic support analysis process objective by program phase.

process comprises a planned series of tasks performed under the direction of the ILS manager. These include examination of all elements of a system to determine the logistics support required to make and keep that system usable for its intended purpose.

Logistics Inputs for Tradeoff Analysis

LSA conducted prior to program initiation identifies constraints and targets for improvement. This early effort provides supportability inputs into systems engineering tradeoffs conducted during the CE phase. Unless timely evaluation of supportability factors is available, the design process will proceed to solidify without logistics inputs.

LSA Task Requirements

LSA requirements, detailed in MIL-STD-1388-1A, consist of five general task sections involving 15 tasks and 77 subtasks. The following paragraphs summarize the five task sections; the MIL-STD should be consulted for details. The time phasing of the total process is shown in Figure 5.1, and an overview of the time phasing and repetitive nature of the individual tasks is provided in Figure 5.2.

Task Section 100— Program Planning and Control. Management of the LSA effort requires the development of a proposed LSA strategy, tailoring, decisions, requirements for the LSA plan, and design reviews, procedures, and schedules. This front end analysis to include LSA planning and management is the responsibility of the program manager. This task section involves the earliest planning activity for an LSA program. Its purpose is to develop a proposed LSA program strategy for use early in the acquisition program and identify the LSA tasks and subtasks that will provide the best return on investment. The LSA plan must effectively document the LSA management structure and authority, what organizational units will be responsible for task accomplishment, and how all tasks are integrated as well as how the results of each task will be used.

Task Section 200—Mission and Support System Definition. The tasks contained in this section identify the operational role and intended use of the new system and establish support resource constraints, readiness and supportability objectives, supportability design requirements, and measures of logistics support. During the early phases of an acquisition program, this analytical task provides the greatest opportunity for the ILS manager to influence the design of the system and its support. The use study is the prerequisite to all other analysis tasks; therefore, it should be initiated in the preconcept phase. In addition, this task section defines the support and support-related design constraints based on support standardization considera-

LSA Task Sections and Tasks	Pre-Concept	CE	DVAL	FSD	Production, Deployment, Post Prod.	Design Changes
Task 100: **Program Planning and Control**						
Early LSA Strategy (101)	X	X	X			
LSA Plan (102)		X	X	X	X	X
Program & Design Reviews (103)		X	X	X	X	X
Task 200: **Mission and Support System Definition**						
Use Study (201)	X	X	X	X		
System Standardization (202)		X	X	X		X
Comparative Analysis (203)	X	X	X	X		
Technological Opportunities (204)		X	X			
Supportability Factors (205)		X	X	X		X
Task 300: **Preparation and Evaluation of Alternatives**						
Functional Requirements Identification (301)		X	X	X		X
Support System Alternatives (302)		X	X	X		
Evaluation of Alternatives & Trade-offs (303)		X	X	X		X
Task 400: **Determination of Logistic Support Resource Requirements**						
Task Analysis (401)			X	X		X
Early Fielding Analysis (402)				X		X
Post Production Support (403)					X	X
Task 500: **Supportablity Assessment**						
Supportability Assessment (Test, Evaluation and Verification) (501)		X	X	X	X	X

Figure 5.2 Acquisition phase timing of LSA subtasks.

tions and provides support-related input to mission hardware and software standardization efforts. This task section also identifies cost drivers and risk involved for new systems. Technological advancements and state-of-the-art design approaches that offer opportunities for achieving new system support improvements are identified. This task also establishes quantitative support characteristics of alternative designs, operational concepts and support related design objectives, goals, and thresholds.

Task Section 300—Preparation and Evaluation of Alternatives. The tasks contained in this section are highly repetitive in nature and are applicable to successive phases of the preproduction part of the life cycle and to production design changes. The tasks are generally performed in sequence, and the process is repeated at increasingly lower levels of the system's WBS as further information is provided by the systems engineering process. This task identifies the operations and support functions that must be performed for each system alternative. It establishes support system alternatives for evaluation, tradeoff analysis, and determination of the optimum system for development.

Task Section 400—Determination of Logistics Support Resource Requirements. This portion of the LSA defines requirements for the ILS elements. The tasks can be general in order to scope requirements or very detailed and produce extensive procedural and parts listing documentation. This task is designed to analyze the required operations and maintenance tasks; highlight resource requirements that are new or critical; define transportability requirements; and identify any support requirements that exceed goals, thresholds, or constraints. This task assesses the impact of new system introduction on existing systems, identifies sources of manpower and personnel skills to meet new system requirements, and determines the impact of failure to obtain the necessary logistics support resources. It also identifies any known or potential postproduction support problems prior to closing down the production lines.

Task Section 500—Supportability Assessment. This task section determines the supportability assessment program to include a test and evaluation program that serves three objectives throughout a program's life cycle: (1) development of logistics test and evaluation requirements as inputs to system test and evaluation plans; (2) demonstration of contractual compliance with design requirements, and (3) identification of supportability problems for corrective action.

5.4   GOVERNMENT AND CONTRACTOR ROLES

There are unique and joint roles for the government (requiring authority) and contractor (performing activity or prime contractor to vendor) and their specialists involved in the LSA tasks. Government management of the LSA process begins in the preconcept phase before the program is formally initiated and continues throughout the life of the system. The preconcept tasks help define initial support criteria and influence efforts of the potential performing activities (competing contractors) through concept exploration (CE) and demonstration and validation (DVAL) and into full-scale development (FSD). The results of early analytical tasks allow consideration of support in the systems engineering definition of system hardware and software, evaluation of alternative designs, and identification of gross resource requirements. Governmental verification tasks begin early in the process using simulation models and baseline comparison systems. Verification tasks continue in conjunction with the contractor throughout the life cycle.

The contractor's LSA tasks are initiated as part of the preproposal effort in preparation for a competitive CE proposal. The contractor's competitive proposal will respond to the specific and tailored request for proposal (RFP) requirements for LSA and will identify the planned approach, key issues to be addressed, and task scope. Therefore, the burden is on the government to accurately describe which ILS issues are to be addressed by LSA. The government must also understand the cost, time, and workload requirements generated by LSA. The integrated logistics support plan (ILSP) is the government's description of the desired logistics program and anticipated maintenance concept, and forms the basis from which the contractor's integrated support plan (ISP) is developed. The ISP may include the contractor's LSA plan.

Following CE contract award, contractor and government logistics management specialists pursue the LSA tasks on a joint basis. The analytical tasks started during the preconcept and CE phases continue and progressively increase in detail as the acquisition program moves into its successive phases in the transition to production and deployment.

The validity of the analysis and the attendant data products must be successfully demonstrated. Results of formal test and evaluation programs and postdevelopment assessments are analyzed by both the contractor and the government, and corrective actions are implemented, as necessary. The process of testing, evaluating, and correcting deficiencies in both the materiel system and its logistics support continues throughout the life cycle. The government ILS manager's supervision of the contractor's LSA role involves assessment of compliance with contractual requirements; conducting periodic reviews; and providing guidance, models and input parameters, government data and/or factors for use studies, and the government-developed joint service LSAR ADP system (or approving an alternative contractor proposed program).

5.5 LSA DOCUMENTATION

LSAR data requirements are detailed in MIL-STD-1388-2A. LSAR data are a subset of the LSA documentation and are generated as a result of performing the LSA tasks specified in MIL-STD-1388-1A. MIL-STD-1388-2A is structured to accommodate the maximum range of data potentially required by all U.S. Military Services in all ILS element functional areas for all types of materiel systems, and throughout the entire acquisition life cycle. This approach permits standardization of formats and data definitions for government-required LSA data. Tailoring the data requirements is a vital part of the ILS manager's role. Some LSA tasks may be recorded through documents such as the contractor's LSA plan (Task 102), alternative support systems (Task 302), and early fielding analysis (Task 402). If task results are to be performed by the contractor for the government, the LSA program statement of work must establish the requirement. ILS managers should be aware of the amount of documentation they may be generating. Only the LSAR data that are required should be ordered by the government. The ILS manager determines what data are needed and when in order to identify the output reports, the LSA data records, and tasks required to meet the program needs and should also ensure that sufficient qualified personnel are available to effectively apply the LSAR data output.

LSAR data may be prepared and maintained manually using the required MIL-STD-1388-2A format and may also be maintained automatically through the use of computer technology or by combining manual and automatic techniques although the preferred method is the use of automated systems. The joint service LSAR ADP system is a standard automated data system developed by the services for use by contractors, if they do not have a validated system of their own. Some contractors who produce commercial products also use the LSAR for their commercial line.

5.6 DATA VERIFICATION

LSA Input Data

LSA is generally performed by a separate ILS group within a contractor's program office or by a supporting activity, and not by the same systems engineering personnel that perform the design, RAM, and so forth. In view of these typical arrangements, the responsibility of ensuring the timely use of appropriate systems engineering input for all analyses falls on the contractor and the government ILS managers. Key personnel in the contractor's ILS activity must be conversant with the language of the associated systems engineering disciplines in order to ensure an effective linkage. The government ILS manager must possess this same capability. All aspects of systems engineering are dynamic and iterative as an ILS and LSA.

LSAR Data Quality Assurance

Quality means accuracy and consistency. Technical accuracy comes from adherence to engineering drawing and specifications. Technical consistency means tasks are described in the same manner throughout the database. LSAR data are employed to define and quantify logistics support resource requirements. The assurance of the qualitative and quantitative validity of these records is required to preclude misidentification and under/overprocurement of support resources.

LSA/LSAR is a conversion process. The LSA input data are converted to detailed LSA records. Some conversions require the application of complex models, for example, repair level analysis and reliability centered maintenance. Others follow detailed procedures prescribed in MIL-STD-1388-2A, such as conversion of reliability estimates of mean time between maintenance actions for spares and repair parts, to estimates of maintenance replacement rates (employed in provisioning computations).

5.7 LSA APPROACH

Initial LSA activity will address primarily the identification of ILS requirements, the development of support concepts, the preparation of preliminary plans, and the establishment of hard-copy top-level data. A preliminary LSAP will be prepared, primarily laying out the government program and establishing a baseline for subsequent automated programs. Actual tailoring and structuring of the LSA will occur later.

5.8 TAILORING LSA/LSAR

Tailoring LSA

The key to a productive and cost-effective LSA program is proper tailoring of the LSA subtasks so that the available resources are concentrated on the tasks that will most benefit the program. Limitations on acquisition funding require that the LSA effort be applied selectively in order to improve hardware design and support concepts, not merely to collect data. The government ILS manager plays a significant role in the tailoring process. Appendix A of MIL-STD-1388-1A provides excellent guidance in tailoring LSA requirements to fit the needs of a specific program. Programs are tailored in several ways. First, they are tailored by task and subtask and by depth of the analysis. This aspect of LSA tailoring involves consideration of amount of new design freedom involved and funds available for investment in tasks, estimated return of investment, schedule constraints, and data and analyses availability and relevancy.

Programs are also tailored in terms of acquisition phase time and required updating. In addition, tailoring can dictate which activity will perform the task or subtask.

Tailoring LSAR

Tailoring LSAR data is mandatory for government programs and ILS managers. The tailoring decisions should be based on (1) the LSA tailoring process, (2) related engineering and ILS element analysis efforts that result in LSAR data, and (3) deliverable logistics products specified by DIDs. In addition, LSAR data records may be tailored to different degrees by hardware level depending on program requirements. Appendix E of MIL-STD-1388-2A provides detailed guidance for tailoring the LSAR.

5.9 LSA/ILS INTERRELATIONSHIPS

The LSA program is interrelated with all the ILS elements and tasks. Figure 5.3 illustrates the interrelationship of ILS elements to LSA and LSAR. The

ILS ELEMENT	BASIC SOURCES	LSA	L S A R	LSA ADP REPORT	OTHER STUDIES	COMPUTATIONAL MODEL
Maintenance Planning	• Service maintenance system • Organizational and operational concepts • Test field and historical data • R&M predictions	• Repair level analysis • FMECA • RCM • Task analyses • Survivability analyses • Analysis of existing manpower sources	A B B1 B2 C D	LSA-003 Maintenance Summary LSA-004 Maintenance Allocation Summary LSA-016 Preliminary Maintenance Allocation Summary Chart LSA-002 Maintenance Plan		Maintenance simulation models
Manpower and Personnel Maintenance Requirements	• R&M predictions and modeling • Test data • Field data • Historical data	• Task analyses • Survivability analyses	A C D G	LSA-001 Direct Annual Maintenance Man-Hours LSA-002 Personnel and Skill Summary	• Available manhours • Indirect productive time • Battle damage simulations • COEA	Manpower models
Supply Support	• Reliability predictions and modeling • Test Data • Field data	• FMECA • RLA • Task analyses • Survivability analyses	B C D H	LSA-036 Provisioning Requirements	• Battle damage simulations • Cannibalization policy	"Spare to availability" models Sortie generation models
Support Equipment	• Lists of standard support and test equipment • GSA/DLA tool specifications	• Task analyses	B C D	LSA ADP REPORTS LSA-005 Support Equipment Utilization Summary LSA-007 Support Equipment Requirements LSA-020 Tools and Test Equipment Requirements		

ILS ELEMENT	BASIC SOURCES	LSA	L S A R	LSA ADP REPORTS		
Technical Data	• System functional requirements • Production documentation • Technical Manual standards and specifications • Descriptions of personnel capabilities (target audience)	• FMECA • RCM RLA • Task analyses • Survivability analyses	B C D H E1	LSA-015 Sequential Task Description LSA-020 Tool and Test Equipment Requirements LSA-029 Repair Parts List LSA-030 Special Tool List LSA-040 Components of End Item List LSA-041 Basic Issue Items List LSA-042 Additional Authorization List LSA-043 Expendable/Durable Supplies & Materials List LSA-050 Reliability Centered Maintenance Summary LSA-055 Failure Mode Detection Summary		
Training and Training Support	• Existing personnel skills capabilities, and programs of instruction • Training devices available	• Task analyses		LSA-01 Requirements for Special Training Devices LSA-014 Training Tasks List		
Computer Resources	• System functional requirements • Test reports • Field reports	• Post production support analysis				
Facilities	• Facilities available • Funding Constraints • Organizational and Operational Maintenance Concept	• Task analyses	C D	LSA-012 Requirements for Facilities		
Packaging, Handling, Storage, and Transportation	• Existing transportation system and capabilities	• Task analyses	H J	LSA-025 Packaging Requirements Data LSA-026 Packaging Developmental Data		

Figure 5.3 Relationships of integrated logistics support elements to LSA and LSAR. (*Source:* MIL-STD-1388-1.)

LSA database is the source data for ILS task accomplishment. The LSA also integrates systems engineering and ILS as shown in Table 5.1.

The systems engineering process encompasses, either directly or indirectly, the planning and integration of many functional areas, activities, and products. The intent of LSA is to bring together design and support concepts during the systems engineering steps of synthesis and tradeoff analysis to influence the design so that the end result will be reduced quantities, size, weight, complexity, and cost of the ILS elements. When design is properly influenced by LSA, one result is a more cost-effective ILS resource as illustrated by Figure 5.4.

Maintenance Planning. This stems from the maintenance concept derived from system specifications, system requirements, and cost economic analysis. The maintenance concept and plan make up the keystone of the ILS hierarchy. It is the principal thread or vehicle trying the LSA together. The maintenance concept delineates:

- System repair policies.
- Repair levels and organizations.
- Maintenance environment (facilities and location).
- Measure of support effectiveness (MTBF, MDT, etc.).

Manpower and Personnel. Manpower levels and personnel skill requirements are directly influenced by the LSA and are based on task analysis, level of repair analysis, and many of the other ILS and nonILS elements. The LSA pulls this data together as part of the system baseline.

Supply Support. The spares quantities, levels, and support procedures are a direct result of the LSA process.

Support and Test Equipment. Requirements for this equipment are identified through systems engineering and the LSA process. Maintenance task analysis and the resulting data are used to develop and document support and test equipment requirements. Once identified, the requirements are documented and data are input into the LSAR data.

Training and Training Devices. Data from the LSAR are used to develop training, training aids, and personnel manning level requirements for operations and maintenance.

Technical Data and Publications. LSA data, as documented in the LSAR, provide essential input in the development of technical data to be used in publications.

TABLE 5.1 LSA Integrates Systems Engineering and ILS

Systems Engineering Activity	Related LSA Task	ILS Product
Design and configuration management	Identification of components, LSA control number, comparative analysis	Maintenance planning, cataloging, provisioning
Reliability data (e.g., component MTBF)	Repair level analysis, maintenance MH requirements, design interface, task and skill analysis	Provisioning, maintenance planning, manpower and personnel, training
Maintainability data (e.g., component MTTR)	Design interface, repair level analysis, task and skill analysis, design support criteria	Scheduled maintenance services, provisioning, manpower and personnel, training, STE
Standardization and parts control	Design interface, technological opportunities, tradeoffs, design support criteria	Parts procurement, requirements determination, technical data PHS&T
Failure-modes effects and criticality analysis	Design interface, reliability centered maintenance, maintenance MH requirements	Maintenance planning, provisioning, parts procurement, technical data
Life-cycle costing	Tradeoffs, repair level analysis, alternate support plans	Provisioning, maintenance planning, requirements determination, STE
Human-factors engineering	Design interface, support system alternatives, task analysis	MANPRINT, manpower and personnel, training and training devices, computer resources support, STE
Safety engineering	Design interface, support system alternatives, tradeoffs	MANPRINT, maintenance planning, facilities, training

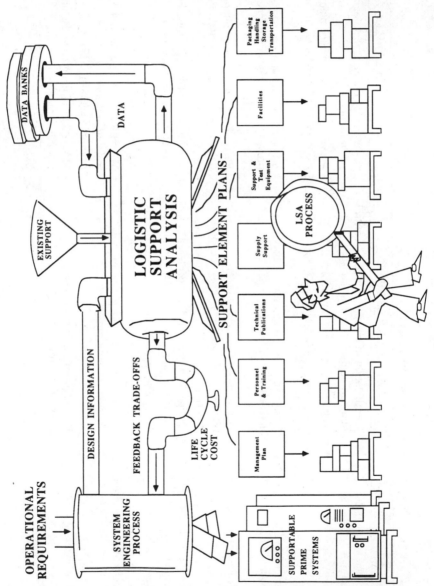

Figure 5.4 The logistics support analysis process.

Computer Resource Support. The computer resources encompass both embedded and stand-alone systems or programs. The ILS elements are applicable to both hardware and software. LSA is performed on the embedded hardware, and the resultant data are entered into the LSAR. The stand-alone resources and the software programs are analyzed for their impact and effect on the system's operation and maintenance procedures. The results of these analyses are applied to the processes used, but there are no provisions for their specific documentation or entry into the LSAR. LSA results can be used to determine computer resource requirements and limitations, to plan for postdeployment software support, and to prepare the computer resources management plan. The ILSP will describe manpower and personnel requirements for developing and fielding computer resources and the training required to operate and maintain these resources, based on LSA data. The plan for acquisition, test, and evaluation of software and software support will also be identified in the ILSP, and the method for detecting and correcting software errors will be identified in the LSA.

Packaging, Handling, and Storage. Specific requirements for packaging, handling, and storage items will be derived from the LSAR. These include equipment physical dimensions, container requirements, coding and labeling, storage and space requirements, preservation and packaging criteria, and handling procedures and constraints. A packaging, handling, and storage plan will be developed using LSAR data and will detail these requirements and their development procedures.

Transportation and Transportability. Like the packaging, handling, and storage area, transportation and transportability requirements will be derived from the LSAR. This data will form the basis for transportation and transportability planning and are used to ensure compatibility with existing systems or modes.

Support Facilities. Information regarding support facilities is derived as a result of the analysis of maintenance, operations, and training tasks, plus space estimation for repair functions, maintenance operations, and spares storage. Justification for support facilities is documented in the LSAR and its input/output records. Facilities engineers must be made cognizant of support facility requirements derived from the LSAR as they occur.

Design Influences. Design influences cross all lines of the LSA process, providing an interface between system design and system requirements—for example, maintainability engineering influences design by ensuring that maintenance support is considered. While often measurable (e.g., mean time to repair), maintainability programs have many subjective factors such as limiting maintenance personnel requirements, special tools, and maintenance skills; standardizing parts; maintenance expenditure limits; and providing for ease of access and fault isolation.

5.10   LSA INTERFACES

Engineering Interface

This serves as a controller of actions to ensure that the following engineering inputs are received and considered:

- Design
- Reliability
- Maintainability
- Human-Factors engineering
- Standardization
- Safety
- Packaging, handling, and storage
- Transportation

Reliability and Maintainability. The contractors's reliability and maintainability programs generate LSA applicable data. This includes the mean time between failure (MTBF) of configured items and piece part failure rates and modes. Maintainability parameters developed are input both qualitatively and quantitatively into the design process for use in tradeoff and risk analyses and in the development of support capabilities. Results of the maintainability program and failure mode and timeline analyses are documented in the LSAR.

Human-Factors Engineering. Data gathered during this function are also fed into the LSA database, and LSAR reports provide a feedback loop to complete the interface. Manpower levels and personnel skill requirements are directly influenced by human-factors engineering. This influence is also reflected in the LSAR reports.

Production. LSA interfaces with production by providing updated information gleaned from field tests, analysis, and other program interfaces concerning design, maintainability, and human-factors–safety engineering. This permits correction of problems noted in prototypes before production runs are made. LSA, through provisioning, also alerts production on the spares quantities needed during production runs. The LSA also furnishes production with the packaging requirements for components. Production data are fed back into the LSA process to update the technical data for publication and training using as-built information.

Support Costs. These costs are derived from the LCC process, interfacing with ILS and LSA activities. These costs are obtainable as the programs become more defined and come on line. Initially these costs will be gross estimates based on LCC model results. LSA data are essential for the costing program, and cost data are used in LSA.

Quality Assurance. LSA interfaces with the quality-assurance function through data exchange and organizational structure. LSA data are fed back to quality assurance concerning failure rates, which, in turn, affects maintainability and provisioning in the LSAR.

Integrated Logistics Support Management Team (ILSMT). The LSA program, as a subset of ILS, interfaces with the ILSMT by using government- and contractor-provided technical data in the form of specifications, guidelines, allocations, and requirements as a baseline for analysis. The LSA program provides the ILSMT with technical data packets at LSA reviews that include training programs, publications, provisioning lists, and maintenance plans resulting from LSA tailored specifically for the program.

System Test–Evaluation. Throughout the system's test and evaluation, LSA data gleaned from the LSAR are used to establish system support packages. Data from the test sites on hardware–software performance, support systems procedures, and training are fed back into the LSA for inclusion, validation, and update of the LSAR.

Subcontractors' and Vendors' LSA Participation. The prime contractors ensure that LSA data on subcontractor or vendor items are included as part of the LSA process. If the subcontractor or vendor is too small to have the expensive LSA process, then the prime contractor still has responsibility for ensuring the integration of data into the LSAR.

Safety. Like human-factors engineering, safety engineering is a discipline that interfaces with the LSA. The results of safety analysis are input into the LSA process and selected data are input into the LSAR; conversely, LSAR data are fed back to the safety engineer for use in analysis and program update.

Deployment–Site Activation. This interface interacts with ILS very heavily and LSA plays a key role. LSA transportation–transportability, packaging, handling, and storage data are critical to the deployment function; facilities, publications, support equipment, spares, training, and maintenance procedures are also critical to site activation.

5.11 LSA AND THE WORK-BREAKDOWN STRUCTURE (WBS)

The WBS is a hardware–software–task-oriented structure. It is the management and technical framework for the project office. Project management updates, maintains, and aligns the WBS on an as-required basis. The WBS and functional group codes are interrelated for rapid transition and parts identification.

The LSA control number uniquely identifies the location of every usage of each part in the end-item breakdown. Experience has shown that the most efficient method of accomplishing this is to use the functional group code numbering system established in the master file during maintenance task analysis. Since the LSA control number and the functional group code (FGC) number both have 11-character field lengths, it is a simple process to use the same numbering system for both applications. Basically, the LSA control number will be the FGC number of the next higher assembly containing the item being identified. The system lends itself to the easy correlation of each part into its next highest assembly and the related LSA data for that item available in the master file as shown in Table 5.2.

5.12 LSA AND THE MANAGERIAL FUNCTIONS

Actions required to manage the LSA process include, to some degree, all the functions of the manager, for example, *planning,* which identifies LSA objectives, and then *scheduling* the actions required to achieve them. Planning is also *decisionmaking* as it involves selecting among alternatives. *Organizing* involves establishing an LSA organizational structure, identifying LSA tasks and/or subtasks; assigning tasks to specific organizational units, delegating authority, and providing for the coordination of authority relationships both within the LSA program structure and with related program structures. *Staffing* involves continuous manning of the technical and management positions identified in the organizational structure. *Directing* provides motivation, communication, and leadership. *Controlling* measures performance, corrects negative deviations, and ensures accomplishment of plans.

5.13 LOGISTICS SUPPORT ANALYSIS RECORD

LSAR Purpose

The purpose of the LSAR is to provide a uniform, organized, yet flexible, technical database that consolidates the engineering and logistics data necessary to identify the detailed logistics support requirements of a system. The LSAR database is used to:

- Determine the impact of specific design features on logistics support.
- Determine how the proposed logistics support system affects system RAM characteristics.

TABLE 5.2 Functional Group Code (LSA) Control Numbering System

Capability	1	2	3	4	5	6	7	8	9	10	11	Level
99[a]	1	6	—	—	—	—	—	—	—	—	—	System
99[a]			0	4	—	—	—	—	—	—	—	Subsystem
24 (10)[b]					A	—	—	—	—	—	—	Major group
24 (10)[b]						A	—	—	—	—	—	Functional group
24 (10)[b]							A	—	—	—	—	Removable assembly
576 (580)[b]								A	A	—	—	Module
576 (580)[b]										A	A	Piece part

FGC (LSA) Structure (Alternate)

Capability	1	2	3	4	5	6	7	8	9	10	11	Level
24 (10)[b]	A	—	—	—	—	—	—	—	—	—	—	System
24 (10)[b]		A	—	—	—	—	—	—	—	—	—	Subsystem
24 (10)[b]			A	—	—	—	—	—	—	—	—	Major group
24 (10)[b]				A	—	—	—	—	—	—	—	Functional group
24 (10)[b]					A	—	—	—	—	—	—	Removable assembly
576 (580)[b]						A	A	—	—	—	—	Module
576 (580)[b]								A	A	—	—	Piece part
576 (580)[b]										A	A	Not used

[a]Uses numeric only and parallels the WBS.
[b]Increased capability if alphanumerics are used in these positions.

- Influence the design.
- Provide input data for tradeoff analyses, life-cycle cost studies, and logistics support modeling.
- Exchange valid data among the functional organizations.
- Provide source data for the preparation of logistics products.

LSAR Data Process

The LSA process is conducted on an iterative basis through all phases of the life cycle. The LSAR, as a subset of the LSA, documents the detailed logistics support data generated by the LSA process. LSAR data resulting from each iteration of the LSA tasks is used as input to follow-on analyses and as an aid in developing logistics products.

The LSAR data may be maintained manually on the LSAR data records or equivalent format approved by the requiring authority. Automation of the LSAR data is not mandatory but should be a consideration in tailoring the LSAR data effort.

Updates

Updates to the LSAR are based on engineering releases, notice of revisions (NORs), and official direction. LSAR data will be revised where discrepancies are submitted by written request, approved by the government, and do not violate the intent of the engineering. When discrepancies are noted in engineering, an engineering change proposal (ECP) must be submitted. This LSA/LSAR update cycle is shown in Figure 5.5.

Cost-Effective LSAR

Two types of features contribute to the cost-effectiveness of an LSAR system: additive features and basic features. An LSAR system must have the basic features of adaptability and growth, increased management control, reduced labor intensity, and system documentation to be cost-effective. It should also have some additive features such as automatic data exclusion, data entry efficiency, engineering logic edits, mass copy, mass delete, model interfaces, optional flow, program key functions, reference library, review cycle support, user access control, word processing, superseded item usage safeguard, and used item delete safeguard. The value of each additive feature depends on how the lack of each affects the LSA program. The value of the additive features is a function of LSA program requirements and changes from program to program. The value of the basic features remains constant for all LSA programs.

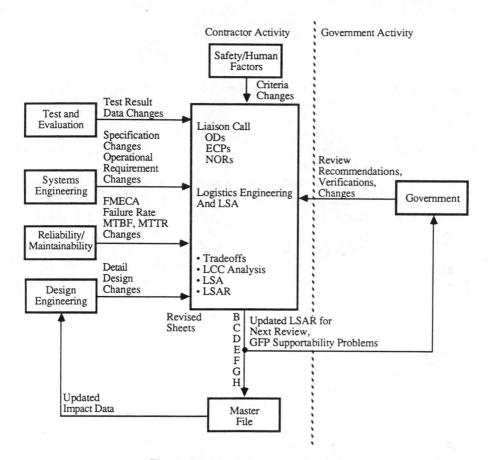

Figure 5.5 LSA/LSAR update cycle

5.14 SUMMARY

Logistics support analysis (LSA) is the iterative analytical process that identifies the logistics requirements necessary for the support of the new system.
The LSA uses quantitative methods to identify the logistics standards that
should be reflected in system design. Quantitative methods are also used in
the evaluation of system design alternatives as well as the needed support
factors provided by the various logistics elements.

The LSA process is used primarily by the government and its contractors.
It is an expensive program that requires financial commitment up front to
achieve reductions in future costs throughout the system life cycle. The primary objective of the LSA is the development of logistics resources that
optimize system support at an affordable cost. The successful implementation
of an LSA program demands and depends on the establishment of a free and

continuing information interchange between the LSA staff and the engineering disciplines. This interchange is essential to the achievement of an optimum balance between the system and its operational capability, logistics requirements, and cost.

In achieving the objective of LSA, the program must exert influence on system design and development that is based on logistics considerations. Initial analysis efforts are directed toward an evaluation of the effects of alternative approaches. Initial logistics requirements or goals are established on the basis of known constraints and logistics risks. Subsequent analysis is directed toward refinement of initial assumptions.

The effective LSA program identifies both quantitative and qualitative logistics support issues and requirements. Alternative concepts and operating parameters are translated into support concepts, support costs, and supportability.

LSA data resulting from the LSA process are stored in the logistics support analysis record (LSAR). The LSAR is the mode for communicating information for risk analysis, tradeoff studies, effectiveness studies, and life-cycle cost projections.

The LSA process can be divided into two general areas: (1) support analysis and (2) supportability assessment and verification analysis. System level support analysis is characterized by (1) the analysis of existing systems that are comparable; (2) the determination and analysis of personnel, cost, and operational readiness drivers; (3) the identification of targets of opportunity in the logistics arena; and (4) the tradeoffs between alternative support concepts. The supportability assessment and verification analysis phase of the LSA process extends throughout the product life cycle. The purpose of this is to demonstrate, within established limits, the validity of support analysis and of the support products derived from that analysis.

The application of LSA is mandatory for all DoD materiel systems. Their applications must be tailored to the requirements of each acquisition to ensure cost-effective implementation. LSA programs for major systems are relatively costly. These costs are most warranted when LSA is used as the integrated source and record for development of ILS planning and definition of ILS products. The program that provides front-end funding for LSA and other ILS activities is more likely to be a successful program.

PART 3

NEW DIMENSIONS IN LOGISTICS

Lifetime expenditures for systems have been continuously increasing, and the availability of those systems has been decreasing. In other words, the government is paying more over the long run for systems that are less reliable and therefore less available. It is widely believed that manufacturers pay inadequate attention to reliability and maintainability because most contracts fail to burden the contractor after delivery of the system. In addition, previous efforts to provide contractors incentives to design-in reliability and low-cost support features have generally been unsuccessful. As a result, the Department of Defense has initiated an effort to address contractor motivation to reduce support costs and improve product reliability through the use of warranties.

A brief history of the warranty program and military warranty policy is included. A detailed discussion of the warranty classifications with supporting data tables provides the reader with a more comprehensive understanding of the government warranty program. There is an extensive discussion of warranty development in the system life cycle and system specifications as well as the how-to of administering the warranty.

Chapter 7 discusses computer resources in the logistics environment. Computer resources have the potential to become the most demanding of all support requirements. Now, most government agencies require that a computer resources plan be developed and maintained during a system's life cycle. Computer resources life-cycle management documentation provides information to allow the acquisition manager to understand and be able to meet sound software development practices for a software system acquisition.

Computer-aided acquisition and logistics support (CALS) is a DoD and industry strategy to enable and accelerate the integration of digital technical information for system acquisition, design, manufacture, and support. To achieve benefits from CALS, a phased strategy has been planned. Phase I will replace paper document transfers with digital file exchanges and begin process integration. In parallel, technology is being developed for Phase II, which involves substantial redesign of current processes to take advantage of a shared database environment beyond the 1990s.

In addition to the CALS effort, the military services are faced with the challenge of developing modern systems that rely on critical computer resources. These computer resources must be managed a part of the total life-cycle process to include software development. This is accomplished through the development of a computer resources management plan, the establishment of a computer resources working group, and the development of a software development plan.

The CALS program has made significant progress. Initial standards have been developed as well as an applications testing program. Both DoD and industry are collaborating in the effort. Ultimately, CALS will result in lower system life-cycle cost, shortened acquisition times, and improvements in reliability, maintainability, and readiness.

Phase II of CALS is the most controversial for it allows the Pentagon to develop standards that allow DoD to delve into contractors' computers. The biggest danger is the military's inclination to attach unrelated pet projects to hot programs. If this happens, and CALS becomes an umbrella for disparate projects, it could be diluted or weakened.

MANPRINT (manpower and personnel integration) is a hot item these days in DoD weapon system acquisitions, although the term "MANPRINT" is used primarily in the U.S. Army. The U.S. Air force calls it IMPACT, and the U.S. Navy is still hanging on to "HARDMAN." It is a comprehensive management and technical program to enhance human performance and reliability in the operation, maintenance, and use of system and equipment. This is accomplished by evaluating human resource goals and constraints in six areas called "domains": manpower, personnel, training, human factors, system safety, and health hazards. The first four domains directly influence human performance and human reliability. The impact of system safety and health hazards is more indirect but can also degrade total system performance if they are overlooked during system development. The impact of considering the soldier in system design and development is far-reaching. If done properly, system performance will be enhanced; manpower, personnel, and training resources will be used more effectively; force capability will improve; and unit readiness will increase.

Chapter 9 covers all aspects of configuration management. Configuration management, an accepted and approved management discipline, identifies the functional and physical characteristics of an item during its life cycle, controls changes to those characteristics, and records and reports change processing and implementation audits; thus traceability is achieved for systems. One of the rewards of effective configuration management is improved supportability, including updated manuals, identified spares, identical and interchangeable equipment, and known configuration.

In Chapter 10, nondevelopmental items are discussed. Nondevelopmental items (NDI) are systems or components available from a variety of sources requiring little or no development effort. The acquisition process for an NDI is not a separate process, but a tailoring of events within the materiel acquisition process and should be one of the first alternatives considered for a solution to a materiel need.

NDI offers several major benefits to include reduced fielding time, reduced R&D costs, and utilization of state-of-the-art technology to satisfy user needs. But there are also major challenges to accomplishing NDI programs. Essential integrated logistics support activities normally accomplished in preproduction phases have to be accelerated and may require increased up-front funding. In addition, the standard logistics systems may have to be supplemented by interim contractor support or other innovative logistics strategies. Safety deficiencies may have to be evaluated to determine whether they pose an acceptable risk. And because of the shorter acquisition cycle, internal service must be expedited or tailored to support the NDI strategy.

6
WARRANTIES IN GOVERNMENT CONTRACTS

In the absence of an offsetting decline in acquisition costs, the lifetime expenditure required by typical modern systems—life-cycle cost—has been rapidly increasing. The availability of crucial systems is poor. Operations and maintenance (O&M) costs are rising. Two important factors contributing to this poor availability and high O&M costs are inadequate system reliability and maintainability. It is widely believed that manufacturers are not sufficiently concerned with reliability and maintainability of their products because most system contracts fail to burden the contractor with a major portion of the life-cycle costs.

Efforts to provide contractors incentives to design-in reliability and low-cost support features generally have not been successful. Previous attempts have included the inclusion of target mean time between failures (MTBF) in performance specifications along with target unit production cost goals and solicitation of life-cycle cost (LCC) estimates from competing contractors to incorporate them into the source selection decision process. The weakness of both methods is the absence of enforcement features.

In general, traditional contractual incentives, such as award fees based on achieving acquisition cost targets, have not had an appreciable effect on contractor motivation. DoD recently began experimenting with several contractual terms that seek to address contractor motivation specifically to reduce support costs and improve product reliability. This is the focus of this chapter on warranties in government contracts.

6.1 A BRIEF HISTORY OF MILITARY WARRANTY

1960 to 1980

In 1964 the Armed Services Procurement Regulation (ASPR) added regulations on the use of warranties. Those regulations, which have been updated periodically, have generally been interpreted to mean that the use of an extensive, long-term warranty was to be the exception rather than the rule.

Early government controls against acquiring defective material included warranty control against latent defects. In the late 1960s and early 1970s, more extensive warranty forms were tried, such as on the Navy F-4 gyro [failure-free warranty (FFW)], and the Air Force ARN-118 TACAN [reliability improvement warranty (RIW)]. Indications of potential success for these selected programs encouraged the Office of the Secretary of Defense and the services to enter into a "trial period" for more extensive warranty forms, particularly RIW and MTBF guarantees. During the mid-1970s, these types of warranties were secured on such equipment as line replaceable units (LRUs) on the Air Force F-16 and the Army ARN-123 radio. In addition, a dialog was started between industry and DoD concerning the warranty issue as newer and more extensive warranty forms were being implemented by all the military services. The services supported research studies to evaluate those warranty applications and to develop analysis and implementation tools.

Warranty Initiatives in the 1980s

The successful use of such warranty forms as MTBF guarantees and RIW during the 1970s provided a basis for extending warranty applications to a broader class of programs. By the beginning of the 1980s, the use of warranties in the acquisition of military systems became a "standard" option, but it was applied only selectively and usually required a special effort on the part of the program office to implement.

In 1980, the Air Force issued the first "Product Performance Agreement Guide," which provided a summary of the features of various forms of warranties that could be used in military procurement. In 1982, the Department of Defense issued a set of initiatives, which became known as the "Carlucci Initiatives," to improve and streamline the acquisition process. The initiatives included warranties as one of the means of achieving desired levels of system reliability and maintainability.

Congressional interest in warranty as a means of ensuring acceptable field performance started with the passage of Public Law 98-212, which was part of the 1984 Defense Appropriations Act. That law, implemented by DoD policy guidance in March 1984, mandated that warranties be included in the production contracts. The law, with some modifications, was made permanent by inclusion of the 1985 warranty law in the 1985 DoD Authorization Act. Passage of these laws led to renewed activity in warranty research.

The key features of Public Law (PL) 98-212 are its emphasis on warranting the entire system, its emphasis on having the prime contractor correct defects, and its specific emphasis on warranting performance. PL 98-212 also involves remedies for corrections, exemptions, and waivers to the law. The remedies for correction are essentially for the prime contractor to replace or repair defective parts or if the prime cannot do so, then the prime must pay someone else to repair and/or replace. Exceptions to PL 98-212 occur only for those weapons systems or components furnished to the prime contractor by the government. Waiver authority can be obtained only under two conditions. First the warranty must be determined not to be cost-effective and both the House and Senate Armed Services and Appropriations Committees must be notified.

This new public law left some areas unaddressed and thus caused some confusion. As a result PL 98-212 was updated through the passage of Public Law 98-525.

PL 98-525 was passed in October 1984 as part of the fiscal year 1985 appropriations act. This law was more specific as to when the systems level warranty applies and when Congress is to be notified of waivers. Weapon system was redefined as "Equipment that can be used directly by armed forces to carry out combat missions." PL 98-525 expanded warranty coverage to include design and manufacturing conformity. This was in addition to the warranties for materials and workmanship and essential performance conformity required by PL 98-212. Table 6.1 compares PL 98-212 and PL 98-525. Table 6.2 summarizes the essential features of the 1985 law.

6.2 MILITARY WARRANTY POLICY

U.S. Army

It is evident from the policy statements that the single most important aspect of the warranty acquisition process is to tailor the warranty so that it has minimal impact on standard army logistical procedures. Army warranty policy is provided in Army Regulation (AR) 700-139, *Army Warranty Program Concepts and Policies*. This regulation, although focused on PL 98-525, also applies to nonstatute warranties. It establishes responsibilities, defines policy and procedures, and standardizes the information, fielding, execution, and compliance for all warranties. Additional policy for acquisition of warranties is contained in Army Materiel Command Supplement 1 to AR 700-139 and AR 702-3, *Reliability Improvement Warranties*.

AR 702-13, *Army Warranty Program*, was published to provide the policies, procedures, and guidance necessary to establish an effective program of warranty management within the U.S. Army. The regulation contains a decision logic tree (Figure 6.1) designed to assist the individual who must decide whether to acquire a warranty. It is basically army policy that warranties will not normally be acquired. AR 702-13 indicates that items for army use should be acquired with warranties only when a warranty is in the army's best interest. Often this is taken to mean that a monetary advantage should accrue to the government. However, cost is not the only element considered when determining what is in the best interest of the government. The procurement team should reject a warranty unless it is demonstrated that it would be cheaper

TABLE 6.1 Comparison of Warranty Laws

Item	Section 794 of '84 Appropriations	Section 2403 of '85 Authorizations
Application	All production	Option—may exempt initial production from performance warranty
Dollar threshold	None	Applies to programs with total cost exceeding $10M or with a unit cost exceeding $100K
Remedy	Manufacturer repairs	Manufacturer may repair or replace or reimburse government for repair or replacement at government election
Material–workmanship warranty	Required	Required
Performance warranty	Required for all mandatory performance requirements	Required for defined performance requirements
Design–manufacturing warranty	Not applicable	Required for all items
Congressional notification of waivers	Required for each waiver	Required for all major programs—all other consolidated

Source: Warranty Handbook, Defense Systems Management College, FT Belvoir VA, June 1986.

TABLE 6.2 Summary of 1985 Warranty Law

Factor	Definition	Description
Coverage	Weapon systems	Used in combat mission; unit cost greater than $100,000, or total procurement exceeds $10,000,000
Warrantor	Prime contractor	Party that enters into direct agreement with U.S. to furnish part or all of a weapon system
Warranties	Design and manufacturing requirements	Item meets structural and engineering plans and manufacturing particulars
	Defects in material and workmanship	Item is free from such defects at the time it is delivered to the government
	Essential performance requirements	Operating capabilities or maintenance and reliability characteristics of items are necessary for fulfilling the military requirements
Exclusions	GFP, GFE, GFM	Items provided to the contractor by the government
	Essential performance requirements for items not in mature full-scale production	The first tenth of the total production quantity or the initial production quantity, whichever is less

TABLE 6.2 (*Continued*)

Factor	Definition	Description
Waivers	Necessary in the interest of national defense; warranty not cost-effective	Assistant Secretary of Defense or Assistant Secretary of the Military Department is lowest authority for granting waiver, prior notification to House and Senate committees required for major weapon system
Remedies	Contractor corrects failure at no additional cost to U.S.; contractor pays for reasonable costs for U.S. to correct	Other remedies may be specified: contract price may be reduced
Tailoring	Exclusions, limitations, and time duration	Specific details to be negotiated
	Dual-source procurements	Relieve second source from guaranteeing essential performance requirements for initial product delivered
	Extensions	Extend coverage and remedies as deemed beneficial

Source: Warranty Handbook, Defense Systems Management College, Ft. Belvoir, VA, June 1986.

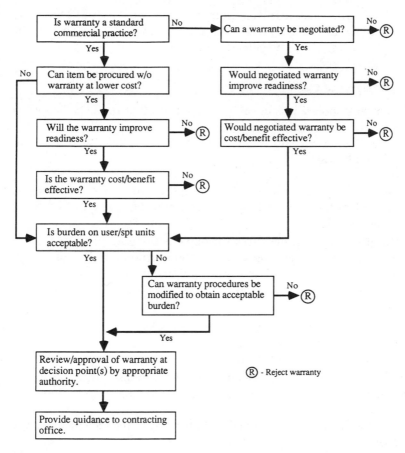

Figure 6.1 Warranty acquisition and fielding logic tree. (*Source:* AR 702-13.)

for the government to procure a warranty than to make the same repairs "in-house," or that the manufacturer–dealer can repair the equipment better, or more rapidly, than the in-house support system.

According to AR 700-139, warranty tailoring is intended to protect the army from the costs and frequency of systemic failures, enact responsive remedies for failures of significant operational impact, minimize or eliminate warranty execution tasks, and become one of the methods used to require the contractor to fulfill the obligation of providing a quality product. Such factors as technical risk, contractor financial risk, and program uncertainties are listed as potential reasons for limiting the contractor's liability under the terms of the warranty. It is also stated that it is not Department of Defense policy to include contractor liability for loss, damage, or injury to third parties.

Warranty coverage for centrally procured equipment should generally include both coverage for failures of individual items and coverage for system defects; the latter may involve a potential redesign liability. The duration of a warranty should be between 10 and 25 percent of the expected life and generally not less than one calendar year of operation.

U.S. Navy

Navy warranty policy was developed to ensure that the U.S. Navy obtained and administered cost-effective warranties and used them to enhance the reliability of systems, subsystems, and materials. Navy warranty policy includes the following:

- Policy on warranty requirements, cost–benefit analysis, acquisition planning, identification marking, failure reporting, period of coverage, and supply policies.
- Development of procedures for implementing warranty terms and conditions, establishment of warranty administration points of contact, and integration of appropriate supply and maintenance regulations.
- A reporting system to ensure proper warranty administration.

U.S. Air Force

With reference to PL 98-525, the U.S. Air Force now requires a warranty plan for each procurement, documenting the responsibilities, decisions, taskings, and strategies for warranties. Specific planning areas include:

1. Brief statement of the need and summary of the technical and warranty history.
2. Membership of the acquisition team.
3. Responsible action point, contracting officer, warranty manager, and other points of contact deemed necessary for warranty administration.
4. Organizational responsibilities for warranty management.
5. Duration, marking, measurement basis, reporting, disposition, material accountability, and other information pertaining to the administration of the warranty.
6. Cost–benefit analysis documentation.
7. Essential performance characteristics that are warranted.
8. FMS coverage and related administrative requirements.
9. Applicability under the law.
10. Procedures for tracking and accumulating warranty costs.

The Air Force Logistics Command has also provided interim guidance on warranty administration. This guidance reflects a position taken by the other services; namely, that warranties should generally be structured to be consistent with current U.S. Military Service procedures. Specifically, the following constraints are included:

- The lowest level of hardware subject to warranty requiring contractor corrective action should be that which can be effectively marked using MIL-STD-130 procedures.
- To the extent possible, warranty duration should be stated as a fixed calendar date and be no longer than that required to identify defects.
- Parameters selected for warranty coverage must be measurable, and the method of measurement must be included in the warranty clause.
- Failure analyses and associated reports should be required for all items returned for correction to provide engineering feedback.
- Generic clauses, tailored to meet specific requirements, should be used to the extent possible, with each procuring activity developing such clauses in coordination with warranty administrative offices.

6.3 WARRANTY CLASSIFICATIONS

Assurance and Incentive Warranties

The term "assurance warranty" is used when the primary intent is to ensure that minimum design, quality, and performance levels are achieved. The government is not seeking anything more than the contract specifics, and the warranty concept and terms and conditions do not provide any incentives for the contractor to do otherwise. This is the type of warranty required by PL 98-525.

The term "incentive warranty" is used for the type of warranty that provides incentives for the contractor to exceed minimum design, quality, or performance levels. For such a warranty, the contractor can adapt a strategy to just meet the minimum performance levels. However, the warranty is structured so that the risks of failing to achieve the minimum levels, or the potential profit associated with exceeding those levels, should normally motivate the contractor to try to exceed minimum levels. This type of warranty may or may not meet the requirements of PL 98-525. Table 6.3 illustrates a comparison between these two major types of warranties.

Assurance Warranty Issues

Assurance forms of warranty have been used in military production contracts for a number of years. Following the passage of PL 98-525, there have been basically only two key changes in warranty practices:

- Application of warranties to weapon systems is mandatory rather than discretionary, as in the past.

TABLE 6.3   Comparison of Assurance and Incentive Types of Warranties

Factor	Assurance Warranty	Incentive Warranty
Basic intent	Meet minimum performance and R&M levels	Exceed minimum level
Warranty price	Expected to be minimal, from 0 up to 1 or 2% per year of hardware price	May be significant, up to 7 or 8% per year of hardware price
Warranty duration	Limited—generally ≤1 year	Can be extensive—≥3 or more years
Technology factors	Warranted item is well within state of the art (SOA), or SOA is so severely "pushed" that only limited warranty protection is realistic	Warranted item pushes SOA, so there is need to protect against failure and there is opportunity for growth
Contractor	Contractor has limited opportunity to control and improve performance prior to and during warranty	Contractor has significant opportunity to control and improve performance
Competition	Should not reduce future competitive climate	May significantly reduce competitive climate
Administration	Generally not a severe burden	May require complex procedures

Source: Warranty Handbook, Defense Systems Management College, Ft. Belvoir, VA, June 1986.

- Of the three types of warranty coverage required under PL 98-525, only the warranty for conformance to "essential performance requirements" reflects a new, postacceptance commitment. (Warranty coverage for conformance to design and manufacturing requirements is traditionally covered under some form of the inspection clause; the warranty for freedom from defects is usually covered under the inspection clause or correction of defects on warranty of supplies.)

The developer of a warranty must be concerned with how best to define and include essential performance requirements and defect controls in a warranty, as well as the form that the warranty will take.

Essential Performance Requirements

The defense industry community recognizes that there could be several hundred performance requirements in a weapon system. Compliance with the majority of stated performance characteristics is determined through an evaluation of the information furnished to support the allocated baseline of the system during design and development. That is, the government implicitly accepts the risk that the contractor's design will achieve specified performance requirements through a review of development specifications and drawings, qualification tests results, and proposed acceptance procedures.

Since a warranty on essential performance requirement survives acceptance of the product, the government, in conjunction with the contractor, must clearly identify those selected performance characteristics that survive the normal acceptance process. Instead of several hundred or so performance characteristics within a weapon system contract, there should be relatively few areas in which the government can clearly describe the compliance and evaluation methods in the operational environment and satisfactorily negotiate any joint evaluation responsibilities with the contractor.

Defects in Materials and Workmanship versus MTBF Requirements

There is a potential conflict between a control on all defects in materials and workmanship and an essential performance requirement on MTBF. If the defects control is limited to those defects that existed "at the time of delivery," then it is fairly clear that the two controls are not in conflict. The defects clause protects against initial quality problems, while the MTBF control is a reliability control for the accepted product. The difficulty in this case is "proving" that the failure was a result of a defect existing at the time of delivery.

If the time of delivery condition is removed, the conflict with an MTBF requirement may surface. This issue should be directly addressed to avoid further problems in implementing the warranty.

Assurance Warranty Forms

Assurance warranty forms have been as simple as a one-paragraph statement and as complex as a set of terms and conditions extending over a number of pages. A warranty that defines a breach only when the number of failures exceeds a stated threshold is a form of assurance warranty. The U.S. Army expected-failure concept is an example of this approach. This concept may be applied to other performance parameters, such as speed, range, power, and accuracy. The product must meet stipulated performance levels, and the warranty does not have a stated or implied incentive to exceed those stated levels.

There is a form of warranty that may have both assurance and incentive features. Consider a warranty that identifies several performance requirements for warranty coverage but has no incentive to exceed minimum levels. There is no direct reliability-related measure. As required by law, the warranty also covers defects in materials and workmanship. The warranty may be worded in such a way that all failures that occur during the warranty period are covered—irrespective of whether the failure exists at the time of delivery, and irrespective of whether the population reliability level exceeds a specified value. For this case, the performance requirements represent an assurance form of warranty, but the defects clause has an inherent incentive in that the contractor's liability is reduced for each failure eliminated. The "strength" of the incentive depends on a number of complex factors, such as the length of the warranty, the contractor's ability to control certain types of defects, and the flexibility and capability to identify problems and institute corrective action.

The more commonly used incentive forms of warranty include:

- Reliability improvement warranty (RIW)
- Mean time between failures guarantee (MTBFG)
- Availability guarantee (AG)
- Logistics support cost guarantee (LSCG)

Table 6.4 summarizes these four forms. They will be discussed in detail in the following subsections.

Reliability Improvement Warranty (RIW)

The objective of an RIW is to motivate a contractor to design and produce equipment that will have a low field failure rate and incur low repair costs in operational use. Under RIW, a predetermined fixed price is agreed on, and contractors repair or replace as necessary all warranted units returned to their repair facilities at no increase in this fixed price. Normally, the RIW contract is awarded to the contractor at the time of the production award; it provides for warranty coverage over an extended period, on the order of 3 to 5 years. Repair of returned items at a fixed price makes contractors thoroughly committed to the operational performance of their equipment and provides incentive for them to improve the reliability of the equipment and reduce repair costs.

Generally, the contractor will repair all returned LRUs or items that are under warranty, and there will be no in-service repair capability. The service organization will simply return an item to the contractor when test equipment or built-in test equipment indicates a failure. Thus, the RIW provision extends the contractor's responsibility for reliability of the equipment into the field, and it is expected that the contractor will consider the impact that design decisions have on field reliability. It is also expected that as a result of repairing field failures, and being given freedom to change the design, the contractor will be in a position to improve the reliability of the equipment. Because the contract is for a fixed price, the fewer returns received by the contractor, the lower the costs. In addition, since the contract price is negotiated on the basis of a specific MTBF, the contractor is motivated to make reliability improvements, at no cost to the customer (government), whenever economically feasible, to reduce subsequent repair costs further.

Under RIW, if a contractor initiated change does not affect form, fit, or function of the equipment, the contractor is authorized to make the change and then is required to document and report it to the government in a timely manner. Changes affecting form, fit, or function still require an engineering change proposal (ECP); however, under RIW, contractors are encouraged

TABLE 6.4 Summary of Four Incentive Forms of Warranty

Incentive Warranty Form	Objective	Approach	Remedies	Application
Reliability improvement warranty (RIW)	Achieve acceptable reliability and motivate contractor to improve	Contractor performs depot maintenance for at least 2 years under a fixed price	Contractor repairs all covered failures and has time option of implementing no-cost ECPs for R&M improvement	Units must be depot-repairable; reduced military self-sufficiency must be tolerable
Mean time between failures guarantee (MTBFG)	Provide assurance that required field MTBF level will be achieved	Contractor guarantees field MTBF; measurements are made and compared with guaranteed value	Contractor must develop and implement solution if guaranteed value is not achieved; contractor may have to provide consignment spares in the interim	MTBF is appropriate reliability parameter and field measurement can be made
Availability guarantee (AG)	Provide assurance that required operational availability will be achieved	System availability is measured in the field or through special test and compared to guaranteed values	Same as for MTBF guarantee	Availability is appropriate readiness parameter, and acceptable measurement methods can be implemented
Logistics support cost guarantee (LSCG)	Control logistics support costs	Contractor "bids" target logistics support cost through use of a model; field parameters are measured, and the same model is used for obtaining measured logistics support costs and compared to target	Contract price is adjusted based on measured vs. target values; a correction of deficiency may be required	Appropriate LSC model exists; generally requires a special test program to obtain measured values

Source: Warranty Handbook, Defense Systems Management College, Ft. Belvoir, VA, June 1986.

to develop and provide no-cost ECPs. This is an essential part of RIW agreements, normally requiring expeditious processing of the ECPs by the government. Experience has shown that contractors will develop such ECPs and will incorporate them into the equipment if the expected savings in maintenance cost will offset the ECP cost. In this way contractor repair costs during the warranty period will be reduced and equipment life-cycle cost–benefits subsequently realized.

The objective of RIW is to achieve acceptable reliability while providing the motivation and mechanism for reliability improvement. This is accomplished through a fixed-price contract provision for the contractor to perform depot repair for all covered failures during the warranty period. The definition of failure is therefore a crucial determinant of the contractor's required performance. The broadest definition of failure includes all units removed from operation because of a determination that they do not perform in accordance with the warranty. An alternative definition of failure includes only those units verified by the contractor to have failed (or to be in nonconformance). If this definition is used, there is usually a provision for adjustment of the warranty price (or for payment on a repair-by-repair basis) if the number of "unverified" returns exceeds a specified percentage of total returns in a reporting period.

The RIW is one of a long line of techniques and methods designed to protect the US Military services from systems with poor reliability. One way to influence an item's reliability is to fashion the design specification so the resulting product is built better or at least less likely to fail. This approach generally increases the cost of producing an item and injects the service deeply into the design process, often overburdening designers with constraining limitations. Enforcement is limited to the requirement that the item pass certain reliability related acceptance tests, which usually mark the end of the contractor responsibility. The critical difference in the warranty concept is the built-in provision for continued contractor responsibility for reliability after delivery and acceptance of the product.

A contractor maintenance arrangement involves contractors in the support of their products, inasmuch as it is usually based on a cost-plus contract, it really does not shift responsibility away from the service. It is too simplistic, however, to limit the difference between a warranty, which is a fixed-price contract, and a conventional contractor maintenance contract to the fixed-price–cost-plus distinction. The RIW concept is as distinctive as a "reliability technique" in that it involves the contractor during the support phase of the product's life, seeks to shift at least some of the reliability risk to the contractor, and is introduced in the transaction early enough to allow responsive design and engineering improvement. It shifts risk so that the service no longer entirely bears the cost of compensating for reliability shortfalls. It also promises that the actual reliability of an item will improve compared with a non-warranty counterpart.

The terms and conditions of an RIW generally include exclusions, failure-verification procedures, turnaround time controls, operating time adjustments, data requirements, and storage and transportation procedures. Each RIW clause specifies the time of repair or replacement. This turnaround time (TAT) usually represents the period the government will be without the item. However, under some RIW contracts, contractors are required to maintain bonded storerooms of units from which they must ship replacement units on notification of a failure. In this situation, TAT represents the time for repair–replacement and return to storage. Failure to meet required TATs is normally excluded if the failure is due to circumstances beyond the contractor's control. Some examples of this are acts of God, acts of a public enemy, acts of the government, fires, floods, epidemics, strikes, freight embargoes, and unusually severe weather. Like the length of warranty coverage, the penalty, if any, for failure to meet TAT requirements is neither uniform nor variable. There are as many different provisions being tested as there are RIW contracts. The RIW is frequently used in conjunction with a MTBF guarantee that requires the contractor to institute corrective action and to provide consignment spares in the event the guaranteed level is not met.

The benefits that the government gets from the RIW contracting procedure include:

- Contractor responsibility for field reliability.
- Life-cycle costs are emphasized to a greater extent.
- Keeping all units in the same configuration is a contractor's responsibility.
- Contractors will be more motivated to introduce changes that will increase the equipment's MTBF, resulting in reliability growth.
- Incentives for contractors to reduce repair costs; such reductions increase their profits or minimize their losses.
- Initial support investment required by the government is minimal.
- Requirement for skilled military maintenance and support manpower may be reduced through RIW usage.

Benefits to the contractors include:

- Increased profit potential when MTBF is improved above the pricing base.
- Multiyear guaranteed business.
- Probability of follow-on contracts.

Negative aspects of RIW contracting involve (1) a high potential for legal disputes, (2) increased system acquisition costs as a result of the contractor's additional effort to achieve good initial reliability and increased spares as a result of sparing at the LRU level rather than at the module level, and (3) an increase in the number of military maintenance technicians required when transition to organic maintenance occurs.

Warranty Pipeline

The prevailing warranty–repair process illustrated in Figure 6.2, comprises the following sequence of events:

1. A warranted unit suspected of failure is tested by military personnel at the using activity to verify failure.
2. If the unit tests "good," it is put back into service or sent to supply as a ready-for-issue spare.
3. If the unit tests "bad," it is shipped, with appropriate data, to the contractor for repair.
4. The contractor receives the unit and verifies the failure and warranty coverage.
5. If the failure is not verified or is not covered by the warranty, corroboration by a government representative is obtained. To cover exclusions, a separate repair contract is usually awarded to the contractor.

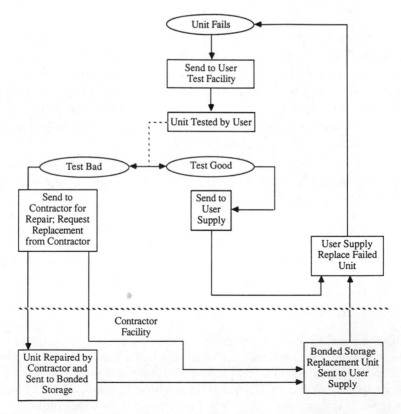

Figure 6.2 RIW pipeline. (*Source:* Air Force Institute of Technology, 1987.)

6. Repair of a covered failure is performed at no additional cost to the government, and required data records are prepared.
7. The repaired unit is usually placed in a secure storeroom maintained by the contractor, pending disposition instructions from the government.

Concurrently with step 3, a notice of failure is sent to the contractor's secure storage area and to the item manager. A requisition is processed to the item manager, who issues a material release order (MRO) to the contractor. The MRO directs that a spare be sent to supply. The spare will normally reach the supply unit before the failed unit physically reaches the contractor. This shortens the equipment pipeline significantly and, for a given mission schedule, reduces the assets needed to support the schedule.

Candidates for RIW Coverage

The following criteria should generally be used for selecting equipment as potential candidates for reliability improvement warranty coverage:

1. A warranty can be obtained at the price commensurate with the contemplated value of the warranty work to be accomplished.
2. Moderate to high initial support costs are expected to occur.
3. The equipment is readily transportable to permit return to the vendor's plant or, alternatively, the equipment is one for which a contractor can provide a service.
4. The equipment is generally self-contained, is generally impervious to failures induced by outside units, and has a readily identifiable failure characteristic.
5. The equipment application, in terms of expected operating time and the use environment, is known.
6. The equipment is susceptible to being contracted for on a fixed-price basis.
7. The contract can be structured to provide a warranty period of several years. This should allow the contractor sufficient time to identify and analyze failures in order to permit reliability and maintainability improvements.
8. The equipment has potential for both reliability growth and reduced repair costs.
9. Potential contractors indicate a comparative attitude toward acceptance of an RIW provision and evaluation of its effectiveness.
10. A sufficient quantity of the equipment is to be procured in order to render the RIW cost-effective.
11. The equipment is of a configuration that discourages unauthorized

    field repair, preferably sealed and capable of containing an elapsed time indicator or some other means of usage control.

12. There is a reason to believe that the equipment will be utilized sufficiently.
13. The equipment is such that the contractor can generate no-cost engineering change proposals subsequent to government approval.
14. Failure data and intended operational use data can be furnished the contractor for the proposed contractual period and updated periodically during the terms of the contract.

Mean Time Between Failures Guarantee (MTBFG)

This guarantee provides a direct means for controlling the operational reliability of fielded equipment. This is accomplished by specifying in the contract the MTBF to be achieved in the field, a means for measuring the operational MTBF, and actions to be taken if the measured MTBF is less than the guaranteed value.

MTBFG Values

Two approaches for determining MTBFG values which have been used are to specify the MTBFG value in the RFP and to have contractors bid an MTBFG value. If contractors are to bid values, the RFP should generally specify a minimum value—one that is consistent with the system specification and development program. The bid value and the MTBFG price are potential source-selection factors.

    A consideration regarding specified MTBF values is to allow for reliability growth. This is generally accomplished by designating an initial period over which no MTBF guarantee is in force. Such a period will allow for stabilization of problems associated with initial installation and operation and for correction of initial production problems. A schedule of guaranteed values may be used to then "increase" the MTBF up to the final desired value. Thus, for the first 6 months of operation, there may not be any guarantee; for the second 6 months, the guaranteed MTBF may be equal to X; and for the next 12 months, the guaranteed value may be $X + Y$, where Y is a positive number.

MTBF Measurement

The contract must specify how MTBF is to be measured. If a current military data system can support such a measurement requirement, that data system may be used. In many cases current data systems may not be adequate, and a special measurement procedure would have to be instituted.

    Generally, MTBF is defined as operating exposures divided by the number of relevant failures. Ideally, the operating exposure is the number of operating hours or cycles of the warranted item. In practice this may be difficult to obtain, and pseudomeasures such as platform hours may be used, and in some cases, a statistical sampling procedure using elapsed-time-indicator readings can be used to calculate operating exposure. In addition, reliability differences may be caused by one or more of the following factors: procurement practices, operational environment, maintenance environment, equipment design and complexity, and data collection and retention procedures. Furthermore, it is not possible to quantify the degree to which these factors affect the observed MTBF.

MTBFG Remedies

In the event a measured MTBF value fails to measure the guaranteed value, the contractor must as a minimum supply the following typical remedies:

- Engineering analyses to determine the cause of MTBF nonconformance.
- Corrective engineering design or production changes.
- Modifications of units as required.
- Pipeline consignment (loaner) spares in accordance with a contractually specified method to support the logistics pipeline pending improvement in the MTBF.

Previous applications of the MTBF guarantee have used a formula for determining the number of consignment spares that reflect the shortfall in pipeline spares as a result of an MTBF lower than expected. Typically, a maximum penalty is specified to limit the contractor's liability. If and when MTBF improves, the government is required to either buy or return the loaners. It is also possible to include a positive incentive if the MTBF exceeds the guarantee value by a certain factor. To date, this approach has generally been used in conjunction with an RIW contract in which there already is an inherent positive profit incentive to exceed the guarantee value.

The MTBF guarantee is best applied when the item is under contractor maintenance, so that problems can be quickly identified and remedies developed. The unit under the MTBF guarantee should be in production if a consignment spares provision is invoked; otherwise, this may not be practical. The MTBF guarantee used in conjunction with an RIW provides a good method of ensuring satisfactory or improved reliability performance. The guarantee obligates the contractor to identify the causes of deficiencies, propose and carry out modifications to correct deficiencies, and provide consignment of spares in the interim, if the guaranteed levels are not met.

Availability Guarantee

An availability guarantee is similar in concept to an MTBF guarantee in that it focuses on a measurable population characteristic rather than on individual system failures. In this case, the characteristic is operational availability, which measures the system readiness state. Availability guarantee is most applicable for systems that are normally dormant or partially dormant, such as missile systems, but that have a high operational availability requirement. A form of an availability guarantee has been used for subsystems of the air-launched cruise missile. The availability guarantee may also be used for continuously operating systems such as a radar warning system.

Availability is influenced by two system characteristics: reliability and restoration capability. The latter characteristic is a function of maintainability and logistics factors.

In practice, an availability guarantee is implemented in a manner similar to an MTBF guarantee. Availability values are specified in the contract. Periodic measurements are made of fielded systems to obtain operational availability statistics. If the measured operational availability is less than the contractually guaranteed value, the warranty remedies are invoked—typically a requirement for the contractor to correct the deficiency and possibly to supply pipeline spares in the interim.

Availability Measurement

The availability measurement process can be somewhat complex and should be tailored to the specific application. For dormant systems, data from periodic checkouts, test launches, built-in test equipment checks, and other sources such as special tests may be combined to yield a measured availability. For continuously operating systems, the ratio of uptime to total time may be measured, and a work sampling approach may be used, or individual measurements of MTBF and mean time to repair (MTTR) may be combined to provide availability statistics.

Logistics Support Cost Guarantee

The logistics support cost guarantee is used when the main focus for control is logistics support costs (LSC). A target logistics support cost (TLSC) is established in the contract, reflecting the costs to support the guaranteed equipment. Appropriate statistics on fielded equipment are collected, usually through a special test, and measured logistics support cost (MLSC) is calculated. The MLSC is compared with the TLSC; if the MLSC is greater, a warranty breach has occurred, and specified remedies must be implemented. For some programs, if the MLSC is less than the TLSC, a positive incentive such as an award fee may be applied.

Target Logistics Support Cost

The TLSC is usually defined through the use of a model that combines acquisition costs, reliability, and maintainability as well as support factors. The

RFP generally provides details on the model to be used to generate these costs. It will include a set of standard factors such as military labor rates and government transportation times, and specifies the size of the population (the number of operational systems) and the number of life-cycle years to consider. Other factors, such as equipment costs and equipment MTBF and MTTR values, are proposed by the contractor and inserted into the model to yield the TLSC. Generally, the contractor does not guarantee the individual proposed values unless special provisions are included.

Measured Logistics Support Cost

Computation of measured logistics support costs usually entails implementing a special data collection system to collect statistics on the values proposed by the contractor that were used to obtain the TLSC. These statistics, together with the same standard (default) values, are then inserted into the LSC model to yield the MLSC.

LSC Remedies

A number of warranty remedies are available. One option is to use a contract price adjustment provision, in which the contract price is reduced by an amount proportional to the estimated support cost overrun. Another option is to invoke a "correction of deficiencies" clause, in which the contractor must identify the causes of the overrun and design and implement a fix. In some cases, a cost-sharing arrangement may be established. To provide positive incentives, there may be a provision that the contractor receives additional monies if the MLSC is less than the TLSC. This may be accomplished by a formula, or, more typically, through an award fee process.

Comparison of Warranty Forms

Table 6.5 summarizes the four incentive warranty forms considered with respect to a number of risk and implementation factors. The table also includes the assurance form of warranty as a point of departure.

Warranty Duration

The period of the warranty is a major element in warranty cost, incentives, administrative factors, investment decisions, risks, and other factors that are all keyed to the duration. The duration of a warranty can be expressed in many ways, including:

- Duration as it applies to individual items versus lots.
- Duration starts with delivery (or acceptance) versus installation versus some other event.

TABLE 6.5 Comparison Warranty Forms

Factor	Assurance Warranty	Reliability Improvement Warranty	Mean Time Between Failures Guarantee	Availability Guarantee	Logistics Support Cost Guarantee
User risk of not achieving objectives	Moderate to high	Moderate	Low to moderate	Moderate	Moderate
Contractor pricing risk	Low	Moderate	High	Moderate to high	Moderate to high
Administration difficulty	Low	Moderate	High	High	Low to moderate
Enforceability Risk	Low to moderate	Moderate	Moderate	High	Moderate to high
Contractor motivation for improvement	Low	Moderate	High	Moderate	Low to moderate
Warranty period	Short	Moderate to long	Moderate	Moderate	Short to moderate
Warranty services provided by contractor	Repair or replace warranty failures; redesign if necessary	Depot maintenance, plus no-cost ECPs	Logistics assets if required, plus no-cost ECPs	Logistics assets if required, plus no-cost ECPs	Logistics assets if required, plus no-cost ECPs

Source: Warranty Handbook, Defense Systems Management College, Ft. Belvoir, VA, June 1986.

- Duration is in terms of calendar time, operating time, or a combination (whichever comes first).
- Warranty period can terminate early or be extended, depending on the item's performance.

The warranty period is the time during which the warranty is in effect. In general, the warranty period should be sufficiently long for the contractor to recover (with profit) any investment in the product. The choice of definition in calendar time or in equipment hours involves a tradeoff between administrative efficiency and accurate reflection of the fact that the "item being purchased is utility and not time." However, the Services must plan for transitioning from warranty coverage to another logistics support concept.

Operating hours is a better measure, but it is more difficult to monitor accurately. A cumulative operating measure is sometimes advocated over a specific unit limitation because fleet operating conditions are such that individual units cannot be easily controlled. Although warranty coverage length should depend on each procurement's circumstances, the current wide variation is not the result of planned experiment. This will make it difficult, if not impossible, to draw any prescriptive guidance for future applications from initial trial warranty outcomes.

When the warranty period is defined strictly in terms of calendar time, how much the government uses the product during the period greatly affects the contractor's performance costs. For initial pricing purposes, programs featuring warranty period definitions of this type include a provision for adjusting the warranty price to actual usage. Table 6.6 summarizes warranty duration alternatives.

6.4   WARRANTY DEVELOPMENT

Warranty and System Life Cycle

This section provides a general overview of warranty-related activities from a system life-cycle perspective. In developing an effective warranty, the program manager needs to plan for the completion of these activities. This section also addresses warranty impacts on the acquisition strategy and procurement plan, the system specification, and the program office organization as key planning factors for the program manager to consider early in the system's life cycle.

Life-Cycle Overview

Figure 6.3 shows how warranty related activities interface with the system life cycle. These activities are summarized by phase as follows:

Concept Exploration. Technical and support concept studies are performed for identifying characteristics to be considered for warranty.

TABLE 6.6 Warranty Duration Alternatives

Warranty Duration	Advantages	Disadvantages
Fixed calendar period for all units: all units are warranted for a fixed calender time, after which all units go off warranty; the actual amount of warranty coverage for individual units will vary, and the user must transition from warranty at a single time; contractor failure and risk exposure will depend on utilization rate	Simplest to administer	Units receive varying amounts of warranty coverage; a sudden shift from contractor to military support could be disruptive; if units are not operated, value will not be received for prepaid warranty expense unless special adjustment provisions are made
Fixed calendar period for successive production lots: the warranty on all units within a production lot expires at a fixed time, but that time varies between production lots—this approach permits an essentially uniform amount of coverage for each unit but results in a situation in which some field units are under warranty and some are not—this may be administratively unacceptable, but it does ease any transition problems; contractor failure and risk exposure will depend on the utilization rate	Permits incremental shift in support; units receive more nearly equal warranty coverage	Confusion may occur regarding disposition of a failed unit; if units are not operated, value will not be received for prepaid warranty expense unless special adjustment provisions are made
Total operating hours for all units: all units are under warranty until a total operating-hour level is reached—this type of coverage reduces uncertainty in pricing the warranty with respect	Assures that the government will receive full value for warranty cost	More difficult to administer than fixed calender period; contractor may be liable for an extended period if operational usage is far below expectation

to failure exposure, but the date of warranty termination is open-ended; coverage on individual units will vary, and a means for measuring total operating hours must be established		
Operate hour or calender time for individual units: the warranty on each expires after a specific number of operating hours or calender time is reached—this type of coverage is similar to the 12,000-mile or 12-month warranties associated with automobiles; this approach provides uniform coverage and the most information for warranty pricing, but it is administratively cumbersome and might be appropriate for only warranty on such items as large, fixed ground equipment	Provides contractor limit on time liability	Requires individual-item operate-time measurement—administration is most complex; value may not be received if time expires; however, coupled with an operate-time adjustment, this problem can be minimized
Total operate-hour or calendar-time coverage for all units: this type of coverage provides for a single end time and limits contractor liability; while time to transition from warranty is not completely specified, it is more predictable than just total operate-hour control	Provides contractor limit on time liability	Administration is complex—value may not be received if time expires; however, coupled with an operate-time adjustment, this problem can be minimized; requires fleet operate-time measurement

Source: Warranty Handbook, Defense Systems Management College, Ft. Belvoir, VA, June 1986.

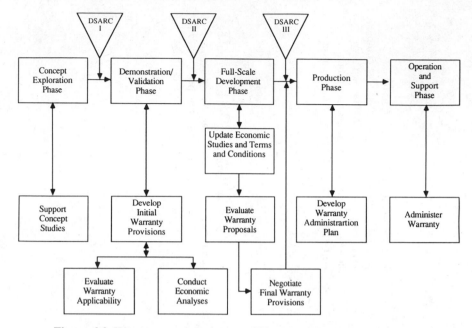

Figure 6.3 Warranty and the system life cycle. (*Source:* AR 702-13.)

Demonstration–Validation. The expected warranty provisions are developed as system requirements to be addressed in the full-scale development phase.

Full-Scale Development. The warranty provisions are updated to reflect better estimates of system RAM, support parameters, and costs, and are included in the production RFP.

Production. A series of tasks is developed to implement, enforce, and manage the warranty provisions.

Operation and Support. The warranty provisions are administered.

Acquisition Strategy and Procurement Plan

To obtain maximum effectiveness from the warranty concept, it is important that the concept be considered early in the system's life cycle, because decisions on the warranty approach can affect equipment configuration and design as well as the planning needed to maintain and support the warranted item. The demonstration–validation RFP may include sample warranty provisions that notify the contractor of the warranty performance requirements being considered for the production system. The sample warranty provisions should be qualitative descriptions of the warranty coverage desired. Actual warranty requirements should be defined only after system performance experience is accumulated and evaluated from analyses and tests performed during demonstration–validation.

The program manager may decide to include a detailed warranty requirement in the RFP for full-scale development (FSD) to indicate the warranty coverage expected for production units. The program manager develops the warranty requirements from the system performance characteristics determined during demonstration–validation as well as further engineering studies and cost–benefit analyses. In addition, the program manager may decide to have the FSD contractor propose alternative forms of warranty that would be more advantageous to the government. Waiting to initiate the warranty agreement until the production contract negotiations would not allow the design to be affected by the warranty considerations. In effect, only materials and workmanship would be warranted at this late stage.

If the FSD contractor is expected to provide production units later, the program manager may decide to include the warranty provisions in the FSD contract, with the options for production already priced. This strategy provides competition for warranty pricing. This is viable if a single production source is expected (the FSD contractor) and if warranty terms and conditions and pricing can be developed at the early date. The government would have the right to change the warranty provisions and negotiate price changes as the system matures and opportunities for a more cost-effective warranty arise.

Table 6.7 presents a general sequence of steps for developing a warranty approach, starting early in the system's life cycle. Steps applicable to the procurement should be included as part of the acquisition strategy for the weapon system.

TABLE 6.7   Warranty Development Strategy

1. Perform studies to identify essential performance characteristics to consider for warranty and identify candidate approaches

2. Develop criteria and models and collect applicable data to perform evaluations to decide between assurance and incentive types of warranty

3. In conjunction with technical, user, logistics, and contractual personnel, develop candidate approaches and assess the feasibility of candidate approaches, including consideration of warranty implementation and administration

4. Develop preliminary clauses or draft provisions for demonstration–validation RFP, or provide "trial balloons" to potential contractors to obtain industry comments

5. Issue an FSD RFP with "expected" warranty provisions for the production contract, or have contractor propose alternative forms of warranty to the government

6. Finalize warranty terms and conditions for the production RFP

7. Develop a warranty selection strategy and decision model

8. Issue an RFP with a warranty option

Source: Warranty Handbook, Defense Systems Management College, Ft. Belvoir, VA, June 1986.

System Specification

A key element in the development of an effective warranty is the system specification, which defines the set of system requirements. It is generally developed prior to the completion of the demonstration–validation phase. The requirements in the system specification are translated to development specifications, generally before or at the beginning of FSD. Requirements in the system specification can be in terms of design details or performance, or, as is most likely, a combination of the two. Performance requirements are preferred in order to interest the largest segment of industry for competitive bidding. Performance requirements also allow greater flexibility in establishing warranty requirements. If the specification establishes detail design requirements, there is a potential for future disputes if the design does not yield the required performance, because the contractor can claim that the design was imposed.

General DoD policy has stated that warranties should not apply to goals or objectives. In addition, qualitative statements cannot be used meaningfully without a potential for dispute. Specific recommendations for including requirements in the specification, giving consideration to warranty development, include:

- Requirements in the system specification and flowdown specifications must be quantitative.
- For requirements to be directly used for warranty coverage, they must clearly refer to the operational environment or special test conditions.
- Methods for measuring conformance to requirements must exist or be amenable to development.
- Only a small subset of specification requirements should be selected for warranty coverage.
- Higher-level, mission-related requirements are generally preferred to sublevel requirements for warranty specification.

Program Office Organization

It is important that the program or system manager plan and coordinate a warranty application early in the system life cycle. The selected warranty approach can affect equipment configuration and design as well as the planning needed to maintain and support the warranted item.

The program office represents the first logical coordination point for ensuring that the warranty is developed and implemented effectively. Program, engineering, logistics, budget, and contract personnel need to know the warranty application at hand and the areas of risk where inconsistency between the warranty and program requirements could void the warranty requirements.

Functional interfaces between the program office, user, and supporting activities are also important in ensuring that the maximum benefit from a warranty application is received. These interfaces identify the multiple features of a warranty application, including the following:

1. Warranted items, coverage, and duration.
2. Maintenance and handling procedures for warranted equipment.
3. Transportation management.
4. Inventory management.
5. Communication of warranty claims.
6. Defense Contract Administration Services responsibilities.
7. Configuration management.
8. Funding.
9. Warranty data reporting.
10. Special training for warranty implementation.

A warranty implementation plan is the program manager's vehicle for describing these features of a warranty application, identifying organizational responsibilities, and establishing procedures and interfaces required for successful implementation and management of the warranty. The program manager can receive assistance from service and DoD activities in planning and developing a warranty application.

Concept Exploration Phase

During concept exploration the program manager evaluates and selects alternative system development concepts for meeting the stated mission need. The concepts should address the functional and performance characteristics necessary to meet the mission need, as well as the necessary interfacing capabilities, and should be accompanied by preliminary life-cycle-cost estimates and logistics supportability plans. Table 6.8 lists the major acquisition activities in this phase and identifies areas that interface with the development and implementation of warranty application.

Although the system is treated in very general terms in this phase, background studies may be conducted in terms of reliable system performance and the expected life-cycle cost. Warranty or other contract methods may be considered as part of the studies as a means of achieving stated goals for reliable performance. The concept exploration phase ends with the development of a system concept paper, which may state the initial requirements for using warranty control techniques.

Demonstration–Validation Phase

The program manager identifies the system development concepts and approaches that have the greatest potential for meeting the mission need in the

TABLE 6.8   Concept Exploration Phase Acquisition Activities and Warranty Interfaces

Acquisition Activity	Warranty Interfaces
Requirements analysis	Identify key parameters as candidates for essential performance requirements coverage
Functional analysis	Relate key performance parameters to applicable hardware–software elements
Trade studies	Analyze various warranty strategies and interfaces as trade studies are conducted in requirements, configuration, and supportability
Technology–risk assessment	Identify potential warranty approaches to addressing risks that are identified
Logistics supportability	Consider impact of various warranty support strategies on overall logistics support structure
LCC assessment	Identify major LCC factors to consider in conducting a warranty cost–benefit analysis
Acquisition strategy–plans	Identify–update major warranty alternatives

Source: Warranty Handbook, Defense Systems Management College, Ft. Belvoir, VA, June 1986.

most cost-effective manner during the demonstration–validation phase. The concepts are verified, and the associated risks and uncertainties are identified and, where possible, resolved, usually through hardware fabrication and demonstration. System and subsystem documents as well as solicitation documents are completed to the extent necessary to support contracting for the full-scale development of the selected concepts. Table 6.9 lists major acquisition activities in this phase and identifies areas of interface with the development and implementation of warranty application.

Although warranty application is generally associated with the production contract, it is important that the system developer understand the warranty requirements, since the requirements may affect design, production processes, parts selection, and quality control in an effort to enhance reliable system performance. The RFP for full-scale development should contain preliminary warranty provisions intended to be used in the production contract.

Program contracting or logistics office personnel perform studies to determine a warranty approach for the weapon system and identify preliminary terms and conditions for the warranty. Major studies related to warranties are summarized as follows:

1. *Initial Screening.* Initial screening is performed in accordance with application criteria to determine whether one or more warranty alternatives are appropriate.

TABLE 6.9   Demonstration–Validation Phase Acquisition Activities and Warranty Interfaces

Acquisition Activity	Warranty Interfaces
Engineering development models	Evaluate technology and performance for identifying key risk factors
Preplanned product improvement (P³I)	Couple warranty alternatives with any P³I alternatives under consideration
Functional baseline	Refine essential performance requirements to be consistent with the functional baseline
LCC update	Establish and/or refine requirements for LCC analysis if LCC is part of warranty acquisition strategy
Test and evaluation master plan (TEMP)	Define any test requirements necessary to implement warranty
Preliminary manufacturing plan	Address design and manufacture warranty requirements
Industrial base issue	Address any potential impacts of warranty on industrial base
Logistics support analysis	Update earlier analyses and define warranty alternatives that are consistent with planned ILS system
Acquisition plans	Update warranty acquisition plans

Source: Warranty Handbook, Defense Systems Management College, Ft. Belvoir, VA, June 1986.

2. *Economic Analysis.* If the results of the initial screening are positive, the candidate warranty alternatives are analyzed to determine the economic feasibility of warranty and the most desirable warranty period.

3. *Development of Provisions.* Initial warranty provisions are developed. The program office should maintain continuous coordination with using commands and support activities.

4. *Incorporation of Provisions in FSD RFP.* After proper initial review with cognizant procurement, legal, and other interested parties, the initial warranty provisions are incorporated into the FSD RFP—primarily for informational purposes, unless a firm warranty commitment is to be made at this time. It may be necessary to prepare special instructions to the bidder to clarify selected points. Additional special briefings with potential contractors may be required to explain the intent of the provisions, since some contractors have had no or only limited experience with these concepts.

5. *Development of Final Preliminary Provisions.* Changes in the initial provisions may be developed as necessary to clarify wording, changes in

coverage, and other areas. In the case of a combined engineering development–production procurement, the final provisions become part of the contract, typically as an option that may be exercised at a later point in engineering development. If it is not a combined procurement, the provision may still undergo additional changes and evaluation as part of the production procurement.

Full-Scale Development Phase

The final products of the FSD phase are product baseline configuration design and a documentation package that reflect the established cost, schedule, logistics supportability, and performance constraints. Table 6.10 lists major acquisition activities in this phase and identifies areas of interface with the development and implementation of warranty application.

During the FSD phase, better estimates of system reliability, maintenance and support parameters, and operating capabilities become available. Warranty applicability and economic studies can be refined and updated, and warranty provisions can also be updated to reflect program or equipment

TABLE 6.10   Full-Scale Development Phase Acquisition Activities and Warranty Interfaces

Acquisition Activity	Warranty Interfaces
Allocated baseline	Define quantitative warranty requirements at appropriate subsystem levels.
System prototype tests	Evaluate data and use to perform warranty analyses, e.g., LCC and R&M
Integrated logistics support	Address warranty implementation and administration.
Quality assurance (QA) plan	Identify approaches to implementing warranty controls on design and manufacture and defects in materials and workmanship.
LCC update	Update LCC model for warranty cost-benefit analysis and refine data base.
TEMP update	Identify and/or update any warranty test requirements.
Acquisition plans	Interface with development and potential production contractors, draft warranty RFP clauses for industry review, and evaluate comments

Source: Warranty Handbook, Defense Systems Management College, Ft. Belvoir, VA, June 1986.

modifications that have occurred during FSD. Major warranty studies in this phase are summarized as follows:

Warranty Feasibility Studies. The initial economic studies performed as part of the demonstration–validation phase may be updated in light of FSD information. If previous studies were not performed, the studies may be initiated.

Development of Final Provisions. If warranty provisions were not finalized as part of the demonstration–validation phase, provisions for the production phase are formulated or refined, with proper coordination between program office and appropriate user and support activities.

Incorporation of Provisions in Production RFP. Provisions are incorporated into the production RFP if they were not incorporated previously. Warranty issues to be addressed in the RFP include warranty management, facilities, in-plant material flow, data, and price. Instructions to bidders regarding required response may be necessary at this point.

Proposal Review. Production proposals must be evaluated with respect to warranty response. The degree to which the full intent of the provisions is adhered to, as well as quoted cost, is of concern. If a warranty price quotation was requested, the economic analysis performed may be repeated, using the quoted warranty cost in lieu of the computed estimates. Any questionable points may be clarified in discussions held with contractors.

Warranty Decisions. On the basis of the economic analysis, as well as mission and logistics factors, the program manager must decide among available warranty options. The decision should be made early enough (ideally at the time of long-lead-item commitment) to permit orderly planning by all affected activities, regardless of the choice made. If a warranty is selected, provisions for funding and for warranty payments must be established.

6.5 IMPLEMENTING AND ADMINISTERING THE WARRANTY

The purpose of an implementation plan is to provide a complete and comprehensive document that describes the features of the warranty, defines the responsibilities for meeting the contractual provisions of the program, identifies the responsible participants, and establishes the procedures and interfaces required for successful implementation and management of the warranty. All three services acknowledge the need for some form of warranty implementation plan. There are basically two different kinds of warranty implementation plans: those prepared by contractors and those prepared by the government. Contractor plans are prepared in response to the contract requirements. The decision as to whether a contractor must submit a warranty implementation plan should be based on the criteria used to determine the need for a government implementation plan.

Depending on the nature of a procurement, warranty contractual provisions may originate in a program office or an item manager's or system manager's office. In the former case, the crafters of the warranty are not necessarily the same people who will have to implement, administer, and evaluate it. In the latter case, the same office will probably develop and manage the warranty to its conclusion. Warranties range in complexity from the very simple to the more complex incentive warranties that may call for protracted contractor participation. If the contractor is required to perform warranty-related tasks for an extended period after the system is fielded, the implementation plan will likely need to include procedures that are workable within the supply support system and the equipment's operating environment.

Some services may require an implementation plan irrespective of the simplicity or the technical needs of the warranty. From a technical viewpoint, the decision as to whether to prepare an implementation plan should be made by the drafters of the warranty contract provisions. For simpler types of warranties that contain no requirement for contractor or government actions to carry out the warranty provisions and require no evaluation of the effectiveness of the warranty, a plan may not be needed. On the other hand, complex, incentive types of warranties may need detailed implementation procedures, depending on how complex the contract provisions are. In general some form of warranty implementation plan will be required if one or more of the following requirements apply:

- The warranty contract provisions require the government to perform actions or tasks.
- The contractor is required to perform actions or tasks that will need government monitoring, inspection, or reaction.
- The contractor is required to submit deliverables related to the warranty.
- There is a requirement to evaluate the effectiveness of the warranty.

6.6 SUMMARY

This chapter has addressed a wide range of topics from warranty acquisition strategy to development of terms and conditions through planning for the operational phases. In the past there were few contractual controls available to the government to ensure that an accepted product maintained its specified characteristics in the user environment. Under the RIW a substantial portion of an item's operation and support costs is fixed for a definite period. The initial requirements for spare parts, manuals, training, and test equipment are reduced, and these costs are deferred until transition to organic maintenance. At transition, the design would be stabilized, leading to better definition of requirements. However, if the RIW concept is appropriate, it should be introduced as early as possible in the equipment's life cycle in order to motivate the contractor to design and produce more reliable equipment.

Although completed DoD warranty programs have exhibited improved reliability, there is no conclusive evidence that the warranty was a major factor. Improvement is traceable not to the warranty but to prewarranty and externally generated component technology advancement. Other improvements could be obtained through better use of testing apart from the warranty program and perhaps at less cost. In addition, methods of reliability measurement and prediction are imprecise. Improvements in these areas would enhance selection, monitoring, and evaluation of warranty programs.

And finally, the warranty requirements imposed by PL 98-212 have generated much controversy within the public sector, DoD, and Congressional and industrial circles. The implications promise to be far-reaching with effects ranging from the health of the defense industrial base to the state of our force readiness.

The next time you buy a product that has a warranty, be sure to read it. You might find that you have limited options should the product fail during the warranty period. Thus, instead of the hassle of getting it repaired, especially if it cost less than $20 to begin with, you just buy another one because you need one now not in a couple of weeks.

7

COMPUTER RESOURCES SUPPORT

The addition of the support of computer resources as a classical element of ILS came as a shock to many who wished this subject would disappear. As embedded computers have become common in every facet of daily life, this area will consume more resources and management time. Computer resource support has the potential to become the most demanding of all support requirements.

Most government agencies require that a computer resource plan be developed and maintained during a system's life cycle. Factors to be considered include: responsibilities for integration of computer resources into the system; personnel requirements for developing and supporting computer resources; and computer programs required to support the acquisition, development, and maintenance of computer equipment and other computer programs. The plan should describe the resources necessary to support the software.

Computer resource life-cycle management documentation provides the acquisition manager with information in order to understand and apply sound software development practices for a software system acquisition. Defense systems are becoming increasingly software-dependent. Software developments have gained great influence because of such factors as cost and schedule overruns, lack of reliability, cost of maintenance, and lack of supportability and interoperability. A number of factors contribute to software problem areas; for instance, widespread lack of properly trained people in key positions because software development is people-intensive; insufficient software management and control; lack of adequate requirements specifications, of research into software development techniques to ensure a better and more maintainable product, of software life-cycle planning, of standardization, and of systems engineers; and duplication in software effort. DoD components are now required to ensure requirements analysis and risk analysis, computer resource life-cycle planning, configuration management of computer resources, software language standardization and control, and delivery of support software are provided for in all contracts.

7.1 COMPUTER-AIDED ACQUISITION AND LOGISTICS SUPPORT (CALS) OVERVIEW

Computer-aided acquisition and logistics support (CALS) is a DoD and industry strategy to enable and accelerate the integration of digital technical information for weapon system acquisition, design, manufacture, and support. CALS has been developed to provide for an effective transition from the current paper-intensive weapon system support processes to an efficient use of digital information technology. The purpose of CALS is to improve industry and DoD productivity and quality. CALS should reduce acquisition and operating costs, shorten leadtimes for acquisition and logistics support, and thus improve military readiness and combat effectiveness. The primary objectives of CALS include (1) accelerating the integration of reliability and maintainability design tools into contractor computer-aided design and engineering systems; (2) encouraging and accelerating the automation and integration of contractor processes for generating weapon system technical data in digital form; and (3) rapidly increasing DoD's capabilities to receive, store, distribute, and use system technical data in digital form to improve life-cycle maintenance, training, and spare parts reprocurement.

In the current environment, a variety of automated systems are utilized by the system contractors working as a production team to enter, update, manage, and retrieve data from databases associated with specific systems. Few of these systems are compatible with each other and/or with similar systems employed by the government to receive, store, process, and use delivered technical data. In addition, the functional capabilities, such as reliability and maintainability design tools, supported by the diverse systems vary greatly. The shorter-term goals of CALS implementation are increased uniform functional capabilities and development of specifications for digital delivery modes of technical data in formats that comply with widely accepted commercial standards developed for that purpose. The longer-term goal is integration of these databases to share common data for system development that would result in a contractor integrated technical information system (CITIS). The data deliverables from, or government access to, specified segments of CITIS data will be explicitly required in future contracts, and developed in accordance with CALS standards and procedures.

Policy guidance issued by the Deputy Secretary of Defense requires acquisition managers to evaluate CALS capabilities in source selection decisions and to implement cost-effective CALS requirements in weapon system contracts. To aid acquisition managers in implementing this policy, a CALS handbook was prepared that provides an overview of the CALS program; a summary of the various ways in which data can be used and the forms in which digital data can be procured or accessed; a set of decision criteria to apply when evaluating alternative digital data delivery and access options; and model contract language for contractor integration of specific functional capabilities, digital delivery of data, and government access to contractor-maintained databases.

7.2 CALS OBJECTIVES AND BENEFITS

The Deputy Secretary of Defense initiated the DoD CALS program in September 1985, with the goal that by 1990 new systems would begin to acquire technical data in digital form (rather than paper) from contractor integrated databases. Substantial quality improvements and cost reductions are expected, including:

- Reduced acquisition and support costs for all programs through elimination of duplicative, manual, error-prone processes.
- Improved quality and timeliness of technical information for support planning, procurement, training, and maintenance, as well as improved reliability and maintainability of weapon system designs through direct coupling to CAD/CAE processes and databases.
- Improved responsiveness of the industrial base by development of integrated design and manufacturing capabilities, and by industry networks to build and support systems based on digital product descriptions.

The Department of Defense and industry are beginning to experience productivity and quality benefits derived from CALS. The momentum and enthusiasm associated with the entire CALS initiative has been effective in getting CALS projects implemented in several major system programs.

Substantial quality improvements and reductions in acquisition and support costs are expected as these programs progress toward fielding in the mid-1990s. For example, industry will eliminate development of duplicative data that drive separate processes in design, manufacturing, support planning, and development of technical manuals, spares provisioning, test equipment, training materials, and other support products. Technical data networks among primes and/or subcontractors and DoD access to industry databases will streamline system acquisition and shorten leadtimes for data delivery and spares procurement. DoD will reduce paper deliverables in contracts and reduce government expenditures for manual processes involving paper handling. Design changes will be consistently promulgated throughout DoD's support structure with assurance that the required technical data will be correctly matched to the weapon system configuration. Most importantly, the design of the major system and its support systems will have high quality by virtue of integrated design processes and a consistent technical database. This ultimately translates into increased system effectiveness and readiness.

7.3 CALS STRATEGY

To achieve CALS benefits, a phased strategy has been planned by a team composed of the Office of the Secretary of Defense (OSD), the U.S. Military Services, the Defense Logistics Agency (DLA), and industry. Phase I will replace paper document transfers with digital file exchanges and begin process integration, and will be implemented between 1985 and the early 1990s. In parallel, technology is being developed for Phase II, which involves substantial redesign of current processes to take advantage of a shared database environment in the early 1990s and beyond. The main roles of DoD in both phases are to:

- Accelerate the development and testing of data interchange and access standards.
- Fund technology development and demonstrations in high-risk areas.
- Encourage industry investment in integrated processes by establishing contract requirements and incentives.
- Implement CALS capabilities in DoD's own extensive automated systems.

To contract for digital data delivery for major systems that are entering development after 1988, DoD receiving systems must be in place between 1990 and 1995. The services and DLA have made a commitment to implement CALS integration requirements and standards in the DoD infrastructure systems. This will provide a common interface to industry. The three highest priority areas are:

- Architectural planning to link DoD islands of automation and interface with industry.
- Equipping automated engineering data repositories with capability for digital input to support spares procurement and sustaining engineering.
- Providing for digital input to automated publishing and paperless technical manual systems.

7.4 CALS SCOPE

CALS encompasses the generation, access, management, maintenance, and distribution of technical data in digital form for the acquisition, design, manufacture, and support processes. Within CALS, the common thread is technical data that include engineering drawings, product definition and logistics support analysis data, technical manuals, training materials, technical plans, reports, and operational feedback data associated with weapon systems and other equipment. The scope of the CALS effort in each of the four main strategic elements is as follows:

1. *Standards*. The standards being developed by CALS are DoD implementations of existing and emerging national and international standards for interchange of text and graphics and for database definition and access. CALS applications will make use of communications standards developed outside the CALS program. Computer hardware, operating programs, and language standards are also outside the scope of CALS.

2. *Technology Development and Demonstration*. The technology associated with CALS includes the development of advanced product data modeling techniques, development and integration of computer-aided reliability and maintainability engineering design techniques, and demonstration of CALS technology in user applications such as electronic technical manuals and parts data access.

3. *Major System Contracts and Incentives*. CALS includes requirements for the integration of major system contractor databases and processes for design, manufacturing, and support; authorized access to contractor maintained databases by government users (and vice versa); and digital delivery of technical information into DoD infrastructure systems.

4. *DoD Systems*. DoD infrastructure systems associated with CALS include those systems for which the primary purpose relates to generating, accessing, managing, maintaining, and/or distributing technical data.

7.5 CALS RELATIONSHIP TO LOGISTICS ADP MODERNIZATION

The DoD "Logistics ADP Modernization Plan" is being prepared by the Office of the Assistant Secretary of Defense (Production and Logistics). This plan will document the overall objectives and management of DoD acquisition and logistics systems. A major subset of this plan will include DoD CALS infrastructure systems that relate to technical data.

7.6 CALS MANAGEMENT

DoD and industry have established an effective organization for planning, managing, and implementing CALS. Key organizations and functional area assignments for CALS are depicted in Figure 7.1. The specific roles of the DoD CALS steering group, DoD CALS office, the DoD "components," and industry are described below.

The DoD CALS steering group serves as the "corporate board of directors"

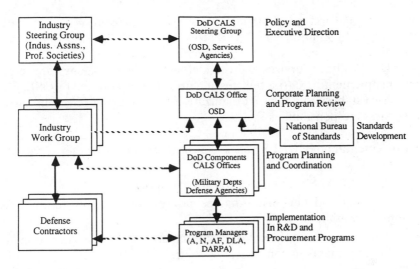

Figure 7.1 CALS management organization. (*Source: Department of Defense, Computer-Aided Acquisition and Logistics* (CALS) *Program Implementation Guide,* Draft, April 1988.)

in formulating CALS policy and implementing the CALS program within DoD. It is chaired by the Assistant Deputy Assistant Secretary Defense (Systems) and composed of senior representatives from each of the U.S. Military Services, Defense Logistics Agency, Defense Communications Agency, and participants from key OSD organizations, including logistics, production support, procurement, information resources management, and command, control, and communication information systems. Work groups are appointed by the steering group to facilitate the coordination process.

The CALS office within OSD provides planning and implementation guidance and performs the following functions:

- Develops DoD CALS corporate plans, policies, and priorities for CALS implementation.
- Accelerates the development, testing, and implementation of standards for digital technical data interchange and product definition.
- Ensures interoperability among the DoD components' technical data systems.
- Sponsors and coordinates technology development and demonstrations for integration of technical data and processes.
- Ensures an effective government–industry interface.

DoD has entered into an interagency agreement with the National Bureau of Standards of the Department of Commerce, to coordinate the development of data exchange and access standards and conformance tests in cooperation with industry. In addition, the CALS office has established a planning group to coordinate DoD plans and oversee architecture design.

The U.S. Military Services and the Defense Logistics Agency are responsible for implementing CALS policies and priorities established by OSD. Each service has established a CALS organization for managing CALS initiatives. The services' responsibilities include:

- Implementation of CALS policies in coordination with other services and agencies.
- Testing of CALS standards in prototype applications.
- Developing and demonstrating technology required for future CALS implementation.
- Incorporating the technical data interchange and integration standards and requirements in contracts.
- Participating in joint programs and serving as lead service–agency as designated by OSD.
- Planning, integrating, and implementing CALS technologies in service modernization projects.

Industry participation is integral to the success of CALS. An industry steering group provides the focal point for CALS planning, technology, and implementation concerns within industry. Their charter includes:

- Coordinating industry planning actions with DoD, industry associations, and professional societies.
- Participating with DoD in CALS standards and technology development and review.
- Providing a forum for promoting industry management awareness and education about CALS.
- Ensuring an effective industry–government interface.
- Recommending acquisition and investment approaches and policy modification.

The industry steering group coordinates activities of the CALS industry task force, which is hosted by the National Security Industrial Association. A complete set of committees and organizational structure is shown in Figure 7.2. The industry task force has grown to over 300 active volunteer members with involvement in a wide variety of technical and business issues.

Most recently, industry has formed an industry-funded cooperative to accelerate the development of the product data exchange specification (PDES), which is central to Phase II CALS. CALS phasing will be discussed later in this chapter. A government PDES users' group has been established under

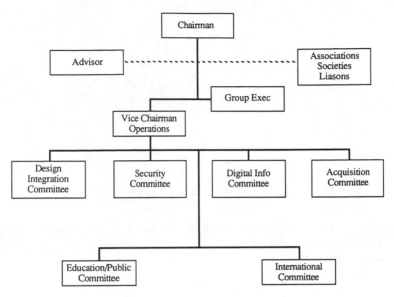

Figure 7.2 Industry Steering Group. [*Source: Department of Defense, Computer-Aided Acquisition and Logistics Support* (CALS) *Program Implementation Guide,* Draft, April 1988.]

the DoD CALS steering group to interface with the cooperative. This users' group has representation from DoD, National Aeronautics and Space Administration, Department of Energy, and Department of Commerce and National Bureau of Standards. In addition to the support being provided through the industry steering group and task force, individual companies are now incorporating CALS objectives and methodologies into their internal system integration and modernization efforts.

7.7   CALS AND THE INTEGRATED WEAPON SYSTEM DATABASE

The long-term goal of CALS is the development of an integrated weapon system database (IWSDB) that incorporates digital engineering product data and logistics support data into a share, distributed database. The IWSDB will provide rapid availability of information to DoD components and industry throughout the lifetime of the weapon system. The two major components of the IWSDB are product data and support data. Their planned transition phases and ultimate culmination in the IWSDB are illustrated in Figure 7.3.

The technical information relating to parts, assemblies, and subsystems used to describe a weapon system for design, analysis, manufacture, test,

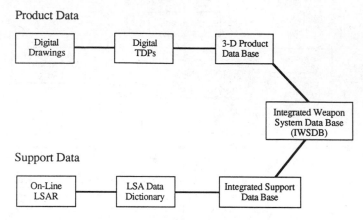

Figure 7.3 Transitional Plan for the IWSDB. (*Source: Department of Defense, Computer-Aided Acquistion and Logistics Support* (CALS) *Program Implementation Guide,* Draft April 1988)

inspection, and reprocurement is called "product data." The three steps involved in the support of the CALS transitional plan for product data are:

- Converting existing engineering drawings from paper to digitized form.
- Developing the capability to transmit an entire technical data package (TDP) (drawings, bill of materials, process specifications, etc.) in a fully computer-interpretable form.
- Developing a three-dimensional product database as specified by the product data exchange specification (PDES).

The evolutionary path for support data begins with the definition of procedures to provide on-line access to existing contractor logistics support analysis record (LSAR) databases defined in MIL-STD-1388-2A, "Logistics Support Analysis Record." The LSAR data element definitions provide the starting point for development of a standard model for database structures and an integrated support database (ISDB) dictionary that can be merged with an integrated product data model that PDES will specify. This will form the specification for the initial ISWDB in the early 1990s. This distributed database will support a full range of life-cycle applications.

The CALS target for the 1990s is for digital data interchange and database access to support a wide range of design, manufacturing, and support applications within both DoD and industry. Examples include:

- Computer-aided design
- Design analysis

- Manufacturing process planning
- Computer-integrated manufacturing
- Supportability analysis
- Maintenance planning
- Technical manual–training material authoring
- Paperless maintenance aids
- On-line provisioning
- Automated spares procurement–reprocurement

The CALS approach is to impose interface and access standards, but to leave development of the applications to the users. Technical manuals are the first application to be addressed. The development of automated technical manuals starts with digital input to DoD automated publishing systems (supported by current CALS standards), followed by paperless presentation to the end user, and ultimately to interactive maintenance aids that provide expert troubleshooting assistance.

7.8 CALS WEAPON SYSTEM CONTRACTS AND INCENTIVES

The incorporation of CALS standards and integration requirements in competitive contracts is viewed by industry as essential to stimulate needed investments. CALS requirements for integration of contractor technical data systems, authorized government access to contractor databases, and digital delivery of data using CALS standards are being implemented in contracts that will deliver data in the early 1990s and beyond.

Recent policy guidance issued by the Deputy Secretary of Defense requires that:

- Systems now in full-scale development or production be reviewed for opportunities to improve quality or reduce costs by changing to digital delivery or access.
- Systems entering development after September 1988 obtain competitive proposals for contractor integration, on-line government access to data, and digital data interchange.
- Services program resources for automated systems to receive, store, distribute, and use digital data for weapon system acquisition and logistic support.
- The Under Secretary of Defense (Acquisition) issue further guidance on contract requirements, application to subcontractors and small business, incentives, and funding mechanisms.

7.9 CALS EXPERIENCE IN EARLY APPLICATION

Implementation of CALS requirements and standards poses a significant technical and cultural change in the manner in which DoD and industry design, manufacture, and support weapon systems. Significant "growing pains" are to be expected during this transition. CALS has recognized the need for a program to provide early feedback on CALS implementation to minimize problems. The results of standards testing in user applications, weapon system demonstration, and early weapon system prototype implementation will provide the necessary feedback to take appropriate corrective actions and ensure effective transition to routine digital delivery in the early 1990s.

7.10 CALS AND DoD SYSTEMS

To receive or access digital technical data and use it efficiently in major system life-cycle support functions, DoD needs to modernize and integrate the systems in its infrastructure. Three planning and management tasks are important to CALS success in this regard:

1. The major groupings of functions, data flows, systems, and interfaces must be identified—that is, defining the major pieces of the process, how they fit together, and where gaps exist. At the top level, this process will be addressed by DoD's "Logistics ADP Modernization Plan," of which CALS is an important subset. Successive levels of analysis and planning are the responsibility of the OSD and U.S. Military Services CALS offices.

2. When existing DoD systems must interface with one another and/or with industry, CALS standards need to be incorporated. The DoD CALS office has the responsibility for providing the interface standards and implementing guidance. The Military Services are responsible for modifications to existing information systems. Progress is reviewed by the DoD CALS steering group.

3. Where gaps in capability are identified, new infrastructure information system acquisitions or major modifications must be undertaken. The DoD CALS steering group is responsible for providing coordination and guidance to ensure that there is no unnecessary duplication among the services' CALS acquisitions. Oversight of individual ADP acquisition programs is provided by the Major Automated Information System Review Council (MAISRC) or the equivalent service review council. The DoD CALS office advocates funding and management priority for these acquisitions, and reviews compliance with CALS interface standards and integration requirements.

7.11 CALS ARCHITECTURE PLANNING

Since 1988, OSD has led the U.S. Military Services in a series of planning sessions to more clearly define the scope of CALS and coordinate the ongoing programs. Planning sessions will continue to define areas where corporate DoD solutions are needed as distinguished from those where service-specific solutions are appropriate. To support this effort, a top-level functional review of each service's programs and processes that relate to CALS has begun, using a formalized systems architecture development approach. This approach will define corporate elements of the CALS "system of systems" in terms of the required data, the functions, and the network architecture. Architectural guidelines based on this structured approach address elements critical to CALS phase II, such as indexing and a locator system for accessing data in a highly integrated, but geographically distributed, database environment. These elements will be corporately developed to ensure consistent CALS implementation within DoD and industry.

Each service and DLA has initiated planning and integration efforts to define the needs and technical approach for connectivity among systems in their separate infrastructures. The U.S. Army CALS (ACALS) program, the largest of these efforts, includes conceptual system design (architecture), technical demonstration, concept development, and fielding of a digital system capable of receiving, storing, processing, disseminating, and using weapon system acquisition and logistic technical information. As the ACALS concept is developed, selective elements within its architecture design will be evaluated as a prototype for CALS architecture throughout DoD.

7.12 CALS AND COMPONENT INFRASTRUCTURE MODERNIZATION

The services have undertaken a wide range of projects for modernizing and improving their infrastructures. Major areas of activity are automation and integration of engineering drawing repositories, computer-aided design systems, technical manuals, and communications systems. These projects are the focus of joint service and DLA reviews and are discussed in more detail below.

The effort to automate all DoD engineering data repositories with interoperable capabilities is progressing rapidly. The jointly developed and acquired U.S. Army–Air Force repository system is now operational at all air force sites and repository installation at army sites has begun. DLA is also developing a DoD-wide system for locating engineering drawings in DoD repositories; their Military Engineering Data Asset Locator System (MEDALS) is expected to be fully operational in 1989. To extract technical data packages from the automated repositories, the army's Technical Data Configuration Management System (TD/CMS) is being explored as the basis for common system procedures to be used by all data repositories.

A computer-aided design (CAD) system acquisition is being planned by the U.S. Navy. Requirements are being developed to procure commercially available work stations that will be capable of interface with the engineering drawing repositories. OSD has directed the other Services to review their requirements for CAD systems and participate in this effort with the navy.

Although each Service has prototype efforts to explore technologies for the generation and maintenance of technical manuals/technical orders and other technical information, only the air force has a programmed comprehensive infrastructure modernization program for electronic technical orders. The Air Force Technical Order Management System (AFTOMS) will provide digital receipt of technical orders, as well as distribution to users and updates as changes are required. This program will build on the current Automated Technical Order System (ATOS) baseline, and provide significantly more capability to the ultimate users of technical orders at base level. Until AFTOMS is implemented, ATOS will continue to accommodate paper-type technical orders through scanning and maintaining digital storage of technical orders.

The AFTOMS concept for wholesale–depot level management of digital receipt, archiving, cataloging, distribution, and change management will also apply to applications for the army and navy. Using the air force concept as a building block, the other services are developing initial concepts to fully integrate the entire technical manual process within their organization.

On-line transmission of the full volume of technical data for major weapon systems is beyond the economical capability of current communication networks in DoD and industry. In the near term, CALS will accomplish bulk data transfers of engineering drawings, technical manuals, and other voluminous documents via physical media, such as tape or optical or optical disk. On-line interaction will be used primarily for lower-volume transaction processing and database access where operational requirements dictate and it is economically prudent. Studies are under way to identify the most effective and efficient means for digital data transmission and communication. Individual CALS program requirements will be thoroughly addressed in design tradeoffs conducted during infrastructure modernization program reviews.

7.13 CALS STANDARDS AND SPECIFICATIONS

The creation of a unified DoD–industry interface, whether through exchange of digital data files between linked data systems, or through access to distributed databases, requires common definition of data interchange and access rules. This is the purpose of the CALS standardization effort. Standards can be divided into several basic categories as shown in Figure 7.4:

Functional Standards. Military standards, military specifications, and data item descriptions (DIDs), which define the data creation procedures and define the content and format of the data products.

Technical Standards. Federal standards, military standards, military specifications, and other relevant conventions for the management, for-

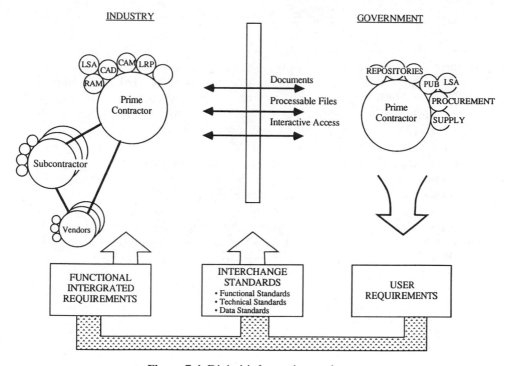

Figure 7.4 Digital information exchange.

matting, and physical or telecommunications exchange of digitized text, graphics, alphanumerics, and other forms of data.

Data Standards. Data dictionaries and sets of rules that govern data integrity and data consistency. The standards also include file structure definitions, index keys, and other descriptive information needed for access to databases.

A major CALS objective is a standardized approach for integrating technical data within a major system program. Functional integration requirements are contractual tasks to be used in statements of work (SOW) articulating the required contractor capabilities for the integration of data systems and processes. These requirements specify the integration, design, manufacture, and support processes for the performance of DoD contracts. They also establish the means by which contractors must demonstrate the capability to access and share databases among and between functional areas. These functional requirements will eventually be incorporated in updates to appropriate military standards and specifications.

As the CALS program evolves, technical data required by the government for a single system will be logically integrated (not necessarily physically integrated) into tightly coupled, controlled, and secure system technical databases, allowing access and transfer of data to those parties with proper authorization and "need to know." The integration of ADP systems and applications that are utilized by the contractor to enter, update, manage, and retrieve data from technical databases for a specific system is called a "contractor integrated technical information system" (CITIS). The required functional integration of those contractor processes necessary to ensure the security, currency, and accuracy of the technical information resident in the system technical databases will be articulated and contractually specified as requirements for the contractor's CITIS. In addition to requiring integration of the contractor's internal data and processes themselves, further integration of internal contractor data and processes with the government furnished information for each weapon system is essential.

The collection of ADP systems and applications that are utilized by the government to enter, update, manage, and retrieve data from technical databases for a specific system, called a "government integrated technical information system" (GITIS), which exists on multiple distributed ADP systems. A GITIS will cross functional boundaries and may combine data from more than one U.S. Military Service to support all requests for data from a single weapon system's user community. This degree of integration will require tight control and coordination of the separate physical databases to allow transparent support to the user. The needed control and coordination will be provided by a logical data structure called the "CALS integrated weapon system database" (IWSDB).

The integration of reliability and maintainability with computer-aided design (CAD) and computer-aided engineering (CAE) is a high-payoff CALS target that will provide significant improvements in the inherent reliability and maintainability characteristics of a system's design. These gains will translate into greater operational effectiveness and decreased life-cycle costs associated with the system when deployed. Integration of reliability and maintainability tools with the CAD/CAE process will require changes to the conventional postdesign analysis processes. These changes will consist primarily of:

1. The development of user-friendly analytical tools that are tightly coupled to the product design database and can be rapidly executed by the designer to provide short-cycle feedback about the efficacy of the design approach during the design process itself.
2. The development of effective means to take advantage of "lessons learned" from prior design experience and field use in the form of design rules, expert systems, and so forth.
3. The development of fully characterized component design, performance, and reliability data in a format readily accessible by these automated tools.

7.14   PHASING OF THE CALS PROGRAM

The CALS program is currently organized into two overlapping phases. The first focuses on converting current paper flows to digital form and the second on replacing the parallel and duplicative requirements imposed by the various acquisition disciplines and functions with the requirement to develop integrated product databases or models. This includes all the information needed for the design, manufacture, and support of a weapon system that can be made available to a variety of industry and DoD users through electronic means.

Phase I of CALS focuses on exploiting current and near-term technology to enhance the highest impact acquisition and logistics functions; specifically:

1. Engineering drawings and other information used to support competitive spares procurement.
2. Technical manuals and other information used to support system maintenance.
3. Logistics support analysis records (LSARs) and other information used to plan logistics support.
4. Life-cycle configuration management of system technical information.
5. Automated interfaces among RAM data, logistics, systems engineering, and CAD.

In the first phase, the interface standards being implemented will permit digital data interchange in neutral format within and among the services, and between DoD and industry. This interchange of technical data without resorting to paper products will result in increased accuracy and timeliness of data transfer at lower costs. Wherever possible, DoD is adopting nationally and internationally approved standards rather than creating unique DoD protocols to define this interface.

The mechanism for implementing these interface standards is a set of "core requirements packages" that will be used by the services in near-term major system and data system acquisitions. The initial Phase I standards were coordinated throughout DoD and the defense industry, and published as MIL-STD-1840A.

While Phase I converts current paper flows to digital electronic flows in a file transfer environment, Phase II is targeted on new functional capabilities through redesign, integration, and consolidation of the numerous parallel, duplicative processes that have evolved in the current paper-based culture.

Phase II CALS will exploit the new power of the computer by redesigning data formats and integrating what are now separate and often redundant collections of data. Phase II will integrate support into the design process and develop a variety of logistics data products from a design database, to produce a large savings in production and improved readiness through improved planning and support. Phase II will be designed to the extent feasible through industry cooperative efforts, because industry must implement most of the systems to create this capability.

The logical collection of shared data for a specific weapon system that supports both CITIS and GITIS users throughout the weapon system life cycle is called an IWSDB. The physical location of the data may be distributed among contractor or government ADP systems. The CALS IWSDB structure is evolving and will be the basis for Phase II environment. The IWSDB will be governed by a data dictionary. This data dictionary will be consistent with CALS data standards, which will include PDES as well as data standards to control support data. The data standards provide the data relationships, data integrity, and data consistency required to accommodate the changes in user requirements and computer technologies that are inevitable throughout the 20- to 40-year life of the major system. DoD expects Phase II to yield the following benefits:

1. More complete integration than is possible in Phase I of contractor design, manufacturing, and support data systems based on advanced product data models.
2. Near real-time updates of technical data to match major system configuration.
3. On-line access by government users to distributed contractor and government databases.
4. Databases owned by DoD, but possessed and maintained either by DoD or by contractors.
5. Automated technical manual authoring and delivery.
6. Automated interfaces of spares procurement with flexible manufacturing systems.
7. Integration of R&M engineering as an on-line part of the CAD/CAE design processes.

7.15 CRITICAL COMPUTER RESOURCES (CCR)

General

The U.S. Military Services are currently faced with the challenge of developing modern systems that rely on critical computer resources (CCR). The critical nature of computer resources in these systems, requires emphasis on the management of computer resources as part of the total life-cycle process. It is imperative that software development is not treated as a discrete activity but rather as an integral part of the broader system acquisition framework.

Critical Computer Resource Characteristics

Critical computer resources have at least one of the following characteristics:

1. Physically a part of, dedicated to, or essential in real-time performance of the mission of the system.
2. Used for systems engineering, integration, specialized training, diagnostic testing and maintenance, simulation, or calibration of the system. This includes system and non-system training devices.
3. Used for research and development of the system.

Life-Cycle Software Engineering Centers (LCSEC)

When computer resources are identified as a necessary element of a proposed system, the program manager (PM) is assisted in acquiring the needed computer resources and in assuring their quality and long-term supportability through support of the life-cycle software engineering centers (LCSECs). The major function of these centers during system development is to ensure the acquisition of high-quality software designed, developed, and documented for long-term supportability. The LCSECs plan for and provide postdevelopment support for major system software, and act as the focal point for dissemination of policy relating to acquisition of critical computer resources. Other functions of LCSECs are:

1. To provide software engineering technical support to the materiel developer through participation in the computer resources working group (CRWG) during acquisition.
2. To provide guidance with respect to software supportability.
3. To participate with development and readiness activities in the management and control of software throughout the life cycle.
4. To evaluate the computer resources management plan (CRMP) for analysis and/or justification for the target system's programming language and determine the system's software supportability for the remainder of its life cycle.
5. To ensure standardization of computer equipment and computer software support facilities wherever practical.

Software Development Cycle

Software for a major system may be developed during any phase of the system acquisition life cycle. Regardless of when in the acquisition cycle the software development occurs, it always entails the same set of activities. It is customary to refer to the set of activities as the "software development cycle" (SDC). The SDC is composed of (1) software requirements analysis, (2) preliminary design, (3) detailed design, (4) coding and computer software unit (CSU) test, (5) computer software component (CSC) integration and test, and (6) computer software configuration item (CSCI) testing.

Each SDC activity has associated software products and documentation describing the work efforts corresponding to that activity. The reviews and baselines relating to the particular documents and activities of the SDC apply regardless of the acquisition life-cycle phase in which the system is developed; in other words, the products and reviews are determined by the software activity, not the system development phase (see Figures 7.5 and 7.6). The SDC will be discussed in more detail later in this chapter. Regardless of the development method used for system acquisition, the software activities must be planned and scheduled in context appropriate to the system development.

Initial Planning for CCR

During the "requirements–technology base activities" phase, which is the initial planning period for the acquisition of a system that will utilize computer resources, initial planning may be directed toward refining proposed solutions or developing alternative concepts to satisfy a required operational capability. The decision to incorporate computer resources in a system should be

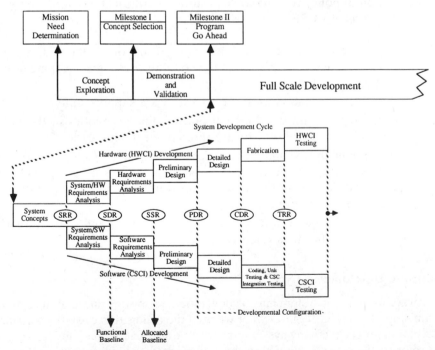

Figure 7.5 System development cycle within the system life cycle. (*Source:* DoD-STD-2167.)

recorded in the operational and organization plan (O&O plan) or mission-need statement (MNS). The O&O or MNS should contain a realistic statement of the threat and identify computer resources as a component of the proposed system to address the threat. Program initiation will occur via the O&O plan or MNS approval.

Planning for the acquisition, development, and long-term support of computer resources is initiated during requirements–technology base or concept exploration activities, and updated throughout the remainder of the system life cycle. This planning is the responsibility of the PM, who establishes a computer resource working group (CRWG) to aid in the management of the system's computer hardware and software. Critical to the successful management of CCR is the need to develop the computer resource management plan (CRMP) at the proper time in the life cycle of a system and to manage CCR in accordance with the approved CRMP.

Computer Resources Working Group (CRWG)

A CRWG is established for each system in which computer resources are likely to be used and will be formally chartered by the PM with the coordi-

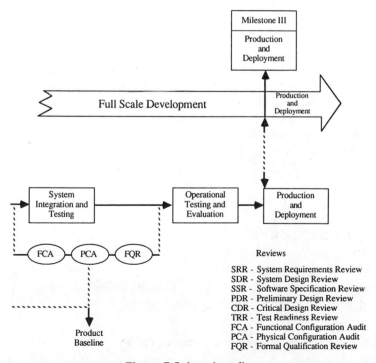

Figure 7.5 (continued).

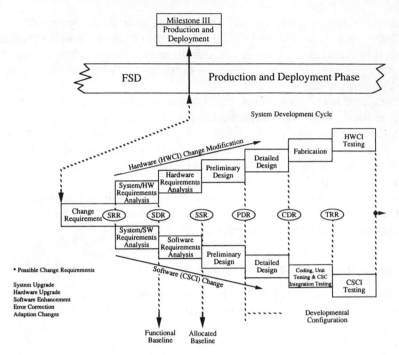

Figure 7.6 System support cycle within the system life cycle. (*Source:* DoD-STD-2167.)

nation of the operating, supporting, and participating commands. For modification programs and those acquisitions closely related to ongoing programs, an existing CRWG may be utilized.

As an advisory board to the PM, CRWG members participate actively in all aspects of the program involving computer resources. Having a major role in the management of computer resources, the CRWG responsibilities include the following:

1. Advising the program or system manager in all areas relating to the acquisition and support of computer resources.
2. Maintaining the CRMP.
3. Monitoring compliance of the program with computer resources policy, plans, procedures, and standards.
4. Integrating software test activities with the overall test program through the test integration working group (TIWG).

Computer Resources Management Plan (CRMP)

In order to initiate the investigations defined in the O&O plan or MNS that lead to the specification of software requirements, the CRWG prepares the

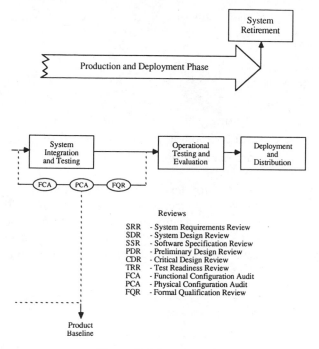

Figure 7.6 (continued).

initial CRMP. The CRMP is the primary dynamic document used to establish the necessary framework and support system for computer resources control during development, production, and postdeployment. The CRMP becomes a part of the required program management documents that are reviewed and updated for each system acquisition milestone.

The CRMP documents the computer resources development strategy and the software support concept and the resources needed to achieve that support posture, identifies the applicable directives for managing computer resources in the system, and defines any changes or new directives needed for operation or support of computer resources in the system and the requirements for software development and test support, and assigns responsibilities for the performance of related tasks during the course of a system's life cycle.

7.16 COMPUTER RESOURCES—SOFTWARE

General

Software development is usually an iterative process, in which an iteration of the software development cycle occurs one or more times during each of the system life-cycle phases is shown in Figure 7.7. The software development cycle includes six phases:

1. Software requirements analysis
2. Preliminary design
3. Detailed design
4. Coding and unit testing
5. Computer software component integration and testing
6. Computer software configuration item testing

Each iteration of the software development cycle, regardless of the system life-cycle phase during which it occurs, is initiated by allocation of system requirements to that software or a subsequent revision to those requirements.

The relationship of the software development cycle phases with the products, reviews and audits, baselines and developmental configurations are

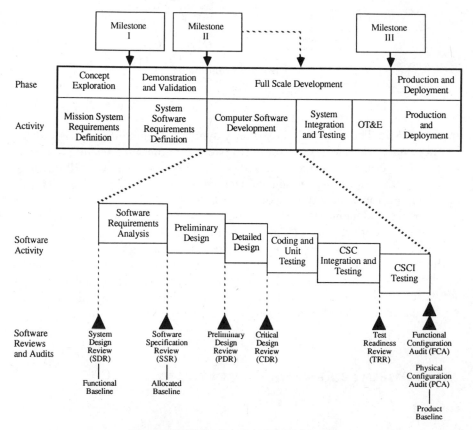

Figure 7.7 Software development cycle.

shown in Figure 7.8, which reflects the sequential phases of a software development cycle, as well as the documentation that typically exists prior to initiating an iteration. During software development, more than one iteration of the software development cycle may be in process at the same time. Each iteration represents a different version of the software. This process may be described as an "evolutionary acquisition" or "incremental build" approach. Within each iteration, the software development phases also typically overlap, rather than form a discrete termination–initiation sequence.

System Life-Cycle Phases and Computer Resources

The concept exploration phase is the initial planning period when the technical, strategic, and economic bases are established through comprehensive studies, experimental development, and concept evaluation. This initial planning may be directed toward refining proposed solutions or developing alternative concepts to satisfy a required operational capability. During this phase, proposed solutions are refined or alternative concepts are developed using feasibility assessments, estimates (cost and schedule, intelligence, logistics, etc.), tradeoff studies, and analyses.

For computer resources, the software development cycle should be tailored for use during this phase and may result in demonstration of critical algorithms, breadboards, and so on.

Major documents resulting from the concept exploration phase include the initial system segment specification (SSS), which documents total system requirements. The SSS may differentiate between the requirements to be met by computer software and those applicable to hardware design. When applicable, definitions of interfaces between computer equipment functions, communication functions, and personnel functions are provided to enable the further definition and management of computer software and computer equipment resources. Normally, this information is derived from systems engineering studies. Deliverable products at the end of the concept exploration phase typically include preliminary SSS(s), preliminary prime item development specifications, software listings, and software test results.

During the demonstration–validation phase major system characteristics are refined through studies, systems engineering, development of preliminary equipment and prototype computer software, and test and evaluation. The objectives are to validate the choice of alternatives and to provide the basis for determining whether or not to proceed into the next phase.

During this phase, system requirements, including requirements for computer resources are further defined, and preferred development methodologies for computer software and databases are selected. The results of validation activities are used to define the system characteristics (performance, cost, and schedule) and to provide confidence that risks have been resolved or minimized. For computer resources, the software development cycle should be tailored for use during this phase, resulting in prototype software items.

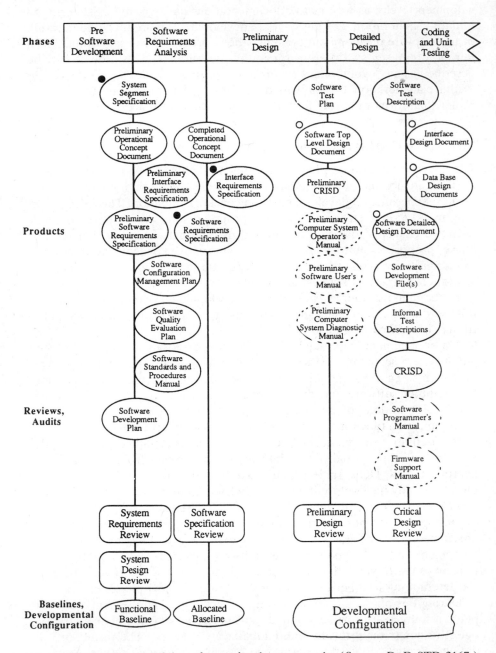

Phases

| Pre Software Development | Software Requirments Analysis | Preliminary Design | Detailed Design | Coding and Unit Testing |

Products

- System Segment Specification
- Preliminary Operational Concept Document
- Completed Operational Concept Document
- Preliminary Interface Requirements Specification
- Interface Requirements Specification
- Preliminary Software Requirements Specification
- Software Requirements Specification
- Software Configuration Management Plan
- Software Quality Evaluation Plan
- Software Standards and Procedures Manual
- Software Test Plan
- Software Top Level Design Document
- Preliminary CRISD
- Preliminary Computer System Operator's Manual
- Preliminary Software User's Manual
- Preliminary Computer System Diagnostic Manual
- Software Test Description
- Interface Design Document
- Data Base Design Documents
- Software Detailed Design Document
- Software Development File(s)
- Informal Test Descriptions
- CRISD
- Software Programmer's Manual
- Firmware Support Manual

Reviews, Audits

- Software Development Plan
- System Requirements Review
- Software Specification Review
- System Design Review
- Preliminary Design Review
- Critical Design Review

Baselines, Developmental Configuration

- Functional Baseline
- Allocated Baseline
- Developmental Configuration

Figure 7.8 Breakdown of the software development cycle. (*Source:* DoD-STD-2167.)

188

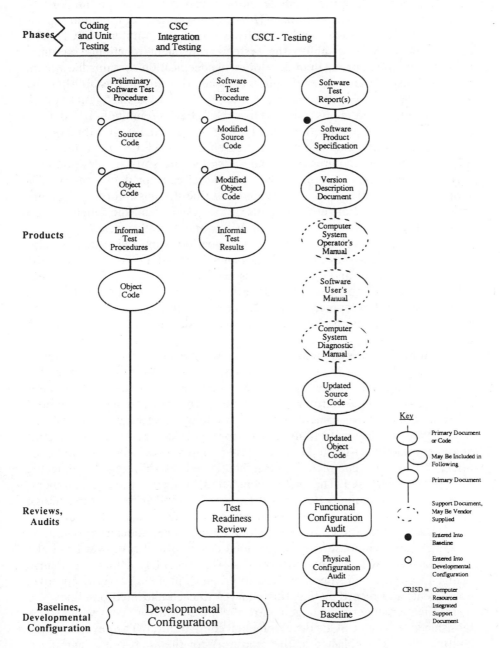

Figure 7.8 (continued).

189

The major documents resulting from this phase are the authenticated SSS(s), authenticated prime item development specifications, preliminary interface requirement specifications (IRSs), and software requirements specifications (SRSs) for each computer software configuration item (CSCI). The authenticated SSS(s) establish the system or segment functional baseline. Each authenticated prime item development specification contains the system requirements allocated to the equipment and software and establishes the allocated baseline for each prime item. Each preliminary SRS contains system or prime item requirements allocated to a CSCI. Each preliminary IRS defines the interfaces and qualification requirements for a CSCI within the system, segment, or prime item. The allocated baseline for each CSCI is established following software requirements analysis within the software development cycle. A preliminary version of the operational concept document (OCD) should also be prepared to identify and describe the mission of the system, operational and support environments of the system, and the functions and characteristics of the computer system within the overall system.

The full-scale development phase is the period when the system, equipment, computer software, facilities, personnel subsystems, training, and the principal equipment and software items necessary for support are designed, fabricated, tested, and evaluated. It includes one or more major iterations of the software development cycle. The intended outputs are a system which closely approximates the production item, the documentation necessary to enter the system's production–deployment phase, and the test results that demonstrate that the system to be produced will meet the stated requirements. During this phase the requirements for additional software items embedded in or associated with the equipment items may be identified. These requirements may encompass firmware, test equipment, environment simulation, mission support, development support, and many other kinds of software.

Software requirements analysis is performed in conjunction with systems engineering activities related to equipment preliminary design. SRSs and IRSs for each CSCI are completed and authenticated, establishing the allocated baseline for each CSCI. The OCD is completed during full-scale development. In addition, a preliminary design effort is accomplished and results in a design approach. For computer software, preliminary design includes the definition of external and internal interfaces, storage and timing allocation, operating sequences, and database design. Detailed design of critical lower-level elements of the CSCI may be performed as well. Formal engineering change control procedures are implemented to prepare, propose, review, approve, implement, and record engineering changes to each allocated baseline.

Informal engineering change control by the contractor starts with the establishment of each CSCI's developmental configuration. The developmental configuration is established as the repository for the approved design documents, software, and software listings. Following successful completion of functional configuration audits (FCAs) and physical configuration audits (PCAs), the documents and listings of the developmental configuration are included in the software product specification (SPS), which establishes the product baseline. This baseline is used to control the software as it is integrated with other CSCIs and hardware configuration items (HWCIs).

Following an acceptable physical design review (PDR) for an item, detailed design of that item begins. During this activity, engineering documentation such as drawings, product specifications, test procedures, and descriptions are produced. For computer software, detailed design is accompanied by detailed design documentation of logical flows, functional sequences and relations, formats, constraints, databases, and incorporation of reused design. The critical design review (CDR) should ensure that the recommended design satisfies the requirements of the SRS. At the CDR, the detailed design documents are reviewed. Equipment–personnel–computer software interfaces should be finalized at this time. A primary product of the CDR for software is government–contractor concurrence on the detailed design documents that will be released for coding and unit testing.

Following CDR, software coding and testing, software integration and testing, software formal testing, system integration and testing, and initial operational test and evaluation are conducted. Software coding is performed in accordance with standards and procedures contained in the approved SDP. Software testing is performed according to test plans submitted for review at PDR, test descriptions submitted for review at CDR, and test procedures submitted for review at the test readiness review (TRR). These activities normally proceed in such a way that testing of selected functions begins early during development and proceeds by adding successive increments to the point where a complete CSCI is subjected to formal testing. Additional test equipment may be required to properly simulate an operational environment to test a CSCI. The scope and realism of software testing may be progressively expanded as additional increments are made available for this purpose. Adequacy of the performance of the software is checked to the maximum extent possible, sometimes through the use of simulation, prior to software installation in a field site or operational computer. Nuclear safety cross-check analysis (NSCCA) is also performed on specified computer resource items during this phase. Satisfactory performance of the software for a large operational system may not be completely demonstrated and assessed until completion of system integration and operational test and evaluation of the equipment or of the system. Software that is relatively insensitive to the system's operational environments may be completely demonstrated earlier.

Functional and physical configuration audits (FCA and PCA) are performed on all items of hardware and software. FCA is conducted on the software at the completion of software formal testing. According to the nature of the software, PCA may be conducted at the completion of software formal testing or after system integration and testing. FCA and PCA may be performed at the system level to authenticate the hardware product specification(s) to establish the system product baseline. This baseline acts as an instrument for use in diagnosing troubles, adapting the computer resources to environmental and operational requirements of specific site locations, and proposing changes or enhancements.

Provisions are made for follow-on support of the equipment and software configuration items and associated documentation. Failure to properly consider these provisions may result in support complications, obsolete documentation, and costly "modernization" programs. This is particularly true where the system is being developed in a phased manner, providing reduced capabilities for early system integration, operation, and evaluation.

The production–deployment phase is the combination of two overlapping periods. The production period is from production approval until the last system item is delivered and accepted. The objective is to efficiently produce and deliver effective and supported systems to the user(s). The deployment period commences with delivery of the first operational system item and terminates when the last system items are removed from the operational inventory. At system transition, the role of the contracting agency normally terminates except for identified residual tasks and phase-out responsibilities. The supporting and using agencies start providing the resources necessary to support the software throughout the deployment phase. Follow-on test and evaluation is performed on operational system items as they are deployed, to assess their operational effectiveness and suitability in a deployed configuration and environment.

After a system is in operational use, various changes may take place on the hardware items, software items, or both. Changes to software items may be necessary to remove latent errors, enhance operations, further system evolution, adapt to changes in mission requirements, or incorporate knowledge gained from operational use. Based on complexity and other factors such as system interfaces, constraints, and priorities, control of the changes may vary from on-site management to complex checks and balances with mandatory security keys and access codes. The authority to change the software must be carefully and specifically delineated, particularly when security, safety, or special nuclear restrictions are involved. The same six phases of the software development cycle are utilized for each change during the production–deployment phase.

Software Testing

Testing is a difficult and time-consuming process. Effective field testing of new or modified applications software is also expensive. As a result, there is a temptation to keep testing to a minimum. In addition, user demand can also result in introduction of new software applications without adequate testing or analysis of system supporting requirements. However, field testing helps ensure that only supportable system applications are introduced since logistics requirements are identified up front. A well-structured software test plan cuts the risk associated with system changes.

A good software test program has several major benefits such as providing data on required user support, obtaining data for life cycle cost analysis for the particular module or information system, allowing resolution of phase-in problems in a controlled environment without disrupting the whole system, and providing proven software and hence ensuring user satisfaction.

Thus testing allows indentification of key logistics drivers critical to determining long-term system support requirements.

Computer Software Work-Breakdown Structure (WBS)

Software WBS elements provide a uniform framework for:

1. Planning, budgeting, and allocating responsibilities within government organizations responsible for the acquisition of defense systems and contractor organizations responsible for their development.
2. Uniform reporting of progress and status on software efforts throughout defense system acquisition programs.
3. Consistent accumulation of resource expenditure data across defense programs that can be used to calibrate and validate software cost estimation models and methods.

MIL-STD-881A establishes criteria for the preparation and use of summary, project summary, contract, and extended contract work-breakdown structures by U.S. Military Services and contractors. In addition, the services will identify the software components in the project summary and contract WBS down to the level at which the software components have been defined prior to the release of the invitation for bids or request for proposal. The services will also negotiate the placement of the software components in the contract WBS with the contractor. The contractor will then extend the contract WBS during the program, as the software system is defined in greater detail.

Support Software

Two types of support software are acquired on defense system acquisition programs: (1) support software that operates as part of the prime mission software (i.e., operating systems, database management systems, on-line diagnostic programs, which execute during system operation) and (2) support software that does not operate as part of the defense system but is used off-line to support the development, test, and maintenance of the prime mission software (i.e., operating systems, compilers, linkers, loaders, simulators, debuggers, off-line diagnostic and utility programs that are used to develop prime mission software during the system development and are delivered to maintain the system during operation, normally as part of a software development–maintenance facility). Support software that executes during any mode of system operation is considered prime mission software.

Contractor Software Development Plans

The contractor's plans for software development describe the resources and organization including contractor facilities, government furnished equipment, software, and services required, and the organizational structure, personnel, and resources for software development, software configuration management, and software quality evaluation. The plan also addresses the development schedule and milestones and risk management. Risk management procedures should be established that identify, assess the probability of occurrence of and potential damage from, and assign appropriate resources for reduction of the risk areas of the project.

Software standards and procedures are also part of the plan. These describe the tools, techniques, and methodologies to be used in software development, criteria for departing from a top–down approach, the software development library and associated access and control procedures, the format and contents of software development files, associated procedures, and organizational responsibilities, design and coding standards, and format and contents of all informal test documentation.

Software configuration management is a key part of the contractor's software development plan. Here configuration identification procedures, configuration control including software problem and change reports, and review boards, configuration status accounting, and configuration audits as well as authentication procedures are addressed. Configuration management must be implemented to provide technical and administrative direction and surveillance in order to (1) identify and document the functional and physical characteristics of each CSCI, (2) control changes to those characteristics, and (3) record and report the processing of changes and the status of implementation. Contractors must form a software configuration control board (SCCB) for control over developmental configuration. Contractors must implement procedures to generate periodic status reports on products in the developmental configuration and in the allocated and product baselines.

Software quality evaluation is an essential and necessary part of the software development plan. This includes the evaluation of development plans, standards, and procedures; evaluation of the contractor's compliance with those plans, standards, and procedures; evaluation of the products of software development; implementation of a quality evaluation reporting system; and implementation of a corrective action system.

Life-Cycle Support for Software

Contractors must define a preliminary version of the information required to perform life-cycle support for the contractually deliverable software. This definition must include the identification of:

1. The support environment, describing support software, equipment, facilities, and personnel.

2. Support operations, including general-usage instructions, administration, software modification, software integration and testing, system and software generation, software quality evaluation, corrective action system, configuration management, simulation, emulation, reproduction, and operational distributions.
3. Training plans and provisions.
4. Predicted level of change to the deliverable software in the support environment.

7.17 SUMMARY

Computer resources are defined as the totality of computer equipment, programs, associated documentation, contractual services, personnel, and supplies. Computer resources acquired as part of a system or equipment acquisition are managed as part of a system or equipment acquisition. The system acquisition life cycle provides a basis for categorizing program management activities for software development.

Requirements for computer resources may be documented during the conceptual phases. An adequate definition of essential system interfaces between computer equipment functions, communications functions, and personnel functions should be established to enable further definition and management of computer software and equipment as configuration items. During the demonstration–validation phase (D&V), computer resources are further defined and preferred development methodologies for computer software are selected and major documents are initiated to include the computer resources management plan. During FSED, the adequacy of computer software is checked to the maximum extent feasible through the careful use of simulation or other suitable techniques before installing the software in a field site or other installation. The overall objective of software testing is to make sure users are provided quality, supportable, automation tools.

The software development cycle includes the analysis phase, the design phase, coding and checkout phase, the integration and test phase, installation phase, and the O&S phase.

The CALS program has made significant progress. The publication of the initial set of technical information exchange standards represents a major CALS milestone accomplishment. A standards application testing program has begun and initial applications to weapon systems are being demonstrated. In addition, the incorporation of CALS concepts into DoD and industrial infrastructures is under way, and advanced technology to meet long-term CALS requirements is being accelerated through close DoD and industry collaboration.

CALS provides a unique opportunity to achieve major productivity and quality improvements through carefully planned and managed investment by both government and industry. Initially, the changes will be gradual, as building blocks are put in place and specific portions of the weapon system life cycle are enhanced.

As the cumulative impact of CALS integration and infrastructure modernization is realized in DoD and industry, more far-reaching changes will occur in the way functions are accomplished, leading to additional major savings. CALS implementation will result in a lower weapon system life-cycle cost, shortened acquisition times, and improvements in reliability, maintainability, and readiness.

The final phase of CALS, and the most controversial, will come in the mid-1990s when the Pentagon develops standards to let it delve into contractor's computers. Defense would be able to view designs in progress or look at technical weaponry information it may have cost or never viewed. If a sole-source supplier goes out of business, the Pentagon will have its blueprints. The biggest danger CALS faces is the military's inclination to attach unrelated pet projects to hot programs. If this happens, and CALS becomes an umbrella for disparate project, it could be diluted and weakened.

8
MANPRINT

The manpower and personnel integration program (MANPRINT) is a comprehensive management and technical program to enhance human performance and reliability in the operation, maintenance, and use primarily of weapon systems and equipment. However, it can be applied to all types of equipment. MANPRINT achieves this objective by focusing attention on human resource goals and constraints during system design, development, production, and upgrade. Human operators and maintainers have a limited range of aptitudes and physical characteristics; as such, they may be regarded early in the decision process as fixed components of a system. Therefore, design must consider the human element. It is one thing to influence design to integrate effectively with personnel characteristics; it is another to evaluate that design to determine how effectively it can be manned. The core of manning evaluation is the ability to predict the number and type of people required and how many would be available.

Human resource goals and constraints are addressed in MANPRINT through six domains: manpower, personnel, training, human factors, system safety, and health hazards. The first four domains directly influence human performance and human reliability. System safety and health hazards impact more indirectly but can also degrade total system performance if overlooked during system development. These domains are discussed later in this chapter.

MANPRINT is a multifaceted program. Individuals and organizations concerned with training, testing and evaluating, logistics, and materiel development have their own perspectives on MANPRINT. However, all benefit from MANPRINT. MANPRINT improves coordination between users and developers of new systems. Consequently, developers, trainers, acquisition specialists, testers and evaluators, logisticians, laboratory scientists, and engineers are all involved in MANPRINT. The term "MANPRINT" is primarily used by the U.S. Army, but the program is DoD-wide.

MANPRINT's success depends on the combined efforts of these specialists in the military and industry. Although each has its own responsibilities, their interests mesh in the communication of MANPRINT requirements and the development of design solutions. The services must establish and maintain a close and continuous relationship with industry, informing industry of both requirements and assessments of industry's responses to requirements, as well as working with industry on a day-to-day basis in a joint effort to ensure that MANPRINT is meaningfully included in system design. Industry must design and fabricate a system that enhances human performance and human reliability and thereby improves total system performance.

Increasingly, the military has found it necessary to rely on engineering and technology to obtain quantum jumps in capability to meet short-term and projected long-term threats. The DoD looks toward technology to replace people whenever possible in the interest of achieving the optimal distribution of manpower. But new system technology is not a solution in itself. If system design is not governed by preestablished MANPRINT goals and constraints, the services will be plagued by mismatches among the equipment, the operators and maintainers, the civilian maintainers at depot level, and the force structure. The central objective of MANPRINT, then, is to influence system design in an effort to optimize total system performance by enhancing human performance.

Advanced equipment may not, in itself, reduce the manpower requirements, because the advanced technology and complexity may require personnel with higher aptitude and skill levels, particularly for maintenance and repair. If personnel lack the requisite aptitudes and skills, then the options to meet the threat are to increase the force structure, manning levels, or training time. But the ceiling on manpower and the demand for more combat power limit increases in force structure. The services might increase training to acquire the needed aptitudes and skills, but increased training time may require more instructors, and removes personnel from the field. This requires more personnel to replace those being trained in order to maintain the same level of readiness. DoD has concluded that, given a declining end strength, advanced systems must be designed for typical military personnel and the projected force structure.

MANPRINT seeks to reduce both the demand for skilled manpower and the operating and support costs associated with the acquisition of new materiel in accordance with DoD Directive 5000.53. By dealing with these issues up front, it ensures that total force structure and military personnel capabilities are reflected in the decisions that affect each individual weapon system development.

8.1   THE TOTAL SYSTEM PERSPECTIVE

The total system includes all of the people, equipment, doctrine, training, and so forth necessary to field and sustain the system in peace time and combat. The total system includes the principal item and the associated support items of equipment, the other support equipment, and training devices. Each has its own respective logistical tails necessary for sustainment.

Historically, requirements documents have not quantified total system performance, constrained the total manpower numbers, limited the personnel skills anticipated for a given level of performance, or constrained training time. This approach has resulted in a "manning the equipment" concept, where the service member is an afterthought, a support item to be married to the equipment after the essential design direction has been determined. Despite the supposed reliance on high technology and/or equipment complexity to achieve system performance, all too often design capability and readiness goals are not achieved. Military personnel frequently cannot properly operate, support, and maintain the equipment.

Force structure and environmental parameters must be provided by the services to ensure that new systems are designed to integrate easily into that structure and environment. Generally, if a new system adds a new or significantly increased manpower requirement to the force structure, the service must man the new structure by degrading (qualitatively or quantitatively) an existing organization. The impact can be far-reaching, affecting the mix of commissioned specialties and enlisted military occupational specialties, grade structure, and authorized level of organization (ALO).

The services continue to face significant manpower–machine interface problems, many of which are the result of unbalanced attention to hardware capability at the expense of total system and human performance. In the past, increased capability achieved with advanced technology, has often been accompanied by increased in soldier task complexity. System design was not impacted by MANPRINT design constraints or by a disciplined process that insisted on putting "the man in the loop." Instead, the system design process was built on the unstated premise that sufficient numbers of skilled service members would always be available to operate, support, and maintain the system.

8.2 ILS/MANPRINT

ILS/MANPRINT is a complete integrated effort identifying support considerations, ensuring effective and economical system support during the product life cycle and is an inseparable part of all aspects of system development, acquisition, and operation. The major ILS/MANPRINT impact on system engineering and design due to logistics support requirements and the system maintenance concept is provided by MANPRINT, maintainability, and logistics support analysis (LSA) programs.

Coordination and integration of ILS/MANPRINT activities into the product program control structure consists of tradeoff decisions and integration of ILS through LSA interfacing with reliability, maintainability, MANPRINT testability, logistics, and design. It also includes participating in design reviews, conducting logistics design appraisal reviews, and evaluating all changes through a logistics change board representative.

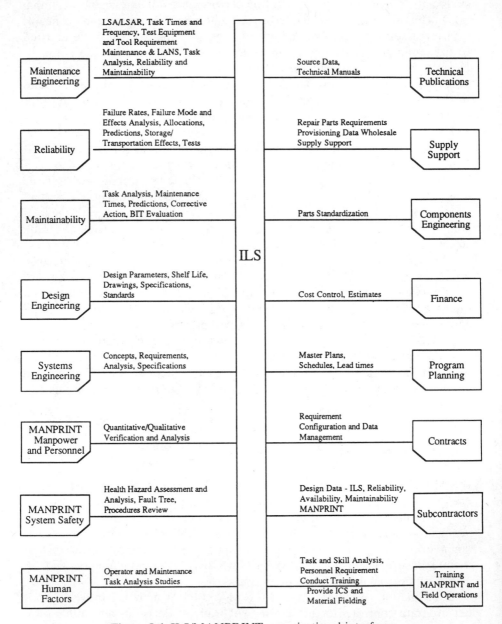

Figure 8.1 ILS/MANPRINT organizational interfaces.

ILS efforts require participation by all functional areas within and some key functional areas outside the logistics organization as shown in Figure 8.1. MANPRINT engineers provide skill capability, personnel–equipment interface analyses, and MANPRINT test and evaluation requirements.

8.3   THE FUNDAMENTAL GOALS OF MANPRINT

The goals of MANPRINT are to improve:

- Total system performance, by including human performance as an integral element. Total system performance is a function of equipment performance and people performance as they each are affected under varying environmental conditions, which include physical, social, and operational conditions.
- Manpower and personnel utilization. All too often personnel requirements are dictated by materiel system design.
- Unit effectiveness by affecting the first two goals. This will enhance the ability of units to perform their mission.

To effectively accomplish these goals, the MANPRINT issues must be addressed early and continuously throughout the materiel acquisition process.

8.4   MANPRINT IN LIFE-CYCLE SYSTEM MANAGEMENT

MANPRINT in the Preconcept Exploration Phase

MANPRINT must be considered and integrated during all phases of the life cycle for all materiel systems. MANPRINT in this phase considers the human element in terms of manpower, capabilities, skills available or achievable, and forecasted training capabilities and training burden.

MANPRINT in the Concept Exploration Phase

MANPRINT analyses must be accomplished in sufficient detail prior to initiation of concept exploration to provide a baseline to which technical approach alternatives and their resulting MANPRINT implications can be compared. MANPRINT requirements and constraints must be established for inclusion in requirements and solicitation documents. MANPRINT data must be developed to determine probable and projected MANPRINT requirements, develop planning for personnel support and training programs, and support operational and organizational concepts. Estimates of manpower and personnel costs, including training costs and projections of the cost of recruiting and retraining service members with the required aptitudes, must be developed and considered in cost effectiveness analyses.

MANPRINT in the Demonstration–Validation Phase

MANPRINT data to support the quantity of systems processed and the identification of personnel requirements are developed during this phase. The development of an initial training strategy, identification of training devices and aids, and update of the system MANPRINT management plan (SMMP) is also accomplished during this phase.

MANPRINT in the Full-Scale Development Phase

All issues initiated previously are continued and updated. In addition, engineering change proposals are reviewed for MANPRINT implications during this phase.

MANPRINT in the Production–Deployment Phase

During this phase, care must be taken to ensure that MANPRINT actions have been completed and the materiel is ready for fielding. New equipment training and institutional training must be ready to prepare service members to operate, maintain, and support the emerging materiel. Manpower spaces must be documented with sufficient leadtime to ensure that personnel with the requisite skills and abilities are available to fill the spaces. And finally, personnel assignment policies must be established to support initial fielding and sustainment.

8.5   HUMAN PERFORMANCE AND RELIABILITY

Although human performance and reliability might be considered to be overlapping terms, it is useful to consider them separately. "Human performance" is the degree to which an individual is able to accomplish a task or series of tasks under specified conditions to meet a specified standard. "Human reliability" refers to the probability that a human will make an error in the operation, maintenance, or support of a system. Equipment reliability, measured by mean time between failure (MTBF) and mean time to first failure (MTFF), affects system results. Human reliability, measured by mean time between human error (MTHE) and mean time to first human error (MTFH), also affects system results. Thus the study of human reliability separately from performance will provide important estimates of the probability of mission success.

8.6   THE DESIGN CHALLENGE

The human contribution to a system must be designed into system effectiveness, system reliability, system durability, and cost-effectiveness. MANPRINT seeks to lessen the impact by improving the system design rather than extending training time or assigning more or higher aptitude personnel. It is imperative that human performance and reliability impediments be eliminated early, beginning with system concept design.

8.7   OBJECTIVES OF THE MANPRINT PROGRAM

The objectives of the MANPRINT program are to optimize total system performance. It will bring options and a better understanding of the impact of hardware–software designs on human performance and reliability. The disciplined approach to achieve the MANPRINT objectives is reflected below:

- Influence materiel system design for optimum total system performance by considering human-factors engineering, manpower, personnel, training, system safety, and health hazards before allocating functions between people, hardware, and software.
- Ensure that materiel systems and concepts for their employment conform to the capabilities and limitations of the fully equipped personnel, operating, and maintaining the materiel in an operational environment.
- Assist the trainer in determining, designing, developing, and conducting sufficient, necessary, and integrated training.
- Improve control of total life-cycle costs of soldier–machine systems by ensuring consideration of the costs of personnel resources and training for alternative systems during the conceptual stages and for the selected system during subsequent stages.
- Develop a unified, integrated MANPRINT database to define ranges of human performance. Compare these ranges against system performance and provide for the timely development of trained personnel.
- Provide MANPRINT data for the development of manuals, training, media, and technical publications. Ensure that the content of these publications matches the personnel capability in aptitude, education, and training to perform the operator–maintainer tasks.
- Apply MANPRINT concepts with current educational technology to design and develop school and unit level training, embedded training, and training devices.
- Influence the manpower, personnel, and training (MPT) related objectives of the ILS process.

- Integrate combat development and technology base information systems with personnel long-range planning.
- Ensure that personnel trained for specific force modernization systems are assigned, in sufficient quantity to support fielding and sustainment, to units and positions for which they are trained.
- Ensure integration of MANPRINT test and evaluation requirements objectives, issues, and criteria into the test and evaluation master plan (TEMP). Ensure that MANPRINT information from tests and evaluations is fully considered in hardware design decisions and the materiel release decision process.

8.8   APPROACH TO MANPRINT ROLES AND RESPONSIBILITIES

Among the key participants in the materiel acquisition process are combat developers, materiel developers, researchers, training developers, logisticians, and industry management. Each must deal with many factors of the program to reach execution and subsequent industry involvement with each program. This requires continual communication and interaction.

MANPRINT does not reduce current mission responsibilities but forms an interactive and mutually interdependent process to improve system performance. As the development process matures, through joint government and industry effort, MANPRINT seeks to ensure that original influences remain despite the potential for changes in the production phase. The MANPRINT initiative provides a variety of options (alternatives for balancing training, force structure, manpower, personnel, and equipment burdens) to key decisionmakers. Active participation by both government and industry is the cornerstone to successful application of MANPRINT, and results in the best designed and most supportable system possible.

Several key issues must be applied and tailored somewhat, depending on acquisition phase, strategy, and system complexity:

1. Anticipate human performance and reliability issues associated with future designs or changes to existing design.
2. Communicate the contract MANPRINT requirements, including goals and constraints, target audience description, and so on to design engineers so as to influence design decisions.
3. Oversee the execution of MANPRINT requirements and advise top management of deficiencies so that corrective actions may be implemented.
4. Participate in design and other reviews to raise MANPRINT issues and to assess MANPRINT progress.
5. Conduct the various MANPRINT analyses called for in the MANPRINT statement of work (SOW) and in specifications or oversee their

conduct if the function is conducted by other than a dedicated MAN-PRINT organization.

6. Examine the interrelationships among the six domains with respect to such design activities as functional allocation.

7. Participate in development of test and evaluation plans to ensure that the validation of MANPRINT requirements is built into the overall test evaluation program.

8. Develop the analytical techniques and employ established analytical techniques as required to conduct the contractually stipulated MAN-PRINT program.

8.9 MANPRINT PROGRAM MANAGEMENT AND ORGANIZATION

MANPRINT requires a comprehensive management and technical effort, starting before program initiation and extending throughout the system life cycle. MANPRINT management and timing emphasizes scheduling activities early in the acquisition process. These activities are documented by the MAN-PRINT joint working group (MJWG), an interdisciplinary organization, in the SMMP activities. The resources required to conduct the MANPRINT-related studies and analyses are identified and become the basis for MAN-PRINT-oriented funding requests.

Of equal importance is the preparation required to permit the MANPRINT element of the material developer's office to provide timely input to industry via the request for proposal (RFP). Industry should be provided draft requirements documents and RFP's for review and comment. Soliciting industry comments early aids in the establishment of reasonable requirements for all materiel acquisition participants.

Just as effective MANPRINT activity is highly dependent on a total understanding of the life-cycle system acquisition events and the ability to provide timely MANPRINT inputs, so, too, is industry's understanding of specific requirements in the design process. Industry should prepare itself organizationally to integrate MANPRINT into the design process. Many companies already have organizations and expertise in place to deal with some of the individual domains. Generally, these include human-factors engineering, training, systems safety, and health hazards. Industry capability to address the manpower and personnel domains is less prevalent.

DoD does not want to dictate how and where industry should integrate MANPRINT. Any number of potential approaches may be taken. MAN-PRINT management may be assigned to the system engineering department, logistics department, training group, human-factors engineering group, mission analysis cell, or another appropriate organization. Indications that the corporate leadership is both aware of MANPRINT requirements and has included them in the technical solution to the requirement is an additional critical factor in the selection process.

The SMMP is the master planning document for MANPRINT activities during system development and is updated as needed throughout the acquisition process. It is the first program management document in the entire acquisition cycle and is initially prepared by the MJWG in the same time frame as the O&O preparation. The SMMP lays out the MANPRINT goals and constraints, issues, areas of concern, data needs, data sources, analyses, tradeoffs, milestones, and decisions that must be made to ensure that MANPRINT is considered in the acquisition process. It is recognized that the SMMP may be rudimentary during its initial phase but will increase in content and specificity as the system development proceeds.

Industry has already accomplished a great deal in developing and engineering new technology. MANPRINT should be considered by contractors in their efforts prior to receiving a contract. The contractor will be required to develop a plan for accomplishing MANPRINT. MANPRINT goals, constraints data, and analyses should be identified in the request for proposal. The industry SMMP should serve as a planning and management guide and an audit trail to identify the tasks, analyses, tradeoffs, deliverables, and recommendations that must be made to address the MANPRINT issues during system design. Industry's version of the MJWG would be responsible for developing the contractor SMMP. The industry SMMP should reflect the contractor's plans and program to accomplish the service's MANPRINT goals and requirements.

8.10 MANPRINT AND PRODUCT IMPROVEMENTS

Product improvements can be very consequential in affecting the MANPRINT domains. In addition, MANPRINT issues must be considered in the configuration decision. The system will normally have been fielded long enough for empirical MANPRINT data to be available. Changes from the original baseline should be identified and assessed. A determination should be made as to whether the original MANPRINT goals and constraints may be breached because of the extent and nature of the product improvement.

When it is agreed that a product improvement should proceed for approval and funding, the product improvement management information report (PRIMIR) is the basic document and report on a product improvement. The PRIMIR is the document used in the prioritization process. MANPRINT issues and concerns should be included in the PRIMIR, particularly where it is judged that MANPRINT issues are a primary reason for desiring the product improvement approval.

8.11 MANPRINT AND NDI PROCUREMENT

Nondevelopment item (NDI) procurement is growing in importance as a method of supplying systems and equipment. NDI procurements can be both large in dollar value and receive service-wide distribution. Market investigation is called for early in the concept formulation phase to assess the availability and viability of NDI as an alternative acquisition strategy preferred over new system development. While MANPRINT in an NDI cannot influence design except where modifications are approved, it can become a critical discriminator in deciding whether an NDI strategy should be approved by the decision body.

An eventual NDI program starts in the same way as any other program, namely, with the identification of a deficiency through the mission area analysis (MAA), followed by the development of the SMMP by the MJWG and an O&O plan. The SMMP MANPRINT goals and constraints, issues, areas of concern, analyses, and tradeoffs are used during the market investigation to determine whether any equipment available in the marketplace meets requirements to include MANPRINT requirements as shown in Figure 8.2.

It should be kept in mind that for commercial, off-the-shelf equipment, the marketplace data and information will have to be expertly assessed to determine if MANPRINT requirements have been met. From a MANPRINT standpoint, a human-factors engineering analysis (HFEA) should be conducted so that experts in each MANPRINT domain will provide an assessment of how well marketplace equipment complies with MANPRINT requirements. For NDI consisting of already developed equipment, a great deal of data and information will have been derived during the development process. Here again the data and information needs of the SMMP should be matched against what is available. If critical MANPRINT issues, questions, and con-

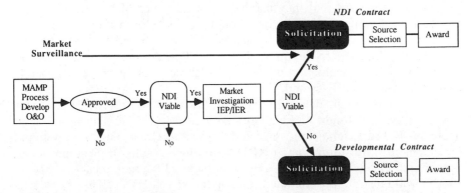

Figure 8.2 The NDI decision process. (*Source: MANPRINT Primer,* HQ Department of the Army, June 1988.)

cerns cannot be addressed because of a lack or data or other reasons, additional effort may be required before the NDI decision is made.

8.12   MANPRINT FUNDING

Implementation of an effective MANPRINT program requires resources—people and money. MANPRINT program management may require more money up front for information that results from analyses, studies, assessments, and evaluations. A key document in determining MANPRINT program management information requirements is the system MANPRINT management plan (SMMP). This document and other studies like human-factors engineering analysis, system safety tasks analysis, and so forth require contractual funding.

The ability to effectively plan and budget for the MANPRINT tasks and analyses is ultimately a matter of identification of requirements, the development of sound cost estimates, and the timely inclusion of these funds into the programming and budgeting process. For those tasks and analyses identified in the SMMP for conduct after program initiation (O&O plan approval), the MANPRINT requirements should be integrated into the tasks and analyses required for the system and funded by the appropriate command or agency. While MANPRINT dollars will not be separately identified, it will be recognized by the materiel developer that, like ILS, MANPRINT dollars are included in the overall funding. This will enhance the MANPRINT community's ability to ensure that the SMMP tasks and analyses are accomplished during contract performance. The MJWG should also be prepared to deal with tradeoffs of tasks and analyses, or work-a-rounds if it develops that all of the desired funding is not obtainable.

From an industry standpoint, there should be an awareness of the types of tasks and analyses that they may come to expect as MANPRINT makes its contractual impact. Contractors should also develop a menu of existing and modified analytical tools to be in a position to respond competently and effectively to MANPRINT requirements in RFP. MANPRINT responses will not stand up in the evaluation process unless these techniques are adequately described and properly costed.

Funds for research, development, test, and evaluation (RDT&E)—also called "Program 6 funds"—are a key source for MANPRINT program management. RDT&E funds are generated through the long-range research and development acquisition plan (LRRDAP), which results in program development increment packages (PDIP) submissions. Planning ahead for MANPRINT will facilitate programmed funding for MANPRINT. This is essential because a key source of MANPRINT funds should be the program manager's budget.

8.13   SYSTEM ANALYSIS AND MANPRINT

Current system analyses address engineering analyses, engineering performance capabilities, and operational performance and force design capabilities. Engineering analyses include item system models—components of a system— and environmental models. In addition to the hardware-oriented analyses, personnel-oriented assessments—such as workload and training analyses— need to be conducted.

Historically, as reflected by each of the ellipses in Figure 8.3, these analyses have been conducted independently and not integrated or linked together. If the system is to be viewed as a whole, these analyses need to be linked and sensitive to all system variables. An integrated system necessitates some degrees of analytical commonality. In addition to linkage, consistent input variables—such as usage rates—should be used. Finally, and particularly important from a MANPRINT perspective, personnel characteristics and performance must be considered in determining force effectiveness.

To accomplish MANPRINT, developers, trainers, acquisition specialists, testers and evaluators, logisticians, laboratory scientists, and engineers must apply the appropriate analytical techniques to predict, understand, and control the relationship between force structure, military personnel, and technology. These analytical techniques can be classified in a number of ways, but it may be most helpful to relate MANPRINT analytical techniques to the hierarchy of effectiveness measures described in TRADOC PAM 11-8 and shown in Figure 8.4, which illustrates that there are multiple levels of measures of performance and effectiveness. Measures at one level are dependent on

MANPRINT

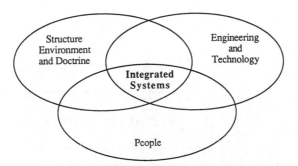

Figure 8.3 Integration of MANPRINT variables.

one or more measures at a lower level. Figure 8.4 shows information that could be used in selecting the best artillery system. The materiel acquisition system has traditionally focused on engineering characteristics (level IV) and hardware performance (level III). MANPRINT requires consideration of the personnel as well as engineering characteristics and performance at all levels. For the model to work, personnel performance must be integrated into the model.

The MANPRINT process, as captured in the SMMP, serves as the basis for identifying information needs and availability. The description identifies:

- System capabilities to include performance requirements.
- Selected acquisition strategy.
- Involved agencies.
- Previous guidance and decisions.

A Way to Look at Effectiveness Modeling

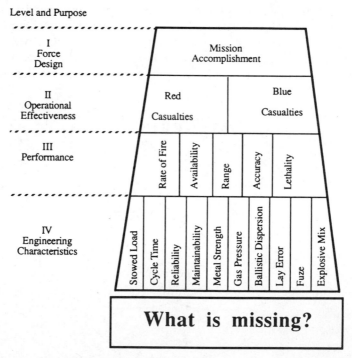

Figure 8.4 A way to look at effectiveness modeling. (*Source: MANPRINT Primer,* HQ Department of the Army, June 1988.)

System capabilities serve as the basis for developing MANPRINT concerns and questions to be resolved. This becomes the information needed that will be further developed as the system evolves. In addition, the acquisition strategy defines the parameters within which the MANPRINT technical effort must plan to operate. Finally, the previous guidance and decisions should contain goals, constraints, and planning factor information to ensure a coordinated and consistent technical effort.

MANPRINT objectives will add to the information needed by establishing requirements that must be researched and analyzed. Information that is most critical to ensuring MANPRINT requirements are identified and integrated into the design process has the highest priority and must be supported with available resources. Failure to develop essential MANPRINT information carries a high risk that performance, supportability, and cost objectives will not be met. Information needs that have a minor impact on MANPRINT objectives should have a lower priority because failure to develop that information is less risky. Figure 8.5 indicates the types of analyses that may support the design process.

8.14 MANPRINT AND READINESS

Personnel performance and reliability are key ingredients of readiness. Equipment readiness is a function of reliability and maintainability. These two attributes are driven by equipment and personnel characteristics, matching equipment to the skills and abilities available in the target audience. Improvements made in this area should result in the equipment compensating for the personnel, resulting in reduced manpower, personnel, and training requirements.

Training status is based on ability to meet performance requirements. This is a function of the training requirement and resources available. The training strategy and concept address both these issues early in an attempt to ensure the training system acquired with the equipment is both effective and supportable.

8.15 MANPRINT DOMAINS—ISSUES AND CONCERNS

Introduction to Interrelationships between Domains

MANPRINT enhances human performance and reliability through the integration of the six domains (manpower, personnel, training, human factors, system safety, and health hazard) throughout the system acquisition process. The quantitative or qualitative altering of one MANPRINT element may have substantive ripple effects across the others.

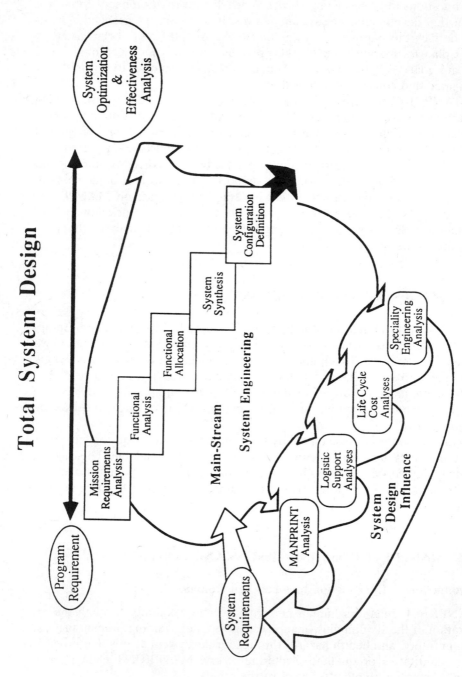

Figure 8.5 Total System Design. (*Source: MANPRINT Primer*, HQ Department of the Army, June 1988.)

Manpower

Manpower focuses on the determination of essential human resource requirements, which requirements will be supported with authorizations, and what the personnel demands associated with those authorizations will be by grade and skill. The manpower and personnel domains interface and overlap at many points. The difference is that manpower deals with defining the human resource demand ("spaces") while personnel focuses on supporting this demand through the acquisition, training, and assignment of people ("faces").

During system development, the concern in the manpower domain is to determine the system's impact on U.S. Army manpower resources and to ensure that each system is optimized from a manpower viewpoint. The force structure implications of the system must be identified. In addition, appropriate goals and constraints regarding the system's human resource demand should be established in terms of affordability and supportability. Early in the development process, based on force structure and organizational design guidance provided, a manpower "footprint" into which the prospective system must fit is developed.

The manufacturer will be furnished this force structure "footprint" in terms of manpower goals and constraints. The manufacturer will be required to demonstrate that these stipulated goals and constraints have not been breached by the system design. And also that the desired total system performance can be achieved with the desired manpower requirement. This will require the manufacturer to consider manpower in the basic design decisions that will impact on task, workload, function allocation between the person, person–machine and machine, and the operational environment projected for the system. Tasks considered must include not only those directly related to the equipment but also the off-equipment tasks that the personnel perform. The operational environment, possibly requiring continuous operations, stress, or extreme climatic conditions, must also be included. Resulting manpower requirements are measured in terms of personnel performance, which allows total system performance to meet the required criteria.

Maintaining manpower requirements within the force structure guidance provided is critical to system development. If increases beyond guidance are required, these increases analysis and programming functions are affordable and will be used to determine whether to ensure that overall DoD end-strength constraints are met. This could create a situation in which the system's manpower requirements are not fully supported by authorizations. This decreased level of manning may degrade actual system performance achieved after fielding to below the level desired. Since early system design decisions will dictate the resulting manpower requirement, early-on manpower analysis and trade-offs are necessary to prevent unanticipated, or unsupportable demands made at system fielding.

Personnel

As indicated above, the manpower process will identify the number of personnel required and authorized. These authorizations will be defined in terms of MOS and skill level (grade). The personnel community must then acquire and assign properly trained, qualified people to fill these established authorizations.

During system development an objective of MANPRINT is to obtain a match between the system requirements and the characteristics of the individual service members and crews who will operate and maintain the system. It must be recognized that individuals vary across many dimensions, including their cognitive, physical, psychomotor and skills and their background and experience as shown in Figure 8.6.

The personnel domain must be concerned with the quality of individuals required by a new system. The services compete with each other in recruiting personnel as well as private industry and institutions of higher learning. The number of quality individuals that can be recruited each year is limited and very costly. These quality individuals must be distributed across all MOSs that make up the force to ensure combat effectiveness in all areas. The aggregate demand for quality must stay in line with what is available. Each new system must be kept within established quality requirement constraints.

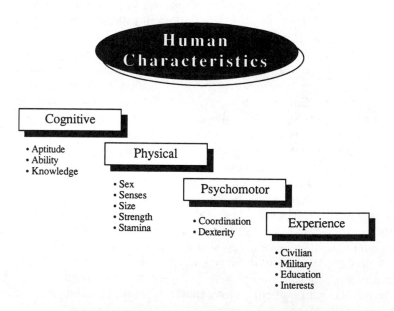

BOTTOM LINE: Access and retain quality people

Figure 8.6 Personnel characteristics. (*Source: MANPRINT Primer,* HQ Department of the Army, June 1988.)

If not, either a disproportionate distribution of quality or a manning shortfall will result.

Training

In the most basic terms, training is the process that prepares personnel to perform specific jobs. Performance standards are established to define how well the service wants the service member to do the task; and finally, performance reflects the service member's ability to accomplish the desired tasks.

As a MANPRINT concern, training goals and constraints must affect system design in a positive way. Traditionally, system designers have not been constrained by the service's training resources. The training community generally faced a completed system design and was asked to structure a training concept that would accommodate the operational and maintenance needs of that design.

The starting point for training is to develop the training *strategy*—who, what, where, and when. The "who" is defined in the target audience description, "where" will be governed by considerations of training transfer and the impact on the operating strength of the MOS, the "what" includes all equipment-related and other personnel tasks, and the "when" is governed by consideration of timing and skill decay. The training *concept* then defines how the training will be accomplished by considering training delivery options such as embedded training, training devices, and resources.

The training strategy and concept (which includes resource consideration) become the basis for developing training goals and constraints. Figure 8.7 shows the training design process in parallel to the equipment design process. By providing training goals and constraints at the start, the training strategy and concept will thus be considered during function allocation. During trade-off analysis, cost, performance, and supportability will be considered. In the process, design decisions will be affected by training considerations.

Human-Factors Engineering (HFE)

It is important to relieve any definitional confusion between HFE and MANPRINT because it is obvious that some of the MANPRINT domains are shown in HFE definitions. First, it should be recognized that HFE was institutionalized as a program long before the advent of MANPRINT. In a real sense, then, HFE has always attempted to address some issues not encompassed by MANPRINT. The MANPRINT program gives a new emphasis to the human resources and human capability areas as related to system performance. MANPRINT relies strongly on the HFE program. In fact, in contracting for system concepts or system development, MANPRINT may employ MIL-H-46855B as the basic document to implement the MANPRINT program. One of the distinctions between MANPRINT and HFE, as currently defined, is

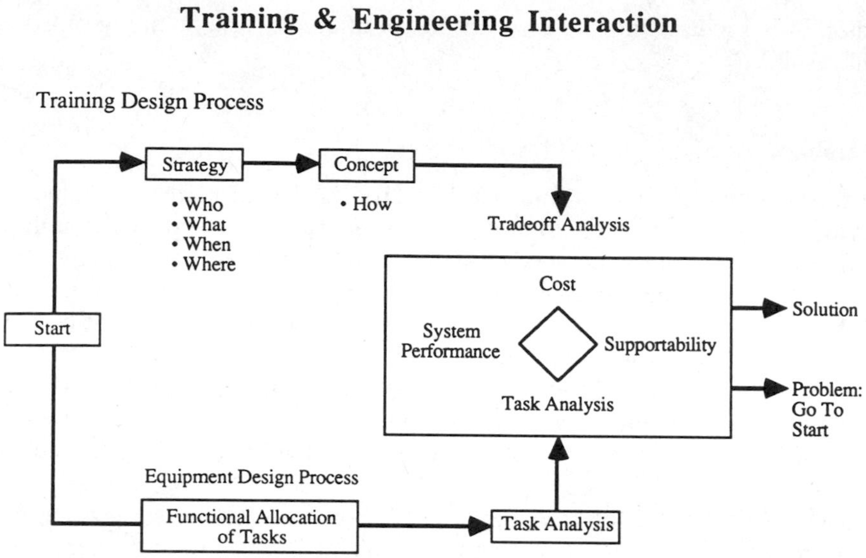

Training & Engineering Interaction

Training Design Process

Figure 8.7 Training design process. (*Source: MANPRINT Primer,* HQ Department of the Army, June 1988.)

that MANPRINT is specifically oriented to integrate all of the MANPRINT domains.

In the broadest sense, HFE is concerned with eliminating, through design, the typical sources of human error depicted in Figure 8.8. HFE deals with determination of whether a system is a manual, semiautomatic, or automatic, and of the functional integration into the system of the soldiers who operate or maintain it. HFE recognizes that human reliability and human performance are integral parts of system reliability and system performance. The development process must adapt to the fact that human performance deteriorates with prolonged stress caused by lack of sleep, heat, noise, fatigue, isolation, overwork, and so on. It must also adjust to such fundamental human reliability–performance influences as motivation, conflict, fear, and so forth.

Human-factors engineering is concerned with the design, development, testing, evaluation, and deployment of manned systems so that soldiers will be able to operate and maintain military systems at their optimum performance levels. This includes the systematic investigation of how the design of the service member's job and the tools that are provided affect that individual's capacity to do the job. The major emphasis is on system reliability and performance, personnel–equipment compatibility, understanding of cost–benefits, and ultimately the achievement of user acceptance. Without proper attention to HFE early in design, system flaws and deficiencies will surface that are difficult, costly, and time-consuming to deal with once the design is relatively frozen.

Typical Sources of Human Error

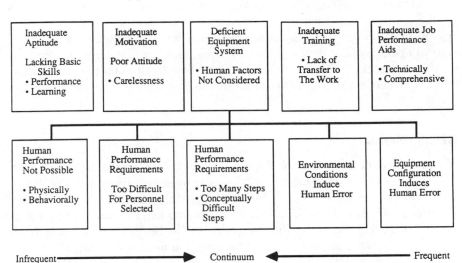

Figure 8.8 Typical sources of human error. (*Source: MANPRINT Primer,* HQ Department of the Army, June 1988.)

System Safety

The services have the responsibility to ensure that hazards to personnel are not system induced. As systems become more complex, and the battlefield reflects the doctrine of continuous and sustained operations, the service member's exposure to system hazards increases. The system safety program is designed to identify and measure safety hazards with the objective of:

- Maximizing operational readiness and mission performance through accident prevention.
- Ensuring safety and health risks are eliminated and residual hazards are formally accepted and documented.
- Minimizing safety retrofits.
- Ensuring that equipment modifications and doctrinal or procedural changes do not lessen safety and health aspects of a system.
- Applying system safety engineering and management principles to developing technology for new systems.

The goal of system safety is to design equipment so that safety considerations do not adversely affect performance or increase demands on manpower, personnel, or training resources. No safety hazards will be accepted without formal documentation of associated risks.

Industry conducts its own safety program that parallels that of the Army MIL-STD-882, which details the tasks and activities that are to be performed by the contractor to identify, evaluate, and eliminate safety hazards of a system or to reduce their associated risks to a level acceptable. Prior to the start of operational or developmental testing, industry is required to produce a safety assessment report that summarizes the hazard potential of a system and recommends procedures to reduce risks to test personnel to an acceptable level.

Health Hazards

Advanced technologies and sophisticated, complex systems have brought with them greater potential of harm due to greater noise, overpressure, shock and vibration, higher levels of toxic fumes, gases, and chemicals, and a myriad of other conditions. These increases in the degree and intensity of hazardous conditions provide major reasons for concern. The heath hazards arising from new technologies, such as lasers and ionizing and nonionizing radiation, give reason for even greater concern. The health hazard assessment (HHA) program meshes with the system safety program in an effort to:

- Preserve and protect the health of all personnel.
- Enhance individual performance and system effectiveness.
- Reduce requirements for system design retrofits needed to eliminate or control health hazards.
- Reduce readiness deficiencies attributable to health hazards that bring about restrictions in training or operational restrictions.
- Reduce personnel compensation by eliminating or reducing injuries attributable to health hazards associated with the use of military systems.

The basic goal of HHA is to identify health hazards as early as possible for elimination and/or control. It is desired that the optimum degree of health features be integrated into a system design within the bounds of cost, operational effectiveness, and time.

The mental as well as physical hazards must be considered in order to minimize potential psychiatric casualties. These casualties can result from a lack of confidence in equipment, organizational or doctrinal isolation, and a nonsupportive social environment.

HHA are not automatically triggered by some activity or even in the materiel acquisition process. These assessments are initiated only on formal request through the Surgeon General's Office (TSG). The formal HHA reports usually become part of the HFEA, which covers all the MANPRINT domains. The HHA is updated based on new or more mature data prior to each Milestone review.

HHA procedures are integrated throughout the materiel acquisition process. In the design process, health hazard analyses are conducted to evaluate hazard severity and probability, to assess risk, and to determine operational constraints. This effort also identifies required precautions, protective devices, and training requirements to minimize potential hazards. Later in the materiel acquisition life cycle, the HHA is used to assess contractor performance and to ensure that health hazard recommendations are incorporated in doctrinal, maintenance, and training publications. Figure 8.9 reflects the general process for influencing the design process.

8.16 MANPRINT INTO THE FUTURE

The full implementation of the MANPRINT program requires the active participation of both the DoD and industry. Its success rests on the combined efforts of all communities involved in the acquisition process. In executing the MANPRINT program, the DoD and industry have discovered confusion and misunderstanding over methods and results of analyses. Generally this is due to a lack of standardization of key analytical techniques such as work-load and task analysis, lack of operational definition of key planning factors such as performance standards, manning levels, personnel cost factors, usage

HOW?

Process Model at the Level of Specific Design Decision Making

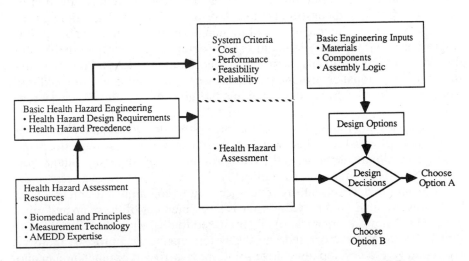

Figure 8.9 Health hazards "HOW?" (*Source: MANPRINT Primer*, HQ Department of the Army, June 1988.)

rates, equipment densities, and replacement schemes, and failure to address force level and contingency analyses.

MANPRINT calls for an expanded focus that not only encompasses the man–machine interface but also addresses the total system–workforce interface and environment. The objective of the MANPRINT program is to optimize total system performance. The success and future of the MANPRINT program is in providing decisionmakers with more complete information and data to assess the numerous variables and alternatives involved in choosing an optimized system design, and in integrating combat, training, and material development with personnel resources and capabilities during all phases of the acquisition process. In the future, to provide guidance and gain management visibility over these issues, the Department of Defense is considering changes to DODD 5000.1, 5000.2, and 5000.39. In addition, research efforts are focused on new analytical tools and techniques for workload and manpower requirements estimation, dynamic anthropometrics modeling, rapid prototyping, and user-rated simulators tied to computer-assisted design systems.

As we look to the future of the MANPRINT program, it is clear that it has started down the road to acceptance. However, it is equally clear that it will not solve all of the system acquisition problems. There will always be some problems that must be managed, however, as the MANPRINT program becomes more fully institutionalized in the acquisition process.

8.17 SUMMARY

The MANPRINT program provides the means through the identification of objectives, goals, and constraints in each of the six MANPRINT domains to effectively manage the integration of human performance and reliability considerations in systems development. The program begins early in the system development process to ensure proper tradeoff and integration of force structure and operational environment, engineering and technology, and soldiers. The orientation is always on achieving the goal of optimum total system performance. To achieve this goal and fulfill the MANPRINT objectives, DoD agencies and industry involved in the system development and materiel acquisition process must be aware of their responsibilities and effectively manage their part of the MANPRINT program. The successful integration of MANPRINT requires a team effort. All players must do their part individually and collectively in order to successfully communicate internally as well as externally to industry. Industry is a key member of the team and must ensure that they are organizationally prepared to deal with MANPRINT and that MANPRINT is integrated into the system engineering process. Industry is encouraged to develop innovative solutions to arrive at optimal human and equipment performance. The total system performance is continuously evaluated throughout the process to ensure MANPRINT goals and constraints are met.

Thus, the keys to an effective MANPRINT program are teamwork, planning, and communications. The MJWG provides the internal mechanism for pulling together the team; industry should have similar organization. The team establishes the requirement and program; in industry it executes the program by affecting design. The plan for executing the program is the SMMP and its industry equivalent. The SMMP serves as the basis for information flow for MANPRINT requirements. The plan must be developed early in the system life cycle and resources requested to execute the plan. The plan becomes the basis for identifying resource requirements and must be tailored on the basis of resource availability. Finally, the key to effectiveness of the MJWG and the SMMP is internal communication as well as coordination between DoD and the contractor throughout the entire developmental process.

The impact of considering military personnel during system design and development is far reaching. If done properly, system performance will be enhanced; manpower, personnel, and training resources will be more effectively used; force capability will improve; and unit readiness will increase. The bottom line of MANPRINT is improved force effectiveness through leveraging technology with the service member in mind to achieve a high return on investment.

9

CONFIGURATION MANAGEMENT

Configuration Management (CM) is an accepted and approved management discipline. It is a system for recording established military requirements for materiel; ensuring that all changes affecting these requirements are reviewed for total impact and cost effectiveness; and maintaining adequate records of the requirements, changes, and hardware status throughout the life cycle of materiel. Configuration management for large and complex systems can be regarded as the means of controlling and documenting the systems development and production descriptions as well as being the source of information for provisioning, manuals, maintenance evaluation, maintenance allocation and other data tasks concerned with insuring system readiness at the time of deployment. Configuration management is equally applicable for the production of consumer goods, although the process may not be as elaborate or burdensome as it is for military hardware contractors. Configuration management practices for the General Dynamics Air Force F-16 FSD and Production program are organized based on MIL-STD-480. The same practices affect many other systems such as Raytheon's PATRIOT missile system, General Dynamics M1 Abrams Tank, and FMC Corp's Bradley Fighting Vehicle.

9.1 CONFIGURATION MANAGEMENT AND SYSTEMS ENGINEERING

Configuration management is the systems engineering management process that identifies the functional and physical characteristics of an item during its life cycle, controls changes to those characteristics, and records and reports change processing and implementation status. Configuration management is thus the means through which the integrity and continuity of the design, engineering, and cost tradeoff decisions made between technical performance, producibility, operability, and supportability are recorded, communicated, and controlled by program and functional managers. One reward of effective configuration management is improved supportability, including updated technical manuals, identified spares, identical–interchangeable equipment, and known configuration.

A fundamental concept associated with system–project development is the use of three baselines to ensure an orderly transition from one major decision point to another. This concept is illustrated in Figure 9.1.

The systems engineering process interfaces with configuration management through technical data that establish baselines to which configuration management procedures are applied throughout the life cycle. This is done by established procedures that identify the complete technical description of the system as it evolves, controls the documents that provide this identification, and continually updates the documentation to reflect the approved configuration of the system. The output of the systems engineering process in the conceptual phase provides the functional configuration identification. This identification translates the required operational capability into performance and design requirements, design constraints, inter- and intrasystem interfaces, test and evaluation requirements, and functional areas of the system that are documented in the system specification at the end of the conceptual phase. During the validation phase, the systems engineering process provides the

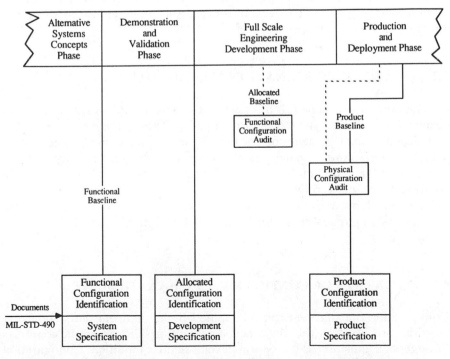

Figure 9.1 Configuration management as related to the acquisition life cycle.

allocated configuration identification, which consists of a series of development specifications that define the functional and test requirements for each major configuration item. During FSD, the systems engineering process provides the product configuration identification, which includes product specifications.

The functional baseline, allocated baseline, and product baseline serve as systems engineering management reference points. They represent the progressive development of specifications, drawings, and associated data. These technical data progress from the general requirements to detailed requirements and provide a level of control that is initially broad in scope. Eventually, they are narrowed to be more restrictive as the design becomes more definitive. A constant closed loop relationship must be maintained between established design requirements and design effort, thereby ensuring that the design effort is at all times directed to meet, rather than exceed or fall short of, total system requirements. These baselines also represent the progressive and evolutionary development of system documentation. System elements developed by the systems engineering process and described in the baselines define product elements of the work breakdown structure.

At any given time, configuration management can supply a current description of a developing hardware unit, software unit, system, and so forth, and provide traceability to previous baseline configurations of that item. Configuration management also contains complete information on the rationale for configuration changes, thus permitting analyses and correction of deficiencies when they arise.

9.2   WHAT IS CONFIGURATION MANAGEMENT?

Configuration management is a collection of tools, techniques, and practices designed to reduce costs and improve quality. These aims are achieved by technical and management control techniques. As an arm of project management, configuration management provides for greater visibility and evidence of order and control. It is a tool that defines the product and then controls the changes to that definition.

Configuration management is concerned with form, fit, and function. Form and fit relate to physical characteristics, while function relates to the functional characteristics.

9.3   INITIATION OF CONFIGURATION MANAGEMENT

Configuration management can be initiated by inputs from the systems engineering process as early as the concept exploration (CE) phase and continues throughout acquisition as the system develops and is modified. Configuration changes occur throughout the life of a system as more knowledge of the system design, operation, and maintenance concept is gained; as mission requirements change; or as nontechnical factors such as cost and schedule influence the design. These changes must be controlled to ensure that they are cost-effective, and that they are properly documented so that all users are aware of the current configuration status.

9.4   ESTABLISHING THE BASELINE CONFIGURATION

One of the more important aspects of configuration is the concept of baseline management, which is formally required at the beginning of the acquisition program. Baseline management controls the definition and identification of baselines. It is a framework for the progressive definition of the product and the establishment of control points. Baseline management plays an important role in the control and change process. It provides a mechanism with which to manage a system by control of progressive definitions, throughout the life cycle of the system. Baseline management also addresses the subject of supportability.

There are three types of baselines: functional baselines, allocated baselines, and product baselines as shown in Figure 9.2, which contains a generalized

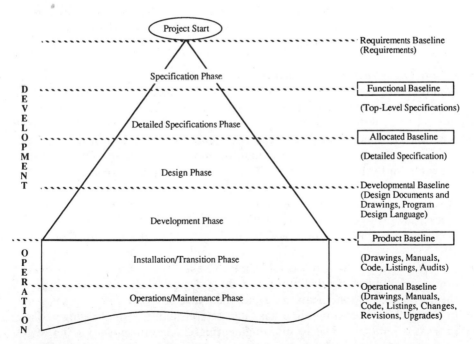

Figure 9.2 Baseline management. (*Source:* By permission from M.A. Daniels, *Principles of Configuration Management,* Advanced Applications Consultants, Rockville, MD, 1985, p. 67.)

baseline management structure. The figure centers on the growth of the system through its simplified life cycle.

Functional Baseline. The functional baseline is ideally established at the end of the CE phase. The system specification (Type A) or development specification (Type B) defines the technical portion of the program requirements. This initial system specification is part of the request for proposal package for demonstration–validation (D&V) and provides the basis for controlling the system design during the D&V phase. Once the system specification has been authenticated, formal configuration control is initiated. The authenticated system specification is the foundation for configuration management during the subsequent phases of the program.

Allocated Baseline. The allocated baseline consists of development specifications (Type B) that define the performance requirements for each configuration item. It is developed and established during the D&V phase, and incorporates the technical approaches developed to satisfy the functional objectives in the functional baseline. During the D&V phase, these objectives are translated through the systems engineering process into subsystem and configuration item performance requirements. Development specifications are usually included in the request for proposal (RFP) for FSD. On authentication, the development specification establishes the allocated baseline for a configuration item. The allocated baseline is the basis for detailed design and development during FSD.

Product Baseline. The product baseline is established by the detailed design document for each configuration item. The product baseline establishes the requirements for hardware fabrication and software coding as the initial article of each item is reviewed and approved by the government as satisfactorily meeting the specification requirements. The baseline will normally include product, process, and material specifications (Type C, D, and E) engineering drawings and other related data. The product baseline is established initially by approval of the product specifications. The configuration item product baseline is the basis of the production RFP and subsequent statements of work. The product baseline is verified by successful completion of the Functional Configuration Audit (FCA) and the Physical Configuration Audit (PCA).

It is important to avoid establishing baselines too early, as this severely limits the contractor's flexibility in making design tradeoffs. The relation of these baselines to specification development and program milestones is shown in Figure 9.3. Once a baseline has been established, a formal change control process is imposed. The formal engineering change proposal (ECP) review process ensures that all aspects of performance, cost, schedule, and interfaces have been considered before any change is implemented.

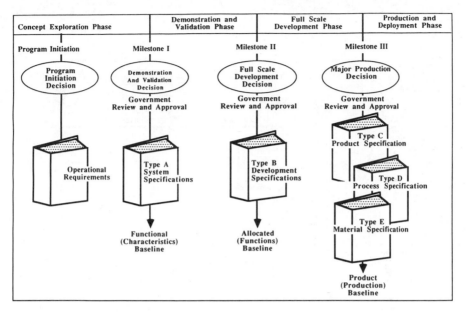

Figure 9.3 Relation of baselines to specification development and program milestones. (*Source: Systems Engineering Management Guide,* DSMC, 1986.)

9.5 THE CHANGE PROCESS

A major element of configuration management is the change process. The process itself consists of the recognition of the need for a change, analysis and documentation of the change proposed, review of the proposal and distribution for assessment of impact of the change, the impact of the assessment and decision, decision documentation and promulgation, and implementation and verification. The change process is a logical sequence of events and consists of three basic building blocks: initiation, decision, and implementation. Figure 9.4 illustrates the traditional configuration control process. However, a new approach is in the work by the U.S. Army.

The objective of this new approach is to integrate materiel change decisions with traditional management and funding mechanisms, obtain higher level visibility and control, allow for flexibility and responsiveness needed to operate on a day-to-day basis, and to vest control and authority at the lowest appropriate level. The scheme developed to accomplish these objectives is illustrated in Figure 9.5. As shown, the process distinguishes between change efforts that can be planned for in advance and those that are basically unanticipated. In the former case, the emphasis is on a development plan that lays out the long-range goals and objectives for the system and using existing planning, approval, funding, and management review methods applied to

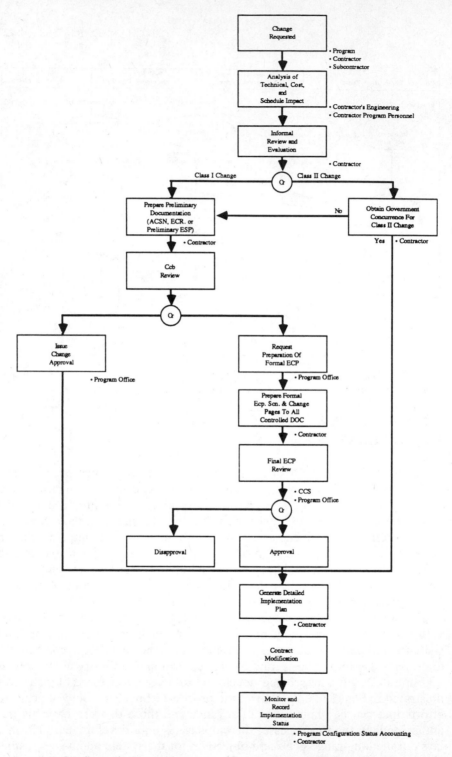

Figure 9.4 Configuration control process. (*Source: Systems Engineering Management Guide,* DSMC, 1986.)

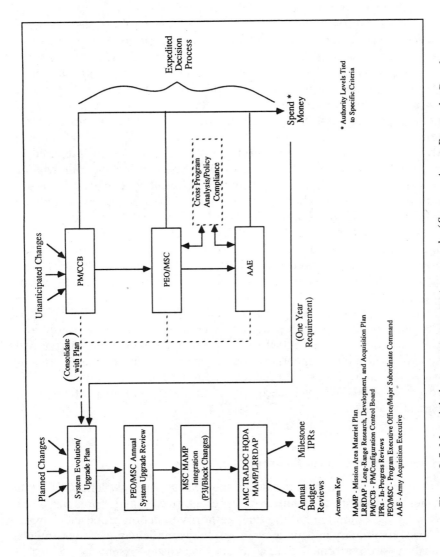

Figure 9.5 Materiel change management approach. (*Source: Army Research, Development and Acquisition Bulletin,* January–February 1988.)

229

block changes. In the latter case—the unanticipated requirements—what is required is an expedited decision process distributing decision authority and control to three different levels, with determination of the appropriate level based on criteria related to the nature and magnitude of the change.

The change process is initiated by government direction or corporate direction, interface working group activity, contractor design or test definition activity, or subcontractor action. A change package includes the following information: statement of the problem and description of the proposed change; alternatives considered; analyses showing that the change will solve the problem; analyses to ensure that the solution will not introduce new problems; verification of interface compatibility, including test, operations, safety, and reliability; estimate of cost and schedule impact; proposed specification or interface control document revision (ICD); and the impact, if not implemented.

After internal review by the contractor, Class I changes are prepared in either an advance change–study notice (ACSN) or an engineering change request (ECR). Occasionally, the change may result in a request for a deviation–waiver. Interface changes are defined on an interface revision notice (IRN). Following approval of the ACSN or ECR by the government, a formal ECP is developed containing the specification change notices (SCNs) and the change pages, together with supporting cost data. On ECP approval by the government, the implementation plan is issued. Implementation status is monitored by both the government and contractor configuration management organizations.

Changes to released ICDs are processed as defined for baseline changes with an additional step. Each proposed ICD change must be reviewed and approved by the interface control working group (ICWG) prior to approval of the ECP, which allows incorporation of the change into the hardware and software. For that purpose, a preliminary interface revision notice (PIRN) approved by the ICWG will accompany the ECP. Approval of the ECP approves the PIRN.

The Configuration Control Board (CCB)

The CCB plays an important role in the change process. The CCB is a permanently established committee of representatives of major organizational elements. This committee has the final authority to act on proposed changes at its level. Its primary mission is to ensure that complete impact assessments and analyses are performed. The CCB also establishes baselines.

During development, the government's CCB is responsible for reviewing and issuing changes to the configuration baseline. The CCB reviews all Class I ECPs to determine whether the change is needed and to evaluate the total effect of the change.

Classes of Changes

Usually there are two classes of changes: Class I and Class II. Class I changes require the approval of a higher authority level than Class II. Class I changes involve changes to form, fit, or function as well as costs, schedule, and safety. The CCB has the authority to approve, disapprove, to defer, or to pass to a higher authority change requests such as the ECP, the RFW (request for waiver) and the RFD (request for deviation). The CCB at every level becomes a focal point for configuration management, while ensuring traceability, feedback, and communications. Changes are prioritized as emergency, urgency, or routine as defined in DoD STD 480A, according to the criticality of the change.

Interface Control

A very important function within configuration control is interface control. The whole point is to orchestrate the interface to say who is responsible for what. Interface control is concerned with the coordination and exchange of data. The ICWG is subordinate to the CCB and somewhat similar to it but is responsible for the detailed technical interfaces only.

The change process and its control require tracking of historical changes to the product, so that problems may be traced to their origin. Configuration control requires several forms and documentation to effect traceability, to authenticate change decisions, and to provide for verification. The use of documentation requires a close interface between configuration management, data acquisition management, logistics publications, and contracts. In a loosely implemented CM system, design changes can occur without proper maintenance of the configuration change documentation after the baseline is established. Lack of good CM systems leads to many pitfalls, including an unknown design baseline, excessive production rework, poor spares effort, stock purging rather than stock control, and inability to resolve field problems. Poor CM is a leading cause of increased program costs and lengthened procurement schedules.

9.6   INTERFACE MANAGEMENT

Functional and performance interface requirements are contained in the appropriate segment or configuration item specifications. When functional requirements are completely allocated, interfaces may be incorporated in the design with full consideration for other design issues such as producibility.

Systems engineering and configuration management personnel must coordinate the large number of contractors and organizations participating in the design effort to ensure compatibility of all interfaces. For this purpose, interfaces are identified and coordinated by the ICWGs, which are generally organized by the prime contractor or system integrator, if the latter is a separate entity. A representative from the government program office may chair the group, although this position is often filled by the prime contractor, with a program office representative as co-chairperson to ensure resolution of any perceived contractual conflicts that may arise between associate contractors.

The ICWGs may be composed of several panels handling specialized interface areas. The chairman organizes these groups, ensuring that the proper specialties are supported by individuals with authority to commit their organizations or to obtain their organizations approval for ICDs developed by the ICWG.

The nature of the ICD varies considerably depending on the interface being documented. It can be physical interface, a radio-frequency interface, or an operational interface. Interface definition takes the forms of drawings, schematics, functions lists, data format diagrams, operational procedures, equations, and other data required by the designers to completely detail their design. The ICD does not duplicate the specification; rather, it describes the design implementation of the requirements in the specification.

The ICD outline is prepared by the prime contractor, and portions are assigned to parties responsible for the agreement of the interface between the system components, including human-factors engineering. Systems engineering must also ensure that various interfaces are compatible or do not force unnecessary costs on interfacing systems. Following completion, the ICD is signed by all parties involved and is placed under configuration management control. The ICD then has some status as a specification in that it represents the baseline configuration, and any changes must be acted on by the appropriate CCB. A number of CCBs may be involved in implementing an ICD change.

9.7 CONFIGURATION MANAGEMENT FUNCTIONS

Configuration management involves four functions:

- Configuration identification
- Configuration control
- Configuration status accounting
- Configuration audits.

The purpose of configuration identification is to document the unique configuration of a system and its components. The objective of configuration control is to ensure that engineering changes are properly defined and presented for management consideration. The purpose of configuration status accounting is to determine the actual configuration of delivered systems affected by retrofit engineering change proposals (ECPs). Configuration audits also allow for traceability.

Configuration Identification

Configuration identification is the process of recording and communicating information about the requirements, the design, or the product itself. This process is accomplished by the use of documentation coupled with naming and numbering. This documentation defines both the original approved configuration known as the "baseline" and the approved changes to the baseline.

Configuration identification involves the physical identification of the hardware parts and software units. The units are further delineated in a hierarchical, top–down manner into smaller detailed units. This process is similar to a work-breakdown structure. By configuration identification, fit, form, and function are defined. The key requirements for configuration identification are (1) identification of all data and all product elements, (2) centralization of the process, (3) well-written procedures for configuration identification (once the procedures are developed, constant reviews are needed to ensure their completeness, adequacy, and applicability), and (4) verification in the development and implementation of an effective configuration identification. Good configuration identification addresses traceability. Traceability is affected by how the configuration identification system operates.

Configuration identification is the family of specifications and drawings that describes the system during the design–development cycle. The identification becomes more precise as the design progresses toward production. This family of documents provides the basis for development, production, testing, delivery, operation, and maintenance throughout the total system life cycle.

9.8 CONFIGURATION CONTROL

Configuration control is the control of changes to hardware, software, firmware, and documentation. It is the mechanism for managing change. One major benefit output of configuration control is the ability to trace the original product through its development and growth. This allows us to manage the product, and its maintenance and operations, upgrades and enhancements, and deployment.

Configuration Status Accounting (CSA)

Configuration status accounting is a management information system that provides traceability of configuration baselines and changes and facilitates the effective implementation of changes. It consists of reports and records documenting actions due to changes that affect the configuration of the system.

CSA has become an essential part of configuration management by providing a process to satisfy the management requirement to know the configuration of the system and the status of actions concerning the configuration. It assures customers that they get what they paid for. Figure 9.6 contains a sample CSA file element structure. As can be seen, CSA establishes the traceability of the progress and changes of systems.

Configuration Audits

In configuration management, the verification process consists of two activities: configuration audits and participation in program reviews. Configuration audits verify conformance to specifications and requirements. The audit function validates the accomplishment of the development and achievement of production status. Audits validate that development requirements are achieved and that system configuration is identified by comparing the configuration item with its technical documentation. Two kinds of audits are performed: functional configuration audits (FCA) and physical configuration audits (PCA).

The FCA is a means of validating that development of a configuration item has been completed satisfactorily and that the item functions as required. It is a prerequisite to the PCA. The FCA is normally performed during the FSD phase just prior to production. The PCA is a means of establishing the product baseline as reflected in the product configuration identification, and is used for the production and acceptance of the system. The PCA may be accomplished during the FSD phase; however, it is usually delayed until the beginning of the production phase and thus may be accomplished on an early representative production unit. A PCA is normally required on the first configuration item (system) to be delivered by a new contractor even though a PCA was previously accomplished on a similar article delivered by a different contractor.

The FCA verifies the functional characteristics, while the PCA verifies the form and fit of the product. The result of the FCA is a formal acknowledgment that the product performs in compliance with requirements and specifications. The PCA results in a formal acknowledgment that the product matches the physical description in its documentation. The FCA is usually conducted after testing and integration, while PCA is done just prior to acceptance. Figure 9.7 illustrates a generic audit process.

9.9 CONFIGURATION MANAGEMENT PLAN

For major systems, a configuration management plan for the acquisition is usually required to be submitted by the contractor either with the FSD phase

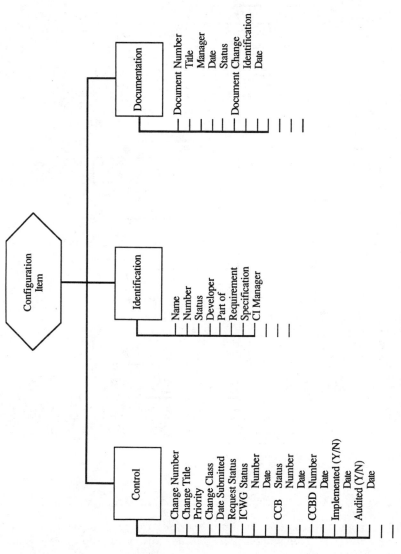

Figure 9.6 Sample status accounting file elements. (*Source:* by permission from M.A. Daniels, *Principles of Configuration Management*, Advanced Applications Consultants, Rockville, MD, 1985, p. 78.)

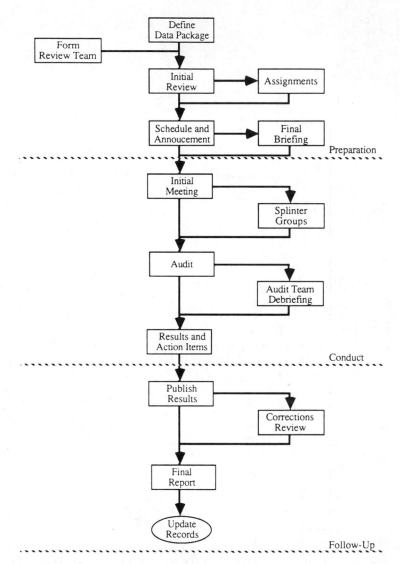

Figure 9.7 Generic configuration audit process. (*Source:* by permission from M.A. Daniels, *Principles of Configuration Management,* Advanced Applications Consultants, Rockville, MD, 1985, p. 87.)

proposal or early in the FSD phase. The configuration management plan is the cornerstone of the configuration management system. The plan should be a vehicle to indicate in detail the "how," "when," "who," "what," and "where" of configuration management activities. The configuration management plan for computer software configuration items is included in the software development plan, the software configuration management plan, or the system configuration management plan.

Once the system is deployed, configuration management becomes the responsibility of the government. The configuration of all units, regardless of location, must be known in order to ensure that changes and modifications can be installed promptly and properly. The government's configuration management plan must make provision for data flow to and from deployed units, to ensure current configuration knowledge.

9.10 SUMMARY

To reiterate, configuration management is an accepted and approved management discipline. It is a process that identifies the functional and physical characteristics of an item during its life cycle, controls changes to those characteristics, and records and reports change processing and implementation status. Through configuration status accounting and configuration audits, traceability is achieved for system configurations. And it is through this traceability that supportability is improved.

10

NONDEVELOPMENTAL ITEMS

A nondevelopmental item (NDI) is a generic term that covers materiel available from a variety of sources with little or no development effort by the service. NDIs are normally selected from:

- Commercial sources (may require ruggedization or militarization).
- Materiel developed and in use by other U.S. Military Services or government agencies.
- Materiel developed and in use by other countries.

The acquisition strategies (AS) available to satisfy requirements cover a full spectrum from traditional full development, to classic "off-the-shelf" NDI. Between the two extremes are "tailored" AS employing a varying degree of NDI. The bottom line is that the Army prefers to buy systems already designed, developed, tested, and in production; or, at least, where principal components are in production as opposed to initiating a new development program.

There are several categories of NDI:

1. Category A—off-the-shelf items (commercial, foreign, or other services) to be used in the same environment for which the items were designed. Research and development (R&D) funds are not required to develop or modify hardware or operational software.
2. Category B—off-the-shelf items (commercial, foreign, or other services) to be used in an environment different from that for which the items were designed. The item may require modification to hardware or operational software.
3. There is a third level of NDI effort. This approach emphasizes the integration of existing componentry and the essential engineering effort to accomplish systems integration. This strategy requires a dedicated

R&D effort to integrate existing proven components into a system configuration, to develop or modify software, and to ensure that the total system meets requirements.

10.1   NDI BENEFITS AND CHALLENGES

NDI offers three major benefits: (1) the time to fielding is greatly reduced and there is a more rapid response to user needs, (2) generally R&D costs are reduced, and (3) the state-of-the-art technology can be used to satisfy the user need.

NDI also presents major challenges. Essential integrated logistics support (ILS) activities normally accomplished in preproduction phases have to be accelerated and may require increased up-front funding. Standard logistics systems may have to be supplemented by interim contractor support (ICS) or other innovative logistics strategies. Contracting strategies must be employed that ensure effective management of hardware and software changes and minimize logistics support, training, and configuration management implications. Safety deficiencies may also need to be evaluated to determine whether they pose an acceptable risk. Procedural safeguards may have to be considered as an alternative to hardware redesign. And because of the shorter NDI acquisition cycle, internal service supporting processes must be expedited or tailored to support the NDI strategy.

10.2   NDI SELECTION CONSIDERATIONS

Requirements Documents. NDI must satisfy the requirements documents.

Life-Cycle Cost (LCC). With NDI, the service expects to field a product faster and cheaper by possibly reevaluating and readjusting user requirements against available marketplace products with possibly a longer interim contractor support period. We must evaluate the total LCC for NDI alternatives, consider risk and cost tradeoffs, and then select the alternative that has the lowest projected LCC within acceptable risks while still meeting user requirements. This is true also for developmental systems.

Operation and Support in Mission Environment. NDI presents special ILS problems because many leadtimes associated with developing elements of organic support exceed the time required to acquire and field NDI. To compensate for this, the service may choose to rely on contractor support on an either interim or permanent basis. This decision should be based on a careful consideration of the risks and costs inherent to this support in relation to the intended use of the equipment.

Availability of the Product and Its Support Items for Purchase throughout the Planned Life Cycle. The market investigation (MI) must assess manufacturer history, production capability, and ability to sustain support over the intended life cycle of the product. The service may choose a one-time buy of spares to ensure support of the product over the entire life cycle.

Safety and Environment. NDI may present special safety and environmental problems due to the possible lack of compliance with normally accepted safety and environmental standards. In cases where standards are not met, a decision must be made at each milestone decision review as to whether the increased risk is acceptable.

Manpower and Personnel Integration (MANPRINT). MANPRINT is the integration of human factors, manpower, personnel, training, health hazard assessments, and system safety considerations into the entire materiel acquisition process. NDI is not exempt from MANPRINT. In fact, NDI requires extra MANPRINT effort. In early proof-of-principle–activities, MANPRINT is a major consideration in determining whether an NDI can be fielded by the service in a strict commercial or off-the-shelf configuration, whether a degree of modification is required, or whether there is no viable NDI solution at all. NDI advantages are manifest, but they cannot overcome the basic requirement to acquire equipment that can be fielded with minimum qualitative and quantitative impact on the personnel inventory and training base.

Nuclear, Biological, and Chemical (NBC) Survivability Requirements. If an item is electrical or electronic in nature, NBC survivability requirements are critical in making a decision to purchase NDI. Electrical items require protection against high- and low-electromagnetic pulse (EMP) damage. This applies to all items, even those in depot reserve, since EMP has such far-reaching effects. Additionally, if radiation hardness is specified, it must be included in the initial development of printed-circuit boards; to attempt to modify these boards later generally requires a total redesign of the board and possibly the entire system. From an NBC survivability standpoint, if the materiel used in the item is vulnerable to the effects of either contaminants or decontaminating agents or if the item is not operable by military personnel wearing NBC protective gear, the equipment may not be suitable for use.

10.3 OVERVIEW OF THE NDI ACQUISITION PROCESS

Requirements–Technology Base Activities and Proof-of-Principle Phase

The NDI acquisition process starts like any other program with establishment of a requirements document. A flowchart of the process for NDI Categories A and B is provided in Figures 10.1 and 10.2, respectively. The O&O plan and subsequent requirements documents must contain a realistic statement

Requirements/Technology Base Activities-Proof of Principle Phase

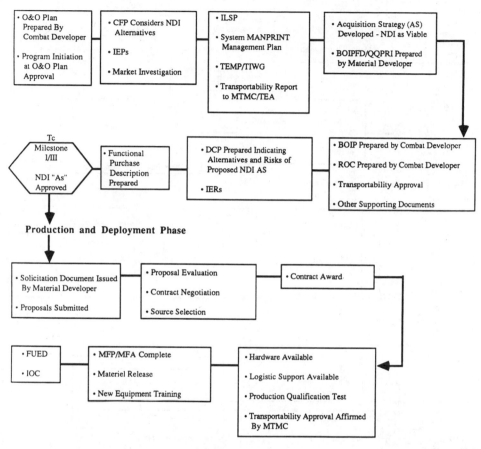

Figure 10.1 Overview of NDI acquisition process (Category A).

of the threat. NDI feasibility surfaces during the normal requirements generation process.

A research, development, test, and evaluation (RDT&E) funded project will be established in this early period to pay any costs generated in preparing the technical rationale in lieu of conducting the MI, testing–evaluating–integrating NDI candidates, ILS, MANPRINT, and so forth. If the results of the MI indicate that an NDI solution is feasible, an NDI acquisition strategy is initiated. This includes the decision to do one of the following:

- Proceed directly to a combined Milestone I/III decision review, which makes the production and type classification decision.
- Proceed to a Milestone I/II decision review and initiate a development

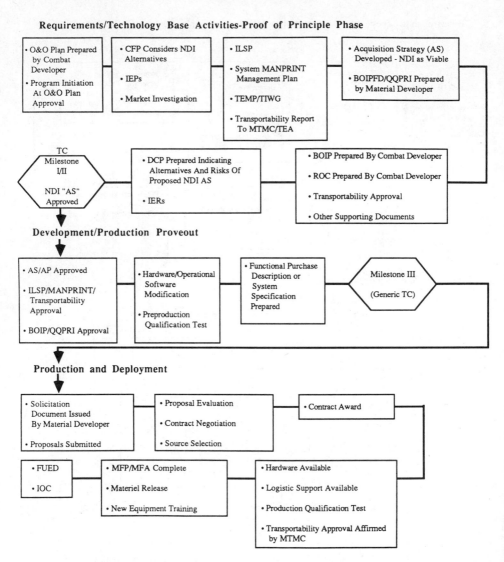

Figure 10.2 Overview of NDI acquisition process (Category B).

proveout phase to conduct a test and evaluation of NDI candidates to determine whether requirements are fully satisfied.

The development proveout phase (including lease, purchase, and/or loan of those NDI candidates and support materiel needed for test, evaluation, or integration purposes) is funded with RDTE funds. The extent of any required modification is one of several factors used to decide whether and how much operational testing is necessary.

The NDI decision is made at the initial milestone decision review. Essential elements in preparation for the review are:

1. Requirements documents.
2. User's statement of the minimal level of support, including computer software support required at initial fielding.
3. Initial personnel requirements identification for submission with requirements documents.
4. Initial ILS plan.
5. Transportability report.
6. Transportability approval.
7. NDI test and evaluation strategy.
8. Preparation of safety and health data sheet.
9. Product quality management plan.
10. Training plans.
11. System MANPRINT management plan.

The initial milestone decision review determines the capability of the marketplace to provide an item for the military, approves the NDI acquisition strategy, approves the issues to be evaluated, and authorizes subsequent testing through the test and evaluation master plan (TEMP).

Development Proveout Phase

Once an NDI solution is authorized, the acquisition plan (AP) is updated to support contracting efforts. There are many contract options available. The AP outlines and justifies the selected option. Additionally, the ILSP is updated with consideration for special factors relating to the approved accelerated acquisition program. After the AP and ILSP have been approved, a formal specification or functional purchase description for the solicitation is prepared. These documents describe the military requirements to industry.

The Milestone III decision approves the AS to support the production–deployment phase and approves and type classifies the item. Type classification (TC) is an integral part of the Milestone III decision review. It is used to implement the Office of the Secretary of Defense requirement that an item must be determined "acceptable for service use" and is supportable prior to expending any procurement funds. Type classification is discussed later in this chapter. The decision authority then issues a "system acquisition decision memorandum" (SADM) revalidating the AS and releasing the solicitation document.

If a particular NDI has been accepted for the operational mission intended, is supportable in its intended environment, possesses a complete technical data package (including ILS and maintenance support), and is acceptable for introduction into the service inventory, the NDI may be designated as "TC standard" (STD). If make and model number of the item are not precisely known, the item may be generically designated TC STD provided the item's functional and physical characteristics are firmly established with completion of (or formal waiver obtained for) all RDTE funded activities. As previously stated, TC activities are conducted prior to production to validate that the item can satisfy the materiel requirement if selected for quantity procurement. TC is then definitized after final product selection when the manufacturer's make and model number are known.

In addition to the TC STD designation, there are several other categories: TC limited procurement test (TC LPT), TC limited procurement urgent (TC LPU), generic TC, contingency (CON), and obsolete (OBS). These will be briefly discussed below.

Type Classification Limited Procurement Test (TC LPT). This designation normally is limited to developmental items. This applies only when the materiel acquisition process decision authority approves additional tests to ensure that developmental items or NDI can be economically produced by establishing a pilot production line and that the mass-produced item is the same as the developmental prototype items or NDI evaluation items. TC LPT items are authorized only to supply hard-tooled models for technical testing and user testing.

Type Classification Limited Procurement Urgent (TC LPU). This designation may be applied to NDI as well as developmental items needed to meet urgent operational requirements that cannot be satisfied by an item presently in the U.S. Army's operational inventory. NDIs designated TC LPU that meet investment criteria may be procured only if they will be issued directly to field elements.

Generically Type Classified Standard. Often when purchasing an NDI, the specific make, model, and component parts are not known at the milestone decision review authorizing procurement. When this situation occurs, TC STD is not possible without significant delays between the times when the successful contractor is known and the contract is awarded. All prerequisites for TC STD must be addressed when requesting generic TC STD. After source selection, the specific data for TC STD must be specified and the required actions taken to attain TC STD subsequent to award.

Contingency (CON). This applies to a system–item that no longer fully satisfies operational requirements but has residual value for use in training or as a mission essential contingency item for reserve components.

Obsolete (OBS). This applies to a system–item that is no longer required or acceptable for service use.

Production–Deployment Phase

Service planners establish the hardware availability date based on both production leadtimes agreed to with the selected contractor and essential support requirements. If required, new equipment training commences and may be performed by the contractor. The ILS strategy varies with the particular NDI and applies to the end item, the line replaceable unit (LRU), or the piece-part level.

10.4   TYPE CLASSIFICATION

Type classification (TC) is the method for identifying the degree of acceptability of a materiel item for its intended mission. TC is an integral part of the Milestone III decision process and as such is not a separate process. It is essentially a certification for Milestone III decisions complying with the requirement for a determination that an item be "acceptable for service use" prior to the expenditure of procurement funds. The decision to TC signifies that a materiel item or system is ready to be produced and fielded. It further signifies that the supportability, reliability, availability, maintainability, and safety of the item meet U.S. Army requirements. TC is applicable to nonexpendable end items which are issued separately for use.

TC Objectives

The objectives of the TC process are to identify the acceptability of a materiel item for its intended mission by the assignment of a TC designation and to provide a guide to authorization, procurement, logistical support, and asset and readiness reporting.

TC Prerequisites

TC is the certification that an item is ready and acceptable for use. The prerequisites for TC are the prerequisites for effective and efficient fielding of a new system. The TC prerequisites include:

1. Approved requirements documents.
2. User test and evaluation determination of acceptability and supportability.
3. Item acceptable and supportable for intended mission or can be made so prior to fielding without development effort.
4. Approved frequency allocation for systems or items that use electromagnetic spectrum.
5. Environmental quality design aspects have been met or regulatory approval for waiver has been received.

6. Request for NSN and LIN submitted.
7. Complete technical data package (TDP) suitable for procurement, if required.
8. Nuclear Regulatory Commission (NRC) license on authorization, as appropriate, for those systems or items containing radioactive materials.
9. Safety and health hazards adequately controlled.
10. Development of a system safety risk assessment and safety and health data sheet.
11. Approval of DoD "Final Hazard Classification for Explosive Items."

Type Classification and Procurement

Items cannot be type-classified unless procurement is planned within the current POM period. An IPR may be held to determine eligibility for TC and to authorize the materiel developer to unilaterally TC when procurement is planned.

Type Classification and Reprocurement–Product Improvement

Reprocured items, specifically, those required after initial purchases and fielding are achieved, will not require separate TC unless they have been so modified or improved as to meet the criteria for a product improvement. An improved or modified item must be separately type-classified when the modification or conversion necessitates special management because it incorporates or requires stockage of major components, such as engines or consumable items that are different from those required for the basic item; or it will not be applied to the total inventory quantity. Separate TC also occurs when there are changes in (1) functional and physical characteristics affecting the quantity of personnel and/or associated support items of equipment required to support the end item, (2) safety and health characteristics, (3) personnel, or (4) mission requirements. The program manager (PM) determines whether separate TC, as a distinct new item, would be required for the product improved item.

Type Classification Exemptions

The following types of items do not require TC:

- Low-density institutional training devices, such as flight simulators that will be fully contractor supported during the entire life cycle.
- Military decorations, medals, and heraldic flags.
- Commercial construction materials, excluding mechanical, electromechanical, electrical, and electronic items.

- Nonmilitary administrative items under General Services Administration (GSA) purview.
- Components of end items, if BOI as separate items is restricted to schools, training centers, laboratories, and depots.
- Expendable items except ammunition, individual equipment, and selected high-density military items.
- Nonstandard materiel for support of allies.
- Items procured for operation and support only by contractors.
- Items procured with nonappropriated funds.
- Items procured only for DoD civil defense, except those used to protect DoD personnel or to quell disturbances.
- Automatic data processing items.
- Locally fabricated training aids less than $300 each.
- Special tools that automatically assume TC of items they support.
- Commercial training devices under a commercial training devices requirement document.
- Sets, kits, outfits (SKO) with a sole basis of issue restricted to schools or training centers, laboratories, or maintenance–test facilities.

Termination of TC

The TC of an item can be terminated when, after an item has been type-classified, a decision is made to terminate the program prior to procurement of any production items. Termination of TC can also occur when the mission assignee agency determines that, under current regulations, an item that has been type-classified is exempt from TC.

10.5   NDI AND INTEGRATED LOGISTICS SUPPORT (ILS)

ILS Importance

ILS is often the most difficult aspect of NDI acquisition because ILS demands day-to-day top management attention. ILS cannot and must not be sacrificed to hardware schedule and cost constraints. If it is known that full logistic support will not be available at the time of fielding, a work-around plan must be developed to provide the best level of logistics support possible until the ILS process provides the requisite support. A successful ILS process can only be achieved through the joint efforts of all involved in the acquisition. This mandates that an ILS plan be developed concurrently with the acquisition strategy (AS). Coordination of the ILSP with all participants is essential, and the formation of an ILS management team (ILSMT) should be considered for this purpose. The ILSP must precisely document overall ILS requirements, the initial support package that will be available during and after fielding, how initial support capability will be achieved, how support will be transitioned, and requirements and detailed plans for each function and element of ILS using information obtained from the market investigation.

Design Influence

During the materiel acquisition process and as part of the market investigation, the design characteristics are evaluated in terms of MANPRINT, supportability issues, costs, and compatibility with support equipment. The design characteristics are included in the functional purchase description or specification and used as source selection criteria, thus addressing the intent of logistics design influence.

Logistics Support Analysis (LSA) and LSA Record (LSAR)

LSA is an integral part of the development of requirements documents and the market investigation. LSA is used in determining initial and life-cycle support concepts and identifying potential support problems and developing solutions. LSA also generates the documentation of ILS element requirements for the LSAR. The LSAR is tailored to provide phased delivery of data required to determine interim and subsequent support resource requirements. LSAR deliverables should be requested to provide timely completion of ILS schedules.

Supportability Test and Evaluation

If commercial marketplace testing does not address the intended military environment and equivalent information cannot be obtained from existing sources, test and evaluation may be required to determine or verify maintenance skill requirements, training requirements, transportability issues, the use of standard support and test equipment, and so on.

Maintenance Planning

The initial maintenance plan generally accepted for most NDIs is to provide the using unit with the capability for troubleshooting to the line replacement unit (LRU), either through the use of built-in-test (BIT) or the use of manual TMDE test procedures. The maintenance personnel in the unit can then remove the LRU and replace it with a working element and evacuate the faulty LRU to the intermediate direct support facility. Intermediate direct support facilities stock LRUs for direct exchange purposes. Depots usually have the capability to repair to the piece–part level. The market investigation should provide information for identifying which maintenance alternatives are viable for the NDI under consideration.

MANPRINT

MANPRINT activities begin during formulation of the O&O plan. Unlike new developments, NDI limits MANPRINT options because we may be starting with a defined end product or component. Because of these limits, early requirements–technology base and proof-of-principle activities must focus on identifying MANPRINT issues and developing accommodations or "work-arounds."

MANPRINT activities predict system demands on future personnel inventories and whether there are unsupported requirements. For strictly off-the-shelf NDI, analysis is required to determine whether the standard NDI configuration meets MANPRINT criteria, goals, and constraints. If it does not, this leads to a reevaluation of the basic NDI decision, a review of other NDI variations, or modification of the initial operations and support concepts. If an NDI strategy is pursued that modifies the equipment or integrates components, then MANPRINT findings might be compensated by simple system design modifications. The results of MANPRINT analysis could dictate modification of commercial equipment, affect source selection, drive contractor logistics support, or eliminate NDI as a solution.

Supply Support

The provisioning strategy follows the maintenance strategy. Actions required to establish interim support are expedited. Phased delivery of tailored LSAR data is used. Normally, intensive management is used to provide for maintenance float and authorized stockage list–prescribed load list (ASL/PLL) items for the interim support concept. Prescreening of the manufacturer's part numbers for national stock numbers (NSNs) by the contractor is required because NSNs must assigned to all spares and repair parts anticipated to require any replacement in the field. If NSN assignment and LSA data collection cannot be completed 240 days prior to first unit equipped data (FUED), then interim contractor support must be obtained. This decision needs to be made early enough to arrange for the correct amount of support for the correct length of time and the proper amount and type of funds to be secured to provide for interim contractor support.

Support Equipment and TMDE

Requirements for test equipment and associated support items of equipment (ASIOE) must be identified as early as possible. Use of government standard test equipment in lieu of contractor-recommended unique test equipment must be determined early.

Training and Training Devices

Overall training requirements have to be determined on an expedited basis. Extensive contractor assistance may be required to initiate new equipment training and establish the institutional training base. If training aids or devices are required, use of contractor-owned or contractor-provided equipment may be necessary.

Transportation and Transportability

Prior to completion of the solicitation documents, requirements for transportation within the Defense Transportation System (DTS) are determined. Commercial standards are used to the extent that they satisfy military requirements; however, any required modifications should be included in the solicitation document. Where necessary, transportability experts participate in precontract award negotiations. The high cost of postproduction modification to enhance transportability characteristics must be avoided. NDI problem items are required to move through a formal approval process at each milestone and require Military Traffic Management Command (MTMC) participation throughout.

Technical Data

Because of the compressed acquisition cycle associated with NDI, emphasis should be placed on supporting training and initial fielding with authenticated commercial manuals. The decision to use commercial manuals, however, must be based on acceptability of the manuals. Supplementation of the commercial manual to add maintenance allocation charts, repair parts lists, and so forth may be necessary. If commercial manuals are not acceptable, then action must be taken to prepare acceptable manuals. In addition to manuals, technical data including drawings, test data, and so forth should not be procured beyond the detail required (i.e., form, fit, and function).

10.6 NDI AND TESTING

Testing requirements will be tailored to each specific system. The following testing guidance by NDI category is not a rigid requirement, but rather general characteristics of testing activities appropriate to each NDI category. The goal of minimum testing still remains regardless of NDI category.

> *Category A.* No testing prior to production qualification testing (PQT) except when the contract is awarded to a contractor who has not previously produced an acceptable finished product and the item is assessed as high-risk.

Category B. Feasibility testing is required in the military environment. PQT is required if feasibility testing results in fixes to the item. Limited user evaluation may occur during feasibility–preproduction tests.

Integration of Components. Hardware–computer software integration tests are required, user testing is required, and a final PQT is required.

Follow-on Evaluation. Testing of the NDI after the first unit is equipped is oriented to the validation and refinement of operating and support cost–data, RAM characteristics, logistics support, training, provisioning planning, and so forth. These tests can materially aid the logisticians in supporting NDI throughout its life cycle.

10.7   NDI AND RAM

Quantitative RAM requirements must be developed for all NDI. Prior to the initial milestone decision review, a tailored RAM rationale report (RRR) is written on the basis of thorough user analyses. These RAM parameters contained in the RRR are considered against what is available in the marketplace as described by the market investigation. The criteria for evaluating RAM for NDI should be the same as for development programs.

Many approaches can be taken to gather valid RAM data during the market investigation. One approach is to request and/or review any RAM analysis that the manufacturer performed in the development of the item. In the market investigation, a range of values limiting RAM requirements may be used as a baseline for the RAM assessment. When quantitative RAM data are not available, it may be possible to assess relative RAM values. These approaches and others should be used to obtain enough RAM data on which to base an NDI decision. If the independent evaluator determines that the market investigation did not provide data adequate to resolve the RAM issues, testing may be required. When market investigation and/or testing demonstrates that commercially available equipment cannot meet the service's RAM requirements, an alternative strategy is to modify existing commercial equipment to meet RAM requirements or modify existing mission profiles to determine whether the commercially demonstrated RAM values are acceptable. Finally, when RAM is an extremely critical design characteristic, or when the commercial RAM parameters are far inferior to the requirements, the NDI strategy may be abandoned.

For NDI Categories A and B, a "Reliability Program" (MIL-STD-785) and "Maintainability Program" (MIL-STD-470) are required and may be tailored for each acquisition. Depending on the information or data gathered during market surveillance and market investigation, reliability and maintainability program tasks may be waived or implemented in part. Reliability and maintainability programs must be implemented for end items assembled from commercial components, unless market surveillance or market investigation information can show that the integration process is of low risk.

10.8 NDI AND PRODUCT IMPROVEMENTS

A product improvement (PI) is a configuration change to an existing weapon system or piece of equipment in response to a user-validated need. The improvement requires testing to ensure that it accomplishes what is intended without jeopardy to any interfacing system and is installed as a modification kit in the field or during production if the weapon system is still in production. Since PI is the preferred method for satisfying materiel requirements, the desirability of making improvements to existing equipment rather than initiating new developments must be carefully weighed. Such improvements may be changes of an evolutionary nature to improve combat effectiveness and extend the useful life of the system or changes for safety, cost reduction, standardization, legislative compliance, energy conservation, deficiency correction, or improvement of RAM. An approved PI will create publication of a modification work order (MWO), a depot maintenance work requirement (DMWR), or other appropriate documents and will cause an engineering change to production packages (i.e., TDPs).

The preplanned product improvement (P^3I) is an acquisition strategy (AS) that can be extremely useful in reducing costs and extending the useful life of a system. This is an extremely complex strategy that must be fully understood to be properly applied. This concept offers an alternative AS that minimizes technological risk and speeds the delivery of a basis system to the user. The objectives of P^3I are to shorten the acquisition and development time for systems; extend the useful military life of systems; reduce technical, cost, and schedule risk; and reduce the requirement for new starts. The initial costs of systems with P^3I will normally be higher than traditional systems because more up-front costs are visible. If the system is in production, retrofit of fielded hardware that upgrades it to the current block configuration, if deemed cost-effective, is funded without going through a separate PI process for approval.

10.9 NDI AND FUNDING

Actions in support of NDI acquisitions that require additional research and development (R&D) effort such as engineering, design, integration, market analysis, identification of available technology, and identification of necessary test and evaluation are RDT&E-funded until all R&D effort is completed, type classification is accomplished, and approval to enter the production phase of the life cycle is received. Actions in support of NDI acquisitions that have been type-classified, given a production go-ahead, and require no additional type R&D effort are procurement-funded. Efforts that are taken to verify or evaluate contractor's claims are considered actions leading directly to the contract award decision, and the use of procurement funds is appropriate.

10.10   NDI AND PPBES

Programming and budgeting for NDI pose special problems. Because of the brevity of the NDI acquisition process, the standard planning, programming, and budgeting execution system (PPBES) leadtimes and funding "windows" may restrict the opportunities for rapid procurement and fielding. This can be minimized through careful advanced planning and, in the case of urgent requirements, reprogramming techniques.

10.11   SUMMARY

NDI systems are available from a variety of sources requiring little or no development effort. The acquisition process for an NDI is not a separate process, but a tailoring of events within the materiel acquisition process and should be one of the first alternatives considered for solution to a materiel need. There are two general categories of NDI: Category A and Category B. There is a third level of effort that emphasizes integration of existing componentry and essential engineering effort to accomplish systems integration. No acquisition, including NDI, is exempt from minimal essential test and evaluation necessary to verify the MANPRINT, quality, safety, reliability, performance, supportability, transportability, and availability characteristics of a system to include life-cycle cost unless previous test and performance data or market analysis are adequate for verifying operational effectiveness and suitability of the system.

The process of assigning a "suitability-for-issue" code to an item is known as "type classification." The basic purpose is to increase the combat effectiveness of the military forces even though it results in conservation of money and materials.

PART 4

INTRODUCTION TO PROVIDING THE SUPPORT

In Chapter 11 the primary purpose of an acquisition is to field materiel systems that not only perform their intended functions but are ready to perform those functions repeatedly without burdensome maintenance. In addition logistics effort is discussed. In the transition to production it is essential that planning for the transition occur early in the development phase. This involves planning for changes to the system, good configuration management practices, establishing change implementation procedures, and planning for product improvements. Transition planning should be completed before entering the initial production phase so that the system support package can be validated prior to the production decision.

The supportability objective during the transition to production is to ensure that earlier predictions and assumptions of support requirements and system performance are verified and validated in the early production articles. Evidence of materiel system problems usually becomes apparent when a program undergoes transition from full-scale development to production. This transition is not a discrete event in time; it occurs over months or even years. The transition process is impacted by design maturing, test stability, and certification of the manufacturing process. The actual manufacturing involves all the processes and procedures designed to transform a set of input elements into a specified product. Failure to consider all production requirements early in the acquisition cycle has resulted in more than acceptable risk in the transition from development to production.

Product design and development cannot logically be separated from the production design. The application of the systems engineering process to production functions can identify the products required to transform the design into a capability for efficient and economical production of equipment. Systems engineering also ensures that production functions support producibility analyses, production engineering inputs to trade studies, life-cycle-cost analyses, and consideration of tools, test equipment, facilities, personnel, software, and procedures that support manufacturing in the D&V and production phases.

Supply support is an essential element of the logistics integration effort and is responsible for the timely provisioning, distribution, and inventory replenishment of spares, repair parts, and special supplies. The computation of requirements for major end items is an important function in the logistics system. It is through this process that the services evaluate readiness position at different points in time, ensure that excesses are not accruing, and provide basic input to support its annual budget request and subsequent apportionment of the appropriation dollars.

Provisioning is one of the most important logistics processes. It means laying-in an adequate supply of material, when and where needed and within monetary constraints, to support a weapon system or end item of equipment during its initial period of operation.

Provisioning is based on reliability data in the LSA/LSAR. Provisioning creates inventories that may be classified as basic demand, in-transit, safety, speculative, or dead depending on the reason for which they exist. There is an in-depth discussion of how each is determined through economic stockage principles, economic order principle, variable safety level principles, and the economic order quantity concept.

Major ILS management risks in the transition to production process include inadequate planning, extensive changes, delayed organic support, delayed completion of testing, and inadequate producibility in design. Intensive ILS management is required to ensure that support items remain compatible with late changes to the materiel system entering production.

Chapter 12 encompasses operational and postproduction support. The overall objective is to maintain a materiel system in a ready condition throughout its operational phase within the operating and support cost program levels established in the program objective memorandum and budget. System readiness objectives established early in development constitute the baseline for planning operational and postproduction support and supportability assessments during the operational phase. However, the first empirical measure of system readiness occurs when the materiel system is deployed in the operational phase. The readiness and RAM experience during the operational phase is used to adjust the logistics support resources that were programmed during the FSD and production phases. Any performance and RAM deficiencies detected must be corrected as early as possible in the operational phase to prevent reduced readiness.

11
SUPPORTABILITY ISSUES IN TRANSITION TO PRODUCTION

11.1 ILS MANAGER'S PRIORITY TASKS DURING THE TRANSITION PHASE

The primary purpose of the acquisition process is to put new products in the field, products that perform their intended functions repeatedly without burdensome maintenance and logistics efforts. The successful fielding of a reliable and supportable product requires that the ILS manager provide strict watchdog management during the transition phase to ensure that adequate technical engineering, manufacturing disciplines, and management systems are applied to the ILS elements and supportability features of the product. Transition-phase ILS priority items include:

1. Providing timely funding for all ILS elements.
2. Involving ILS specialists in the preparation of comprehensive hardware and software specifications and data descriptions.
3. Continuing an active LSA process.
4. Establishing adequate funding for initial spares and support equipment.
5. Ensuring ILS inputs to configuration control and the comprehensive assessment of the impact of changes on all support elements.
6. Establishing a technical management system for tracking support equipment reliability, configuration control, and compatibility with end-item hardware, firmware, and software.
7. Funding and scheduling of manuals and other support documentation.

The Transition Plan

Transition plans, which are detailed accounting of the items and issues to checkoff in "readiness" reviews, are primarily a management tool for ensuring that adequate risk-handling measures have been taken. They must be initiated and tailored to the need of the program by the program manager and ILS manager.

Management of Changes

The management of changes to existing products requires special emphasis on integration, implementation, and baseline control. Modifications are used to correct product deficiencies to provide increased performance, lower life-cycle costs, extend the product's useful life, or remove obsolete capabilities.

Planning for future modifications starts during the development of the system. Design decisions made in development affect the flexibility of the product to incorporate changes throughout its life cycle. Once the product is produced, the decisions are costly to reverse. Especially during the transition from development to production, ease of future modification must be considered in conjunction with performance, cost, and manufacturing and producibility considerations.

Modifications sometimes occur after production has started, but, owing to the length of the production phase, it may be possible to incorporate the modifications into some of the items during their assembly–fabrication instead of waiting until after the items have been "produced." Incorporating changes into the produced items is complex since the produced items may be distributed in many locations with multiple agencies responsible for their operation, support, maintenance, and repair. Managing this aspect of modifications tests the best management talents and requires considerable effort to realize success.

During production, successful modification programs are dependent on a highly disciplined configuration management system. This is especially true when large numbers of items are built over a prolonged period of time and are widely deployed. The ability to make even minor corrections at remote facilities is often limited. The use of a configuration management system is therefore mandatory throughout the life of the system. Adequate communication channels are essential for transferring modification data from the organization installing the modification to the configuration management center, so that up-to-date configuration status accounting records can be maintained.

There is a clear distinction between postproduction modifications and engineering change proposals (ECPs). When products receive changes during production, only those products that have been distributed can be modified through the product improvement program. Research, development, test, and evaluation (RDTE) funds are used to develop modifications that increase the operational envelope of a product; operation and support funds are applied to modification kits. When product improvements are very large, they are budgeted and funded as if they were a new development effort.

If a modification must be incorporated into a number of different products, this multiplies the complexity of the modification. Tailored installation instructions and kit interface components are required for each application. Communications with agencies and users and modification management problems increase dramatically.

Even with a good configuration management system, the impact of testing results can overwhelm the best logistics support planning in the transition to production. Because of the "reality" of professional specialization and organizational compartmentalization in both government and industry, each support discipline is considered a specialty unto itself and is often isolated at the expense of coordination and integration. Experience has amply demonstrated that the traditional approach results in an inability to obtain optimum support in the field following delivery of a product. Ideally and properly implemented, the systems engineering and LSA processes would cure the lack of adequate integration between design engineering and logistics elements.

Change Proposal Preparation

The starting point in change preparation is recognition of a deficiency and a decision to employ a design solution. As shown in Figure 11.1, the request to change production—and possible retrofit distributed equipment—may be originated by the government, the contractor, or commercial manufacturer. The top half of Figure 11.1 illustrates one approach to contractor preparation of an engineering change proposal (ECP). The contractor ILS manager must be actively involved in determining the impact of the ECP on affected ILS elements, developing requirements and schedules for required changes to affected ILS elements, and participating in engineering review board and change review board meetings. The government ILS manager must be involved in the government review and approval process and ensure that the impact of ILS elements have been fully evaluated, ECPs for associated changes to support equipment and training devices are available for concurrent review and approval, leadtimes for changes to ILS elements are compatible with the planned implementation of the ECP on the production line, and changes to ILS elements are funded.

Change Implementation

After approval, the contractor initiates action to finalize the change for production and/or retrofit and the concurrent modification of the affected ILS elements (bottom half of Figure 11.1). The government ILS manager normally is responsible for the application of retrofit kits and must ensure that the required changes to logistics support of deployed systems are applied or are available concurrent with the application of retrofit kits to the systems. This latter requirement can be facilitated by grouping retrofit kits into block modifications and applying them to complete production lots.

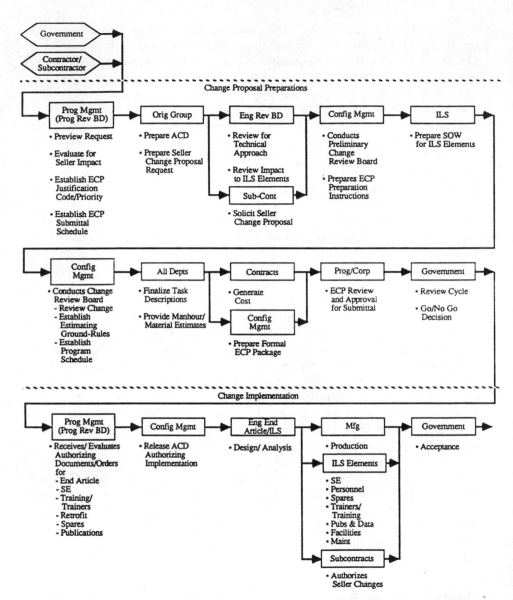

Figure 11.1 One variation of ECP preparation–implementation by a system contractor.

The DoD acquisition cycle generally involves the following steps for modification programs: need, design, test, manufacture, and operate and support. The need and test steps are like those for other programs and require no additional explanation here. During the design step, the modification and modification kits are designed. The modification kit is the collection of hardware, software, data, and instructions that incorporates the modification into the existing system. The modification should be designed so that it can be incorporated in a produced item without degrading its performance. Integrated logistics support planning for the modified system must be done during development.

The manufacture step involves manufacturing and assembling the modification kit. The kit may also contain unique tools, spares, and other items necessary to incorporate and support the change. Kit production and delivery schedules must be fully coordinated with the installation schedule.

Implementation of the installation schedule becomes more complex as the number of products that will receive the modification increases and the number of modifications to an existing product system increases. Nonstandard configurations are a "known unknown" and should be considered in modification program planning. Successful modification programs proceed on the assumption that each product's configuration may vary from that documented.

Planning and budgeting for the modification must consider any required production tooling, test equipment, support equipment, simulators, and trainers. It also includes documentation, training, and operational validation for the modified product.

Of particular importance is compatibility between different modifications, especially when incorporated separately. Systems engineers should analyze all proposed modifications and recommend or disapprove implementation to the appropriate configuration control board or decision authority.

Product Improvements

A product improvement (PI) is a configuration change to an existing system or piece of equipment in response to a user-validated need. The improvement requires testing to ensure that it accomplishes what is intended without jeopardy to any interfacing system and is installed as a modification kit in the field or during production if the system is still in production. Since PI is the preferred method for satisfying materiel requirements, the desirability of making improvements to existing equipment rather than initiating new developments must be carefully weighed. Such improvements may be changes of an evolutionary nature to improve and extend the useful life of the system, or changes for safety, cost reduction, standardization, legislative compliance, energy conservation, deficiency correction, or improvement of RAM.

The product improvement proposal (PIP) and supporting documentation provide a visible audit trail. The PIP will also substantiate the need for the improvement, identify all of the resources required, and provide the plan, including schedules and milestones, for developing and applying the modification or making changes during production.

Special Applications: Preplanned Product Improvements (P³I)

There is no single DoD-specified approach to modification management. However, the Deputy Secretary of Defense, in DoDD 5000.1, directed the implementation of P^3I in major Department of Defense programs. The primary objectives of P^3I are:

- Introduction of higher technological performance during the system's lifetime through more rapid fielding of technological advances.
- Shortening of the acquisition and deployment times.
- Extension of the system's useful life (before obsolescence).
- Reduction of system technical, cost, and schedule risk.
- Reduction of requirements for major system new starts.
- Higher operational readiness during the system's lifetime.

The P^3I concept cannot be applied to all new system developments but should be considered when:

- A short-term need exists to build a system with current technology.
- There is high risk that current technology will not meet a projected future threat and a low risk that future technology will not meet such a threat.
- The system can be designed to incorporate planned technology development.
- P^3I can be an effective means of meeting overall long-term program objectives (based on threat, development risk, and total life-cycle cost). It may not be cost-effective for low-cost, low-technology systems.
- A long-term military need exists for the system. (P^3I can shorten the development time for the basic system; however, evolutionary changes will normally lengthen the total development period.)
- The U.S. Military Services, DoD, and Congress demonstrate a commitment to acquire the system under the P^3I concept, including acceptance of initially higher costs.

P^3I represents a separate and specific acquisition strategy to acquire clearly stated requirements on an incremental basis. It is not a process where the program manager or the user attempts to guess future requirement and configure the basic system to accommodate these guesses. P^3I is evolutionary development or incrementalism. The mission area analysis process results in the identification of a deficiency. The program manager translates this deficiency into a specified materiel solution. When the specified solution is evaluated as "high risk," short-term, reduced-risk technologies are selected for the basic system in favor of parallel development of the deferred technology. Growth provisions are then designed into the basic system to accommodate future application of the higher technology. Thus further requirements must be clearly stated in the requirements documents and validated by the user.

Specific incremental capability improvement must be known in order to design appropriate growth provisions into the system. It is also needed to support the parallel development effort and the continuation of RDTE funds past the production decision of the basic system. When this concept is not followed or is poorly understood, then gold-plating results and sunk costs are built into systems because of poor guesses. When the deferred capability is applied to the basic system, it is accomplished under the rules of standard product improvement. In addition, during the upgrade process, growth provisions are also installed to accommodate the next incremental upgrade as necessary in response to changes in the validated new future requirement. This process continues until the basic design constraints dictate a replacement system.

Table 11.1 provides suggested criteria for choosing P^3I over a new start or a product improvement. It should be noted that no single, standardized approach can be used for every system development.

The decision to use P^3I should be made as early in the concept exploration phase as possible and no later than Milestone I. In order to make P^3I effective, the design strategy should include (1) modular design, (2) a careful architectural interface system, and (3) provisions for space, weight, cooling, power, and so forth. A development process must be established to communicate system growth requirements and identify new technological opportunities. Implementation begins shortly after the design strategy is developed, so that P^3I is incorporated into the acquisition strategy at the outset. The P^3I acquisition strategy should be communicated to industry early in the program, and industry should be included in the process of developing the strategy. The integration of P^3I into the acquisition cycle is shown in Figure 11.2.

Designing for P^3I may increase the initial acquisition costs of the program. The costs of designing in the flexibility P^3I should be documented and analyzed in terms of net utility for meeting long-term requirements.

11.2   DESIGN INFLUENCE

Design influence is an intangible ILS element but significantly affects overall system readiness, supportability, and affordability. As the cost of acquiring a new system gains in importance, the intangibility of ILS influence on design may shift toward very tangible efforts with a very real effect on system development decisions. Management concepts and technical processes are being established that affect both the ILS management capability and technical methodology where the concerned program and logistics managers can, in fact, influence system design to provide for more effective logistics support.

LSA/LSAR and Design Influence

LSA provides uniform methods to (1) apply ILS and MANPRINT influence in system design and selection, (2) develop the required support system, (3) provide the logistics and MANPRINT analyses required for major decision

TABLE 11.1 Criteria for P³I

	P³I	PI	New Start
Motivation for change	Planned for each upgrade	Forecast life of entire system	In reaction to events
Preparation for change	R&D on selected components	R&D on entire system	Serendipity technology base breakthrough, new threat, or deficiency
Organization for change	Replace module	Replace entire system	Complex interfaces must be resolved
Design lifetime	Different for each module	Maximum feasible for all components	Whatever is available at time of PI

264

Characteristic			
Performance relative to SOA through system life	On average closest available SOA	High at start, erodes after design freeze	Catch-up mode farthest from SOA
Procurement plan	For defined modules	For entire system	As needed
Confidence in meeting cost and schedule	High due to manageable number of changes	Poor yield due to large number of subsystem changes	Is often in response to previous failure to meet goals
Confidence in threat prediction	Higher due to shorter time frame	Poor because of long-range of projection	In reaction to threat changes
Budgeting approach	Funding wedge provided at early time	Specific actions funded in advance	No advance funding provisions

Source: Systems Engineering Management Guide, Defense Systems Management College, Ft. Belvoir, VA, December 1986.

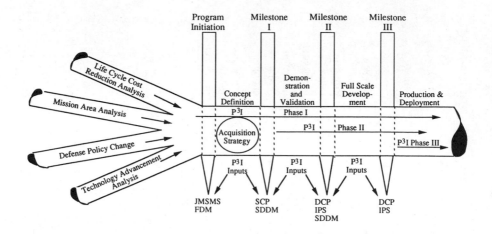

P³I Phase I   • Planning & Research to determine how system will evolve in response to emerging technology and projected need

P³I Phase II   • Incorporation of design or procurement considerations to facilitate system evolution

P³I Phase III   • Application of product improvements which take advantage of Phase I & II. Should also apply additional Phase II considerations

Figure 11.2 Preplanned product improvement (P³I) in the system acquisition cycle.

points, and (4) permit comparison of acquisition phase estimates with the system operational results.

LSA participation in the design reviews will include a report of supportability and readiness design requirements established by LSA, accounting of the LSA recommendation and review process, recommended corrective actions and their status in accordance with MIL-STD-1388 and a statement of the progress of those LSA/LSAR functions and tasks that directly affect design. A typical system design evaluation is shown in Figure 11.3. LSA/LSAR activities will be scheduled with specific quantitative and qualitative milestones that can be measured in objective terms.

LSA is required in all acquisition programs. There are no exceptions. LSA is tailored to the requirements of a specific acquisition program. LSA is the analysis that supports the decisionmaking process concerning the scope and level of logistics support requirements.

Design can be influenced through the use of MIL-STD-1388. However, several major cautions must be observed:

1. Design influence happens very early on in a system development program—in the preconcept and concept–exploration phases. By the time the system reaches the demonstration–validation phase, system design, which influences logistics, is nearly fixed.

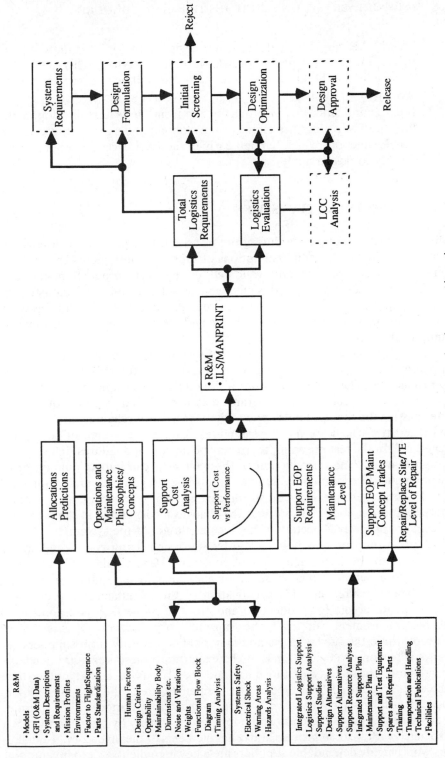

Figure 11.3 Typical system design evaluation as an iterative process.

267

2. LSA/LSAR can effectively influence design only if it is acquired either by contract or through in-house government services. This has to be done deliberately. Otherwise, the logistician will be put into an expensive catch-up mode.

3. If design influence occurs primarily "up front" in system development, then the PM or logistics manager must obtain and exercise those LSA/ LSAR functions which happen early in the LSA cycle and relate more directly to design analysis.

11.3   SUPPORTABILITY ISSUES

The supportability objective during the transition to production is to ensure that earlier predictions and assumptions of support requirements and system performance are verified and validated in the early production articles. Among the evidence that the ILS manager should insist on are demonstrated reliability, a producible design, proven repeatability of manufacturing procedures and processes, certified hardware and software, and verified support equipment.

In the acquisition process, evidence of system problems usually becomes apparent when a program undergoes transition from full-scale development to production. This transition is not a discrete event in time; it occurs over months or even years. Some programs may not succeed in production even after passing the required milestone reviews because of reliability and support characteristics which are not "designed-in," cannot be "tested-in" or "produced-in." In the test program, there may be unexpected failures that require design changes. The introduction of these changes can impact quality, producibility, and supportability and can result in a program schedule slippage. The ILS manager must exercise strong change management discipline during this transition period to ensure that the changes incorporated in the materiel system are properly reflected in the documentation of the system.

The transition process is impacted by (1) design maturity—a qualitative assessment of the implementation of contractor design policy, (2) test stability—the absence or near absence of anomalies in the failure data from development testing, and (3) certification of the manufacturing process— includes both design for production and proof of process. In addition, just-in-time procurement can aid in the transition to production. This is a new concept where the flow of parts and material goes directly from suppliers to workstations on production lines at the exact times these items are needed. In addition, the quality of the parts and material is such that no further inspection is required. This type of procurement is often called a "pull" rather than a "push" system and is synonymous with "zero inventory." Just-in-time procurement reduces the costs and improves the quality by reducing the cost associated with work in process and other elements of the inventory. To make this work, highly dependable suppliers are a key ingredient. "Zero inventory" combined with just-in-time procurement and just-in-time manufacturing is just now becoming a key factor among American businesses. However, it has been popular for some time in Japan with Toyota.

11.4   READINESS AND SUPPORTABILITY

The overall objective for any new system is to provide a needed capability at an affordable cost. Achievement of readiness objectives is essential to attainment of capability. Supportability objectives and supportability design factors are formulated to attain the specified readiness levels within life-cycle cost targets and in compliance with logistics constraints.

In order to influence rapidly evolving system design, readiness and supportability objectives, thresholds, and design requirements must be established. Figure 11.4 identifies program requirements and corresponding logistics support analysis tasks.

Support Readiness Reviews

Support readiness reviews should be initiated and scheduled by the program manager or the ILS manager. To be most effective, support readiness reviews should precede preliminary design reviews (PDR) and critical design reviews (CDR), wherein the ILS manager has an active role. Logistics-related issues from earlier PDRs and CDRs should be prime considerations during later support readiness reviews. The ILS manager should maintain and track support related action items.

Readiness

Readiness of a system is a future oriented attribute. It represents the system's ability to deliver the output for which is was designed. The system readiness objectives are the criteria used in assessing the ability of a system to function properly. There is no universal measure of readiness that is applicable to all systems. Expressions of readiness assume forms that are dependent on the system and its design and use conditions. The program manager must choose a means of defining system readiness that is quantifiable, measurable, and precisely defined by readiness criteria.

Support Requirements Review during the Transition Phase

The ILS manager should take stock of the lessons learned from the results of the development program phase by conducting a support requirements

Acquisition Phase	Program Requirements (DoDD 5000.39)	Logistic Support Analysis Tasks (MIL-STD-1388-1A)
Pre-Concept	• Identify support resource constraints (mission area analysis)	• Perform mission area analyses • Analyze intended use; identify supportability factors <u>Use Study (LSA Task 201)</u> • Select and analyze baseline comparison system <u>Comparative Analysis (LSA Task 203)</u>
Concept Exploration	• Define baseline operational scenarios for system alternatives	• Identify peacetime and wartime employment <u>Use Study (LSA Task 201)</u>
	• Identify support cost drivers and targets for improvement	• Develop a baseline comparison system; determine supportability, cost and readiness drivers <u>Comparative Analysis (LSA Task 203)</u>
	• Identify and estimate achievable values of logistics and R&M parameters	• Identify design opportunities for improved supportability <u>Technological Opportunities (LSA Task 204)</u>
		• Define supportability related design constraints <u>Mission Hardware, Software, and Support System Standardization (LSA Task 202)</u>
		• Update Manpower, Personnel, and Training (MPT) constraints <u>Comparative Analysis (LSA Task 203)</u>
	• Establish system readiness objectives and tentative thresholds	• Establish R&S Objectives <u>(LSA Task 205.2.2)</u>
Demonstration and Validation	• Establish a consistent set of objectives for readiness, R&M, and logistic parameters	• Establish supportability characteristics and supportability related design factors <u>(LSA Task 205)</u>
	• Conduct trade-offs among design, support concepts, and support resource requirements	• Perform evaluations of alternatives and trade-off analyses <u>(LSA Task 303)</u>

Figure 11.4 Development of R&S objectives and supportability design factors.

review prior to recommending that the program proceed to the production phase. Some of the review considerations are:

1. Have the supportability parameters required to satisfy the operational requirements of readiness, mission duration, turnaround times, and

support base interface goals been identified, tracked, and verified in the preceding phases?

2. Have critical supportability design deficiencies identified during testing been corrected or have solutions been identified that can be applied prior to distribution?
3. Have ILS elements been fully evaluated in a representative operational environment? Have deficiencies been corrected, or can they be corrected prior to distribution?
4. Have quantitative requirements for ILS elements been determined?
5. Is there sufficient funding?
6. Can the manpower required to support the system be satisfied by the manpower projections?
7. Will production leadtimes for the ILS elements support the planned production and distribution schedules?
8. Have simulations confirmed the attainability of system readiness thresholds within the target levels for operating and support costs?
9. Have plans for interim contractor support, if applicable, and transition to organic support been prepared?

If these issues have not been resolved, then the ILS manager should develop a recovery plan or recommend further system development.

Supportability

Ultimately, supportability is the degree to which system design characteristics and planned logistics resources, including manpower, meet system readiness and utilization requirements. Each program activity of the ILS manager should:

- Define supportability objectives that are optimally related to system design and to each other.
- Cause supportability objectives to be an integral part of system requirements and the resulting design.

Supportability objectives prescribe conditions and constraints guiding the development of system design and logistics support. These objectives are related to the planned operational role and utilization rates of the system and overall support capability.

Readiness and Supportability Objectives

The procedure employed to develop readiness and supportability objectives requires evaluation in the areas of system mission requirements, deficiencies of current systems employed in the mission area, technological opportunities, and logistics constraints and limitations. During the CE phase, studies based on mission area and materiel system analyses are employed to quantify relationships among the conceptual hardware, mission, and supportability parameters. Numerous studies and analyses are also part of the LSA process and can be found in MIL-STD-1388 in detail.

11.5 MANUFACTURING MANAGEMENT

Manufacturing management is a technique for the proper and efficient use of people, money, machines, materials, and processes to economically generate goods and services from the program start through the production phase. Manufacturing involves all the processes and procedures designed to transform a set of input elements into a specified element. Major functions include design, producibility, manufacturing planning, production control, demonstration and testing, manufacturing methods development, fabrication, assembly, installation, checkout, scheduling, and manufacturing surveillance.

The basic objectives of manufacturing management functions are to (1) do manufacturing planning during the development cycle in acquisition programs, (2) formally document and review pertinent manufacturing criteria before the decision to produce, and (3) properly monitor manufacturing considerations after the decision to enter the production phase is made. Meeting these objectives requires integrating manufacturing throughout the program life cycle, reviewing manufacturing throughout the program life cycle, and reviewing manufacturing capability, feasibility, producibility and readiness preceding major program decision points as shown in Figure 11.5.

Manufacturing management should ensure schedule attainment at the lowest possible cost in agreement with the design and operational requirements. Industrial processes, techniques, and controls involved in manufacturing and delivery should be surveyed continually to determine whether the program plan and milestones are being achieved, to anticipate potential problems, and to take action to prevent or minimize adverse impacts.

Manufacturing management personnel should take part throughout program and project planning, design, development, and production to make sure that (1) manufacturing feasibility is properly assessed; (2) program management has visibility of manufacturing costs and potential schedule impacts; (3) plans for quantity effect the most economical and efficient use of manpower, materials, machines, facilities, and methods. This participation includes an active role in production design reviews and other program reviews during the design and development phases of a program.

Manufacturing management personnel should also take an active role throughout development to ensure that designs are capable of effective and economical manufacture under quantity manufacturing conditions. Manufacturing risks must be detected and resolved to minimize the financial com-

Concept Exploration Phase
- Evaluate Production Feasibility
- Assess Production Risk
- Identify Manufacturing Technology Needs
- Identify Manufacturing Cost
- Develop Manufacturing Strategy
- Identify Deficiencies in U.S. Industrial Base
- Determine Availability of Critical Materials
- Develop Contract Requirements for D/V Phase

Demonstration/Validation Phase
- Assess Producibility of Competitive Designs
- Accomplish Production Risk Resolution
- Reassess Production Transition Risk
- Evaluate Producibility Criteria
- Plan for Achieving Producibility
- Assess Production Feasibility
- Complete Manufacturing Technology Developments
- Plan for Use of Competition in Production
- Develop Initial Manufacturing Plan
- Evaluate Long Lead Procurement Requirements
- Develop Initial Manufacturing Cost Estimate
- Develop Production Readiness Review Plan
- Develop Contract Requirements for FSD Phase

Full Scale Development Phase
- Evaluate Producibility of Design
- Revise Production Risk Evaluations
- Define Required Manufacturing Resources
- Develop Detailed Production Design
- Define and Proof Manufacturing Processes and Equipment
- Accomplish Producibility Engineering
- Accomplish Production Planning
- Develop Production Work Breakdown Structure
- Develop Manufacturing Cost Estimates
- Complete Manufacturing Plan
- Plan for and Accomplish System Transition
- Accomplish Production Readiness Reviews
- Develop Contract Requirements for Production Phase
- Complete Initial Production Facilities

Production and Deployment Phase
- Execute Manufacturing Program
- Integrate Spares Production
- Maintain Production Surveillance
- Implement Product Improvements
- Provide and Support Government-Furnished Property
- Accomplish Value Engineering
- Accomplish Second Sourcing/Component Break-Out
- Complete Industrial Preparedness Planning

Figure 11.5 Manufacturing activities in the system acquistion process.

mitment associated with the decision to enter the production phase. Personnel involvement begins with determining leadtimes, schedule requirements, and reporting manufacturing requirements, analyzing contractor responses, and providing an active manufacturing interface with other functional specialists. Manufacturing elements should implement liaison and operating procedures to ensure effective government involvement with the contractor, to a depth sufficient to establish a strong and mutually knowledgeable relationship among the procuring activity, contract administration activity, and the contractor.

Manufacturing management seeks to reduce logistics costs by providing recommendations to industry and the program office regarding (1) translating fabricating, testing, and failure modes into quantitative requirements for field testing; (2) a responsive repair capability; (3) rapid transportation; (4) use of standard items; (5) diversion of items from production contract; (6) minimizing investment in high-cost items; (7) greater flexibility with subcontractors and vendors; (8) developing requirements forecasting techniques that can be logically explained and demonstrated to the contracting authority; and (9) a higher degree of risk and penalty when the manufacture of spares and repair parts exceed use. This is done by reducing the incentive and cost-sharing arrangement of items produced and not used.

11.6 PRODUCIBILITY

Failure to consider production requirements early in the acquisition cycle has resulted in more-than-acceptable risk in the transition from development to production. Product design and development cannot be logically separated from the production process. The approach taken to design most often radically constrains the producibility of a product. It is for this reason that producibility attributes should be an integral part of all configuration item trade studies.

Once production is viewed as simply another life-cycle function, not unlike companion operation and support functions, then systems engineering tools can be used to manage the production process. The timely application of systems engineering tools and the systems engineering process can contribute to assisting industry in achieving higher levels of productivity.

The capability to produce a hardware item that satisfies mission objectives is as essential to the systems engineering process as other functions such as operations or support. The application of the systems engineering process to production functions will identify the products required to transform design into a capability for efficient and economical production of equipment and facility elements of the system. Systems engineering also ensures that production capabilities are constantly used as design selection criteria. Typical production functions that are analyzed during a fully integrated design effort include such actions as material ordering, material handling, fabrication, processing, quality assurance, process control, assembly, inspection, test, preservation, packaging, storage, shipping, and disposition of scrap, salvage, and waste materials. The systems engineering process applied to production functions supports producibility analyses, production engineering inputs to trade studies, life-cycle-cost analyses, and consideration of materials, tools, test equipment, facilities, personnel, software, and procedures that support manufacturing.

Approach

Manufacturing activities begin early on when a system concept has been defined. Initial activities are concerned with production feasibility, costs, and risks. As development proceeds, trade studies are conducted to establish the most cost-effective methods for manufacturing items, and detailed plans are developed for the production phase. During production, extensive controls are implemented at both prime contractor and subcontractor facilities to ensure that the product will meet the specifications. Manufacturing activities during the system acquisition process are shown in Figure 11.5. Producibility analyses may generate the need for a requirements scrub effort by the program manager and user.

The requirements for contractor production management, are given DoDD 4245.6, *Defense Production Management*, which defines the need to establish:

- Industrial resource analyses.
- Production readiness reviews (PRRs).
- Production risk analysis.
- Manufacturing strategy.
- Comprehensive producibility engineering and planning program.
- Effective integration with the quality program.
- Independent assessment of production readiness.
- Planning for post production activity.
- Incorporation of a variety of cost avoidance and/or reduction techniques.
- Emphasis on life-cycle cost.

During the early program phases, the contractor's production engineering personnel must be integrated into the systems engineering organization to ensure that producibility requirements are incorporated into basic systems engineering documentation, specifications, and plans. The production engineers review conceptual designs together with other engineering specialists, conduct manufacturing trade studies to establish the most producible design, identify required production resources, and prepare the production plan. The basic systems engineering process applied to production functions is illustrated in Figure 11.6.

During full-scale development (FSD) phase, a manufacturing organization is established. The elements of production engineering, product assurance, planning, facilities, and production equipment are often integrated into the organization. Product assurance provides the process control of manufacturing and subcontractor operations. Prototype and qualification articles are produced and tested to demonstrate that the system meets its specification requirements. Prior to the start of full-scale production, PRRs are conducted to assure that all necessary resources and controls are established. Depending

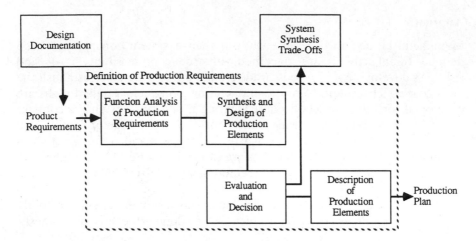

Figure 11.6 The basic production systems engineering process. (*Source: Systems Engineering Management Guide,* DSMC, December 1986.)

on the size and complexity of the system, PRR may be held as a single review or as a series of reviews.

Production Engineering Analysis

Production engineering analysis begins in the CE phase, as stated in MIL-STD-499A, MIL-STD-1528, and DoDD 4245.6. This analysis requires rate and quantity inputs that may themselves be the products of major tradeoff analysis efforts. Typically, the production engineering analysis is performed to:

1. Establish estimates of the production capability required.
2. Assess previous production experience and problems encountered on similar programs in conjunction with cost/schedule-control system reporting.
3. Identify, develop, and document new technology or special processes.
4. Assess production feasibility and identify risk areas.
5. Develop production costs and schedules.
6. Define production risk mitigation approach and associated milestones.
7. Define tooling requirements.
8. Define a production test plan.
9. Establish inspection requirements.
10. Establish personnel skills and training requirements.
11. Evaluate existing facilities and equipment to establish any modifications or new resources for manufacturing.

12. Develop a manufacturing assembly sequence chart.
13. Define producibility criteria.
14. Identify trade areas to reduce risk or cost.

The objective of the production engineering analysis, considered as an integral part of the systems engineering process, should be to permit the production of a quality system on time, at the lowest possible cost. During the early program phases, production engineers work with systems engineers to define the impact on existing resources and provide data on manufacturing alternatives to proposed designs using the basic systems engineering process.

Producibility Trade Studies

Manufacturing trade study areas include engineering design, reliability, maintainability, program schedules, life-cycle cost, effectiveness, producibility, supportability, and other factors affecting overall program objectives. Trade studies are conducted to evaluate the most cost-effective manufacturing process to be employed within program constraints. The trade study process involves the identification of alternate candidates, definition of evaluation criteria, weighing and scoring of the candidates, and examination of adverse consequences.

Manufacturing Planning Support

The results of the production engineering analysis are documented in the production plan, which defines manufacturing concepts and methods. The production plan provides sufficient information to supporting organizations to ensure a timely, coordinated approach to the production process. As the detailed design is completed and prototype hardware is developed, production engineering supports planning by continuing to refine its analyses to more detailed levels and by developing requirements for items not visible in earlier phases. After the baseline design is established, engineering change proposals are evaluated by production engineering as part of the configuration management process to provide manufacturing inputs on cost and schedule impacts.

The PRRs are conducted to establish that the system is ready for efficient and economical quantity production, that adequate test planning has been accomplished, and that problems encountered have been resolved.

Production Strategy–Plan Development

A production strategy is developed as part of the overall program acquisition strategy. This strategy is a comprehensive assessment of the production issues that form the foundation for a formal production plan.

The production planning review is an integral part of the overall acquisition review process. An acquisition may not proceed into production until it is determined that the principal contractors have the physical, financial, and managerial capacities to meet the cost and schedule commitments of the proposed procurement. Competition, value engineering, tailoring of specifications and standards, design-to-cost goals, cost–benefit and tradeoff assessments, preplanned product improvements, multiyear procurements, industrial modernization incentives, and other techniques are used to reduce production, operating, and support costs. Standardization, commonality, and interchangeability must be promoted throughout the acquisition cycle to reduce leadtime and life-cycle cost.

Producibility Engineering and Planning (PEP)

The term "producibility engineering and planning" as used in DoD is identical to the term "production planning" in the academic and industrial worlds. PEP includes all those design activities and disciplines necessary to design a product that is producible, design the processes and tooling, set up the manufacturing facility, and prove the processes and facilities, before entering production.

The PEP program extends throughout the life cycle. It includes actions required to maintain a capability to produce material for equipment operation and maintenance after the production plan is complete. The planning for these postproduction activities starts in the development of the initial production strategy.

11.7 STANDARDIZATION AND INTEROPERABILITY

Standardization is a prime factor in optimizing system design and therefore must be integrated with other design criteria. A design standardization program serves to better reliability, maintainability, and cost-effectiveness. Military standards are delineated in directives to provide commonality and interchangeability among items in the subsystems of a new product. MIL-STD-680 requires the development of a plan to define the management of standardization programs during development.

Standardization affects the type and quantity of test and support equipment, type and quantity of spare–repair parts, personnel quantities and training requirements, the extent of coverage in maintenance procedures, computer software language requirements and production lines, system design, interchangeability, calibration and reliability, PHST requirements, testing, and facility space for storage of spare–repair parts. Interoperability affects interchangeability, PHST, testing, maintenance procedures, software requirements, reliability, and calibration requirements.

Why Have Component Standardization Program?

Why should DoD or commercial manufacturing be so concerned with common parts such as fasteners, resistors, semiconductors, electrical connectors, valves, bearings, and so forth? Management decisions involving the selection and acquisition of new equipment and weapons systems affect the introduction of new support parts in many ways. There are several reasons why component standardization is critical in any acquisition strategy and must be used as early in design as possible.

Life-Cycle Considerations. Each time a new system enters the inventory, it brings with it thousands of new items of spares and support equipment. Any serious attempt to reduce the number of different items in the logistics system, and thereby reduce costs for the life cycle of a system, must begin in the design and development stage.

Government and Industry Studies and Reports. For many years, studies and reports within both government and industry indicated a consensus that component standardization must be effected at that point in the life cycle of equipment or system acquisition where savings related to cost factors can be maximized. This has been defined as the "point of decision" when a design engineer recognizes a functional requirement and needs an item to fulfill this requirement. These studies also emphasize that the item intelligence required by the design engineer differs from that required by the logistician in that the design engineer is primarily interested in the technical characteristics and qualifications of an item, while the logistician is interested in its utilization in a predetermined application. Industrial and productive organizations have developed listings of selected items for the purpose of applying intensified management and control. These listings are known as "company design standards" or "preferred parts lists." The DoD Component Standardization Program builds on this concept and uses the DoD contractor as the first barrier to parts proliferation. This is achieved through common program parts selection lists developed in concert with a centralized body of DoD parts standardization experts via the DoD Parts Control Program imposed on all contracts.

Early Opportunities. Current DoD policies require a maximum degree of standardization during engineering and operational system development, without causing unacceptable compromise of performance, reliability, availability, or cost of systems. The U.S. Military Services have developed implementing regulations reflecting those policies for additional emphasis on standardization during design. This additional emphasis broadly reflects that standardized parts selection during design is, in effect, an optimum compromise where, within a single operation, the many requirements are equated in relation to each other. The selection process thus also integrates considerations of technical characteristics, testing, maintainability, safety, human factors, sources of supply, and previous usage history. The task of selecting, specifying, ensuring proper design application, and controlling parts used in complex systems is a major engineering task. Piece parts constitute the building blocks from which systems are created and greatly impact hardware dependability and readiness. Since the reliability and maintainability of the end item is dependent on these building blocks, the importance of selecting and applying the most effective parts cannot be overemphasized.

Increased Productivity and Enhanced Competition. The opportunity for mass production through standardization has been well known since the time of Eli Whitney. Increased productivity through the use of standards is still valid today. The value of standardization for spare parts competition has been cited time and again by competition advocates throughout DoD.

Military Service Resources. Prior to the establishment of an integrated DoD Parts Control Program in the mid-1970s, parts control during the design phase was usually effected on a U.S. Military Service or Command basis. Limited coordination, between and among the services often resulted in redocumentation of the same item, utilization or substandard parts, and insufficient distribution of intelligence concerning parts of proven reliability. Within the Services, resources devoted to component parts standardization were scarce and the few parts personnel were often concerned with higher-priority tasks than standardization and related problems.

Documentation Currency. Industry groups have repeatedly alleged that the DoD baseline of standardization documentation in terms of coordinated standards and specifications is either lacking, incomplete, or not current with the state of the art. As a result, there has been a strong tendency by both government and industry to develop limited use and specification–source control documentation to meet specific system design needs. The net result has been parts proliferation, sole-source buys, and a multi-billion-dollar investment in data for duplicate nonstandard items. The DoD Standardization Program now serves as a DoD-wide focal point where data for "advanced technology parts" can be organized for updating of the design baseline on a commodity basis. The active involvement of standards engineers in parts selection at the design stage permits an analysis of application trends and provides potential for design documentation standardization that was previously impossible to obtain.

The DoD Parts Control Program. This is an integrated effort by the military services and the Defense Logistics Agency to streamline the selection of preferred standard parts during the design of military weapon systems and equipment as well as in the production and fielded stage. The program promotes the use of quality standard parts in the electrical–electronic and mechanical parts categories. Such common parts represent about 50 percent of the DoD items stocked. The idea of the parts control program is to minimize the variety of parts by relying on defense contractors to serve as the first barrier to the proliferation of unneeded nonstandard items. The DoD Parts Control Program–System has as its objectives the control of item proliferation within DoD and to achieve design-to-cost and LCC savings and avoidances by promoting the use of parts of proven performance during the design of equipments and weapons systems. This objective can be achieved by applying techniques that (1) assist system or equipment acquisition managers and their contractors in selecting parts to equal contractual requirements, (2) minimize the variety of parts used in new designs, (3) enhance inter–intradepartment and systems standardization interchangeability and RAM, and (4) conserve resources.

11.8   QUALITY ASSURANCE

Quality assurance (QA) is a planned and systematic pattern of all actions needed to provide confidence that adequate technical requirements are established, products and services conform to established technical requirements, and satisfactory performance is achieved. Provisions for QA involvement should begin in the earliest program phase and continue throughout the system or equipment life cycle. In system design reviews, QA participation must be part of the systems engineering process. QA participation is also required as part of manufacturing management, production capability, and production readiness reviews.

Designed-in quality must be stressed in the early phases of the acquisition cycle. The contractors must describe their approach for preventing deficiencies in product design, development, and manufacture. Before completing the development that results in the production decision, specifications standards final inspections, final tests and evaluations needed to control the level of quality must be developed. In addition, early in the development, new supplies, materials, and processes must be identified that necessitate technical advances in the development of measuring and test equipment measuring standards, or state-of-the-art measurement techniques so that adequate QA provisions can be made prior to production.

When government source inspection or acceptance is required, the responsibility for QA enforcement of contracts, including special inspection requirements, is assigned to the appropriate contract administration activity. Qualification approval, first-article testing, preproduction samples, and initial production evaluation are usually required to determine whether quality standards have been met. Research, contract administration, logistical support, and using activities must continuously exchange information on the quality of products and services during each acquisition phase.

Quality must be an integral part of the design and manufacturing process. Quality does not cost; it pays.

Total quality management is the application of methods and human resources to control the processes that produce our defense materiel, with the objective of achieving continuous improvement in the quality as illustrated in Figure 11.7. The DoD total quality management strategy also addresses the concurrent need to motivate United States industry to greater productivity. It is a strategy for improving the quality of DoD processes and products and achieving substantial reductions in the cost of ownership throughout the systems' life cycle.

The concept embraces the effective integration of existing management initiatives and initiation of new techniques that have a positive impact on quality. Examples are acquisition streamlining, competition for quality, statistical process control and continuous process improvement, value engi-

---

DoD POSTURE ON QUALITY

- Quality is absolutely vital to our defense, and requires a commitment to continuous improvement.

- A quality and productivity oriented Defense Industry with its underlying industrial base is the key to our ability to maintain a superior level of readiness.

- Improvements in quality provide an excellent return on investment and, therefore, must be pursued to achieve productivity gains.

- Technology, being one of our greatest assets, must be widely used to improve continuously the quality of Defense systems, equipments and services.

- Quality must be a key element of competition.

- Acquisition strategies must include requirements for continuous improvement of quality and reduced ownership costs.

- Managers and personnel at all levels must be held accountable for the quality of their efforts.

- Competent, dedicated employees make the greatest contributions to quality and productivity. They must be recognized and rewarded accordingly.

- Quality concepts must be ingrained throughout every organization with the proper training at each level, starting with top management.

- Principles of quality improvement must involve all personnel and products, including the generation of products in paper and data form.

- Sustained DoD wide emphasis and concern with respect to high quality and productivity must be an integral part of our daily activities.

Figure 11.7 Quality in DoD.

neering, transition from development to production, warranties, and gain sharing. The concept also recognizes that quality extends well beyond the domain of the inspector, and that it must be an integral part of system requirements, engineering, and design, as well as the manufacturing process.

Within DoD, quality has traditionally been defined as "conformance to contractual requirements." This definition has been the crutch for fulfilling the legal requirements in the administration of contracts, while disclaiming responsibility for actual results. This often results in enforcement of contractual requirements regardless of their validity. Many contractors have been satisfied with short-term profits and neglected long-term consequences.

Recognizing that the designer and the manufacturer, as well as the ultimate user of the products, have a key role in the quality equation, the DoD and a select number of industry associations have agreed on a new definition for quality: "Conformance to correctly defined requirements satisfying customer needs." This new definition does not, in itself, resolve the problems; it does, however, provide the correct perception of quality, which expands its domain throughout the product life cycle and involves everyone in implementation to assure success. The DoD "posture statement on quality" captures the essence of the strategy. The traditional view of quality is compared with the current posture in Figure 11.8.

The DoD uses product specifications and standards to impose contractual requirements. These documents are essential to the acquisition process because they provide the baseline for the bidding process, as well as providing the legal basis to determine contractual compliance. One requirement found in these documents is "acceptable quality level" (AQL) or the "lot tolerance percent defective" (LTPD). These provisions were originally intended to institute standard sampling procedures to ensure quality integrity of large production lots. Such numerical values, however, have been used by many manufacturers to justify lack of action in instituting effective process controls to improve quality. These contractors have become complacent with the "good enough for the government" concept, and lost sight of good business practices aimed at customer satisfaction and a lasting relationship based on integrity. Allowing a persistent level of errors as a way of life has contributed to unacceptable failure rates in defense equipment and to the escalating cost of maintenance and logistics support.

The DoD, to rectify the perception of allowable defects and stimulate changes to improve product quality, has recently directed its specification preparing activities to remove AQLs and LTPDs as fixed requirements in military product specifications. This action will provide opportunities to improve quality to the maximum extent possible by promoting competition based on excellence. In the past, the quality effort emphasized final inspection to detect defects after they had been produced in order to determine compliance with the required AQL or LTPD. Intricate sampling plans based on prescribed AQLs required the inspection of products to determine acceptance, thereby relieving the contractor of further responsibility for quality. The new approach

TRADITIONAL VIEW	CURRENT POSTURE
• Productivity and quality are conflicting goals.	• Productivity gains are achieved through quality improvements.
• Quality defined as conformance to specifications or standards.	• Quality is correctly defined requirements satisfying user needs.
• Quality measured by degree of nonconformance.	• Quality is measured by continuous process/product improvement and user satisfaction.
• Quality is achieved through intensive product inspection.	• Quality is determined by product design and is achieved by effective process controls.
• Some defects are allowed if product meets minimum quality standards.	• Defects are prevented through processes control techniques.
• Quality is a separate function and focused on evaluating production.	• Quality is a part of every function in all phases of the product life cycle.
• Workers are blamed for poor quality.	• Management is responsible for quality.
• Supplier relationships are short termed and cost oriented.	• Supplier relationships are long term and quality oriented.

Figure 11.8 Two views of quality. (*Source: Army Research, Development and Acquisition Bulletin,* March–April 1989.)

recognizes the value of sampling inspection techniques as a quality assurance tool. It removes, however, the inference that a predetermined amount of defects is expected and allowable. It enforces the concept that all delivered products are expected to comply with the established technical requirements.

QA Program

An effective and economical program, planned and developed in consonance with the contractor's other administrative and technical programs, is essential. Such a program should ensure adequate quality throughout all areas of contract performance—for example, design, development, fabrication, processing, assembly, inspection, test, maintenance, packaging, shipping, storage, and site installation. Facilities and standards such as drawings, engineering changes, measuring equipment, and the like as necessary for the creation of the required quality must be effectively managed. The quality program must be planned and used in a manner to support reliability effectively.

Contractors must institute effective process controls and in-process inspection techniques that preclude out-of-tolerance conditions during manufacturing in order to achieve continuous improvement and be able to compete on the basis of quality. By stabilizing the process well within acceptable limits, the "defect-detection" (troubleshooting) approach is replaced with the "defect-prevention' technique. The latter does not leave the process to change and then require screening of the good from the bad at the end of the process, nor does it rely exclusively on a sampling inspection that offers a measure of the degree of noncompliance.

11.9 SUPPLY SUPPORT

The efficiency of the supply system is measured by its effectiveness in supporting the major system or commercial product. The recent trend in the development of military supply systems has been toward greater centralization and integration. Some aspects of logistics support lend themselves to centralization and DoD control, while others are accomplished best under control of the U.S. Military Services. Some military supply management problems arise directly from the multiplicity of items that are stocked; others, from the size of the inventories. Many items are low in cost and are easily acquired through normal commercial channels; others can be obtained only through cooperation between the services and industry. Some items are issued in large volume on a recurring basis with demand relatively easy to forecast, while other items such as special types of equipment never have been issued and demand is difficult to forecast. Military supply management involves the largest inventories and the greatest diversity of items to be found in any organization in the world.

In addition to volume and variety, military inventories are characterized by rapid and unpredictable changes in makeup as the result of technological advances that have revolutionized the art of war. Many items become obsolete in less than 5 years.

The inventory managers are responsible for ensuring that items are properly cataloged. The individual inventory managers have responsibility for initiation of cataloging actions for items they manage and for changes to the catalog data. The inventory managers are responsible for planning and computing requirements for assigned items.

Supply support is an essential element of the logistics integration effort and is responsible for the timely provisioning, distribution, and inventory replenishment of spares, repair parts, and special supplies as shown in Figure 11.9. Supply planning for spares and repair parts must be based on technical inputs from maintenance planners and engineering for attaining a predetermined state of supply readiness. Factors to be considered include planning inputs (level of repair, repair–discard criteria, maintenance level facilities); isolation of utilization rates, operating hours, and failure rates and repairables program planning; requirements for provisioning technical documentation

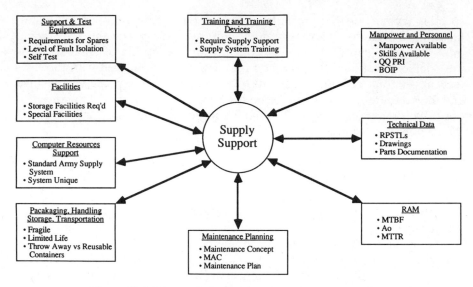

Figure 11.9 Integrating considerations for supply support.

and maintenance plan inputs; inventory management factors; and inventory management controls. It includes all consumables, special supplies, and related inventories needed to support the system. It covers provisioning documentation, procurement functions, warehousing, distribution of material, and the personnel associated with the acquisition and maintenance of the inventories at all support locations. Initial requirements for supply support are determined through the LSA process. Supply support encompasses all management actions, procedures, and techniques used to determine requirements to acquire, catalog, receive, store, transfer, issue, and dispose of inventory. As a result, supply support activities can be designed to meet the requirements of the system.

11.10 SUPPLY MANAGEMENT

The computation of requirements for major end items is an important function in the logistics system. Secondary items are all items not defined as principal items and include reparable and nonreparable components, subsystems, and assemblies, consumable repair parts, bulk items and materiel, subsistence, and expendable–durable end items. Secondary items include end items as well as spares and repair parts.

The U.S. Military Services and the DLA are required to minimize the total variable cost relative to ordering and holding inventory at inventory control points and their stock points. This minimization of the total variable costs is subject to a constraint on the average number of days forecast for delay in the availability of materiel, in terms of requisitions, for release by item managers or by the automatic data processing systems supporting the item managers.

Grouping of Secondary Items for Management Emphasis

Secondary Items are grouped into segments to be managed with different degrees of thoroughness depending on essentiality, investment in inventory, and the desired degree of protection against supply failure. The item groupings are based primarily on annual dollar value of demand or procurement with emphasis on essentiality and criticality. If the status of an item changes, the degree of management attention may be changed to accommodate it.

Very-High-Dollar-Value Items. Very-high-dollar-value items are accorded the highest possible degree of management emphasis. They are intensively reviewed and analyzed by item managers to include all facets of requirements forecasting and supply control operations.

High-Dollar-Value Items. High-dollar items receive less management attention than do very-high-dollar items but are still carefully controlled by item managers. Requirements are reviewed at least quarterly. Maximum use of the computer is made to determine requirements, but computer output is carefully reviewed and validated by managers. The system is a consolidation of the line item stratification and supply control study into one fully automated management system for secondary items.

The Item Management Plan. Very high, high, and medium management intensity items use the item management plan. A requirements objective is computed, assets are compared to those requirements, and the resulting supply position is used to determine whether to buy or repair, cut back due-in receipts, and declare or recall excess.

Routine Supply Control Study. This supply management routine is used for all low-management-intensity items. This routine accomplishes the same actions as the item management plan except that it contains streamlined computations and uses only wholesale-level demand, return, and asset data.

Secondary Item Demand Forecasting

Forecasting demands is one of the most important functions in the inventory management and requirements process. This is essential in the transition to production. Demand forecasting may be based on demand history or, when the history is inadequate, on engineering or mathematical estimates of these demands. Another technique is to base forecasts on the planned use of equipment. The techniques used most widely in demand forecasting are averages, projection of trends, and exponential smoothing.

Moving Averages. The moving-averages method is a demand forecasting technique in which a variable number of periodic records of quantities of items requisitioned are averaged for an estimate of the demands for the next period. Demand data accumulated on a monthly or quarterly basis are usually used for developing averages.

Least Squares. When the examination of demand data accumulated over a period of time indicates a changing demand pattern, demands may be forecast by constructing, graphically or mathematically, the line or cure that represents the demand trend and extending this line or curve into the future. The mathematical technique applicable for lines is called the "method of least squares." The method of least squares is designed to estimate the beginning demand level and the rate of change of a straight line that best represents recorded demand over a period of time. When the plotted data follow a uniformly changing curved path, a curve can be fitted to the data using advanced mathematical techniques.

Exponential Smoothing. Exponential smoothing is a statistical technique that is a refinement of, and is based on, the moving-average technique. While the moving-average technique places equal weight on grouped data in a given base period, the exponential smoothing method gives increasing weight to more recent data in a given base period. Three figures are required for exponential smoothing: the demand that was predicted for the last period, the actual recorded demand for the same period, and a factor called the "smoothing constant." The predicted demand for the last period usually was obtained at the beginning of the period by similar exponential smoothing computation or by the moving-average technique.

11.11 PROVISIONING

Provisioning is one of the most important logistics processes. Broadly defined, provisioning means laying-in an adequate supply of materiel, when and where needed and within monetary constraints, to support a system or end item of equipment during its initial period of operation. Provisioning is a management process for determining and acquiring the range and quantity of support items needed to operate and maintain an end item of material for an initial period of service. The objective of provisioning is to ensure the timely availability of minimum initial stocks of support items at using organizations and at the wholesale-level maintenance and supply activities to sustain the programmed operation of end items until normal replenishment can be effected. The ideal situation is one in which replacement and spare items match failures and there is neither a shortage nor a surplus of spares. Some amount of safety stock is normally incorporated in the recommended spares quantity. The amount of safety stock is a function of both financial constraints and a degree of acceptable risk.

The basis for determining the quantity of spares is the reliability data expressed in the LSA. These data require modification before they can be used in spares computation, because reliability figures reflect primary failures only and not failures caused by other parts in the system or by external conditions such as human error. These types of failure create a demand on the spares inventory and must be considered when determining spares quantities. This is accomplished by developing a mean time between demand (MTBD) figure.

The first step in developing an MTBD figure is to reduce the MTBF to a figure known as the "mean time between unscheduled maintenance actions" (MTBUMA). This is accomplished by the incorporation of secondary failures. Secondary failure rates are derived through judgment and the analysis of historical data on similar products or equipments.

Provisioning Concepts

The phased provisioning concept permits the acquisition of all or part of the initial quantity of selected support items to be deferred until later stages of production. The item(s) selected will normally be in the high-unit-cost category, will be unstable in design, and may not have been determined to be a required spare. The phased provisioning procedures involve establishing a buffer stock at the contractor's facility. Contractors maintain items in buffer stock of the latest configuration.

Spares Acquisition Integrated with Production. This is an effort to incorporate orders for spare recoverable assemblies with production installation. Thus, costs of spares and installs should be reduced. It also ensures timely availability of spares.

Interim Release. The prime provisioning activity controls the "interim release" concept. This permits the contractor to release to production those items having a production leadtimes greater than the leadtimes determined at the provisioning conference. This type of item is normally recoverable with a high dollar value and will likely be required by the operational need date for repair of the next higher assembly or replacement of an installed item.

11.12   INVENTORIES

Inventories may be classified as either basic demand, in-transit, safety, speculative, or dead, depending on why the inventory was held. Basic demand stocks are those required for filling orders under conditions of certainty where demand and replenishment time can be predicted accurately. In-transit stocks are en route from one inventory location to another. Safety stocks are held in excess of the basic inventory because of uncertainties in either demand or the replenishment time or for disaster or security reasons. Safety stock is inventory set aside to protect against short-term variations in either demand or supply. It is a direct result of the risk and uncertainty that characterizes inventory planning. Speculative stocks are created to balance cyclical production. Dead stocks are those for which demand no longer exists.

An inventory acts to achieve a balance between supply and demand. Success here requires regularity in the demand cycle. The interval between a fully stocked inventory and the time it is again fully stocked through replenishment is the inventory performance cycle. Changing an order cycle impacts both inventory performance and average inventory. Average inventory is a measure of inventory contents over time. Placing orders less frequently, with a constant consumption rate, means that each order must be for a larger quantity of inventory stock. The opposite is also true in that inventory stock must be larger because there is a longer period of time before a new order is received.

The initial acquisition cost of inventory items represents only one part of the cost of inventory. Transportation and storage costs must be considered. Replenishment stocks, for example, are affected by transportation times. Obsolescence costs as well as order costs must also be reflected in inventory management.

Stockage Concepts

Stocks on hand at the retail level are kept to a minimum. Normally, stockage will not include supplies for which there are no anticipated requirements.

Economic Inventory Policy

The economic inventory policy consists of three principles: the economic stockage principle, the economic order principle, and the variable safety level principle.

1. *Economic Stockage Principle.* The economic stockage principle governs which items are selected for stockage. This principle first considers the quantities of an item needed to attain a specific economic order frequency and needed for safety stock to ensure a high supply rate for demands. Then it compares the cost of stockage against costs of nonstockage. If the cost of stockage is less than the cost of nonstockage, the item is selected for stockage. If the reverse is true, the item is handled as a nonstockage item. Generally, the stockage criterion under economic inventory policy is more liberal for items with a unit cost of $25 or less, and is more restrictive for items costing

more than $25. The economic stockage principle extends the range of stockage for low-cost items with low dollar investment and limits the range above this cost to the faster moving items. This increases supply effectiveness and, at the same time, reduces the cost involved by less frequent ordering.

2. *Economic Order Principle.* The economic order principle provides specific rules on how frequently and in what quantities stocks should be replenished. It considers the cost of ordering and maintaining replenishment stocks. As larger quantities are ordered, the cost of maintaining inventories increases. However, since fewer orders are placed, the ordering costs for a given period are decreased. The basic principle used is that an item is ordered in such quantities that the annual cost of maintaining the inventory is equal to the cumulative cost of placing replenishment orders for the item over the same period of time. Order frequency tables reduce this principle to a series of varying operating level factors or order frequencies based on the total dollar value of annual demands. Use of order frequency tables has materially reduced the workload in review and replenishment actions, particularly for those items having annual demand of little dollar value. Furthermore, greater quantities of operating stocks of low-cost, fast-moving items, procured as a result of decreased investment in the high-dollar-value items, are resulting in improved supply efficiency. These tables are likewise entered into automatic data processing applications.

3. *Variable Safety Level Principle.* The variable safety level principle considers the operating level factors, the number of demands, item essentiality, and the order and shipping time for the item. Since the number of demands and the time required to order and receive items vary, the method for computing a safety level must be flexible enough to allow for the deviations. Reorder point factor tables incorporate factors used to allow for these deviations. In addition, the reorder point factor tables distinguish between the demands for nonreparable and demands for reparable items. This distinction is made because many of the demands for reparable items may be satisfied by the repair and return to stock of unserviceable items. These tables are also computerized.

Noneconomic Inventory Policy

Under the noneconomic inventory policy concept of inventory management, requisitioning objectives are based on a fixed amount in terms of days of supply that may be on hand or on order at any one time.

Economic Order Quantity Concept

The economic order quantity concept is used to compute replenishment orders and relates the cost of ordering to the cost of the item. Larger quantities of items qualifying for stockage under this concept are requisitioned less frequently if demands are not subject to excessive fluctuations.

The requisitioning objective for economic order quantity items is the sum of the operating level quantity and the reorder point quantity. This concept materially reduces the number of requisitions and the cost of processing low-dollar-value stockage list items.

The EOQ concept is based on the fact that total inventory costs are minimized at some definable purchase quantity. Total inventory costs are a function of the number of orders that are processed per unit of time and the costs of maintaining an inventory over and above the cost of items included in the inventory. The EOQ concept ignores transportation costs as well as the effects of quantity or volume discounts and assumes linear usage with independent demand.

Order points include leadtime, demand, and safety stock in a relationship that reduces reordering to a simple mechanical process. Order points assume that leadtimes can be accurately specified and are fixed. The order point also assumes that due dates do not change once they have been established. However, neither assumption is valid in the real world.

Inventory control rarely reflects the uniform usage rates that are characteristic of EOQ recorder situations. Lot sizing can help here. Lot-for-lot sizing is the most basic form. It is a simple one-for-one approach that matches the order with demand over a specified interval of time. No consideration is given to quantity discounts, transportation charges, or order processing costs. The period–order–quantity (POQ) builds on the EOQ to derive order placement intervals. Time-series sizing combines requirements for several periods to develop a uniform order quantity. Time-series sizing, although more dynamic than EOQ, fails to meet the requirements caused by rapid and frequent changes in the level of demand. These requirements have been met by a process called "material requirements planning" (MRP), which is primarily production oriented inventory planning.

11.13   SPARES AND REPAIR PARTS

Initial sparing of parts for new systems is always a challenge because of the dynamic changes caused by introductions to the field, uncertainties of demand and usage, a high rate of revision, retrofit and modification, inadequate documentation, manufacturing startup problems, and a general lack of experience. The key elements in establishing demand are the usage rate, system usage, and quantity per system to determine total usage quantities. During the early stages of a program, additional adjustment factors may be used, including reliability growth, usage profiles, secondary failures, user responsible failures, learning curves, and wearout. If items are to be replaced during scheduled maintenance, those needs must be added to the unscheduled usage. The usage rate is based on failures plus expanding modification factors to consider replacement level, repair or scrap, level of repair, location, and essentiality.

Whether a part is to be scrapped or repaired has a major influence on the number of spares required. If a part is scrapped, then a spare is necessary to replace every one to be used. If a replaced part is returned for repair, then the repair and recycle time determines the quantity. Low-volume, low-cost parts should be designed for throwaway maintenance.

Spares location also interacts with organization, structure, facility planning, item selection, and operations, which have an influence because the more places there are to stock parts, the more parts that are required. But it does not make sense to stock all parts at every level.

Spares and repair parts, special tools, and test and support equipment are selected from the provisioning list of all replaceable parts and special tools for the major end item by determining those that are required in the performance of maintenance operations.

11.14   MAINTENANCE ALLOCATION

To provide uniformity in maintenance planning and as a basis for selection of spares and repair parts and maintenance coding, a maintenance allocation chart is prepared. The maintenance allocation chart shows by functional description the maintenance operations assigned to each level of maintenance. Take your car, for example; certain services are required after 6,000 miles, more after 12,000, and so forth. Consideration is given to the level of maintenance and supply support provided; the complexity of the repair operation on the specific major end item; the availability of tools, test and support equipment, skills, and facilities; the capability to store and transport authorized spares and repair parts, tools, and test and support equipment; and the time required to effect repairs under combat conditions. Maintenance factors for spares and repair parts are based on anticipated replacement rates. Maintenance factors on new items or items having new application are initially estimated using all available data. Maintenance factors are continually being refined by analyzing data collected from all available sources. Allowance quantities are computed for each spare or repair part allocated for use at each category of maintenance. This computed quantity represents the number of repair parts necessary to fill the supply system and to provide adequate support for the specific weapon system or end item over an initial period of service. It is recognized that a maintenance factor is an "average" factor, and that the factor can be used to accurately forecast usage only when end-item density is large enough to provide for statistical uncertainty.

11.15   REPAIR–DISCARD ANALYSIS

Level of repair analysis (LORA), also known as "level of repair" (LOR) and "optimum-repair level analysis," (ORLA) is a well-established methodology.

The maximum benefit in performing level of repair analysis is obtained by doing so early in the life cycle and designing the equipment according to analysis results. The results of a LOR analysis are used to prepare maintenance plans and determine logistics resource allocations. The analysis will delineate as outputs resources of workers, material, and money. The LOR analytic techniques can also be used to evaluate alternative design and support proposals. Repair–discard tradeoff decisions may be classified into two types: repair level and design-oriented.

The Navy Electronics Systems Command utilizes LORA to identify requirements by year for support equipment, trained personnel, and spares for each repair level. LORA also identifies costs. LORA, however, makes no decision as to how an assembly will be supported. This decision must be made by the LSA.

The repair level decision is useful for developing logistic support concepts during the system planning phases for use during the operational phase after the design has been completed. The decision must optimize the maintenance and support levels at which repairs are most economical to effect. The second type of repair–discard decision is design-oriented for application during the late planning and design phases of the system life cycle.

In a study entitled *The Repair versus Discard Decisions*, the Defense Systems Management College (DSMC) noted five major decision points in the system life cycle where repair–discard decisions might logically be made. The first of the decision points, development of design specifications, occurs during the concept formulation and system definition phases. It depends on operational maintenance and logistics support policies as well as cost-effectiveness. At this level repair–discard decisions are primarily broad policy decisions that become part of the maintenance and logistics support concept. They result in the establishment of both qualitative and quantitative criteria in system development specifications to guide system–equipment design engineers during the development and design phases. The second point, initial design or item selection, occurs during engineering development. The policies and criteria previously established are now applied to assemblies, subassemblies, and modules based on analysis using quantitative repair–discard cost models. The third decision point, initial source coding for provisioning, occurs during the late design and early production phases. The decisions at this point are primarily logistics support decisions, such as range and depth of spares, effect on operational readiness of maintenance and supply delays, transportation and pipeline effects, and numbers and locations of test and repair stations. The fourth decision point, coding–design review, occurs during the operations and support phases of system deployment. The fifth decision point in the repair action, relative to whether a reparable item is still economically worth fixing after a failure has occurred, on schedule maintenance action, as a result of damage, age of the item, wear, or other condition.

Impact on Maintainability

Equipment must be designed for discard instead of repair. Many benefits are possible if failed parts can be discarded, including reduced requirements for accessibility to lower equipment levels, reduced need for skilled personnel and test equipment, fewer test points, and possibly the elimination of an entire level of maintenance. Time to repair will usually be improved with resulting improvement on operational readiness. In addition, the number of individual line items in spare stocks will be reduced.

11.16 SUMMARY

Major ILS management risks in the transition to the production process include inadequate planning, extensive changes, delayed organic support, delayed completion of testing phase, and inadequate producibility in design. Major support problems first become evident when the system is transitioned to production. Transition planning should, therefore, be completed before entering the initial production phase so that the system support package can be validated prior to the production decision. Thus, intensive ILS management is required to ensure that support items remain compatible with late changes to the materiel system.

12

OPERATIONAL AND POST-PRODUCTION SUPPORT

The overall objective of operational and postproduction support is to maintain the system within the operating and support cost program levels established. System readiness objectives established early in development constitute the baseline for planning for operational and postproduction support.

Prior to deployment–distribution, success in achieving system readiness objectives is evaluated by modeling or other estimation techniques employing input data obtained in development and operational testing. The first opportunity to directly measure readiness occurs when the system is initially distributed with its planned logistics support structure. Operational support planning and postproduction support planning are performed early in the acquisition cycle and serve a twofold purpose: (1) to ensure that readiness objectives are met and sustained and (2) to provide advance planning for corrective actions if required.

Logistics support problems increase with the age of the system and the rate of obsolescence of the technology employed in its manufacture. While problems may be encountered in a number of support elements (such as retaining manpower skills and replacing support equipment), the loss of production sources for spares and repair parts presents the greatest difficulties. Each system has unique postproduction support problems, and the success of postproduction support will depend on the logistics manager's ability to anticipate problems and find cost-effective solutions before they reduce readiness and/or increase support costs.

12.1 MAINTAINING READINESS

Assessing Performance

Although adequate development testing and operational testing, with their inherent data feedback, are critical to the success of a system, they do not fully measure the experiences that occur once that system has been distributed. Existing data collection systems provide coverage for many general applications; however, their output may not be sufficiently timely or detailed to support the reliability and maintainability analysis needed while the system is still in production. Supplemental data collection may be necessary and should be considered to provide timely corrections to design and quality assurance deficiencies that would be reflected in high failure rates, poor training that would be reflected in a high false removal rate, or poor technology. The earlier these problems are detected in the operational environment, the less costly the retrofit and the more effective the system will be.

Adjusting the Support

The initial corrective reaction to a readiness shortfall is to draw more extensively on existing logistics support resources. Responsive actions might include accelerating delivery of critical parts, raising stockage levels, modifying training procedures and technical manuals, changing operational or maintenance procedures or concepts, and increasing technical assistance to user personnel.

Initial estimates of requirements for ILS elements are based on anticipated failure rates, maintenance times, and other input factors. Logistics support resources must be recomputed as required based on updated values of RAM and other parameters measured during the operational phase.

Correcting the Design and Specifications

Manufacturing drawings of a system are modified either to correct performance and operational RAM deficiencies or improve and maintain the producibility of major components and spares to reflect changes in specifications and standard components that evolve over time. Relative to the first issue, it is important to detect design deficiencies as early as possible while the system is still in production. Procurement and application of modifications are much more expensive than a production engineering change. Drawing obsolescence, the second issue, occurs primarily in the postproduction period and becomes apparent when components can no longer be procured with the outdated drawing. Inability to obtain components incorporated in the original design can also necessitate modifications.

Updating the Software

Electronic circuitry is finding increased use in a variety of commodity groups. This growth has brought with it increased requirements to develop, test, and maintain the software used to control the system and the software employed with the automatic test equipment (ATE) to test replaceable units. Rapid growth and expanding technology have brought two problem areas:

1. Software programs exhibit a greater tendency for latent defects than hardware design.
2. System developers have encountered difficulties developing and maintaining ATE software compatible with system design during full-scale development (FSD) and production phases.

Responsibility for initial establishment of a complete and tested software capability remains with the system developer. However, it must also be recognized that there will be a continuing need for software maintenance during the operational phase and the postproduction period. ILS managers must establish the funding and the organization required to update the software to correct deficiencies and reflect the design changes.

12.2   EVALUATION OF MAINTENANCE AFFECTING OPERATIONAL AND POSTPRODUCTION SUPPORT

Maintenance Concept

The maintenance concept is one of the first policies that should be planned and published for the guidance of designers, logisticians, and all involved in design and development of the product. In the real world, the maintenance concept usually comes about after the fact and reacts rather than leads the design.

Maintenance Planning and Plan

Maintenance planning involves the definition of the maintenance concept, performing LSA, provisioning, assessing and evaluating the overall support capability, and designing a control mechanism for the corrective action and modification. The lifeline of the acquisition process is maintenance planning, which continues throughout the life cycle of the equipment. There are three important aspects of maintenance: documentation; basic factors of time, skills, and resources; and analysis. During the early phases of a program, the maintenance planning activity is crucial. Each logistics element is dependent on the maintenance plan. This emphasis decreases during the middle of the development process, but increases again as the operational phase nears.

The principal tool for guiding and recording maintenance analysis is contained in MIL-STD-1388-1A, which standardizes the approach to maintenance engineering analysis and incorporates operational considerations. However, the depth and detail of any analysis is governed by the design complexity and the resources available.

The maintenance plan documents the concepts and requirements for each level of equipment maintenance to be performed during the system's useful life. The factors in maintenance planning include (1) definition of actions required to maintain the designed system in a prescribed state of operational readiness; and (2) determination of maintenance functions, including checkout, servicing, inspection, fault isolation, replacement, and repair.

The role of LSA in maintenance planning evolves around the identification of specific maintenance actions to be performed; the systematic application of analysis to identify and describe tools, test equipment, personnel spaces, and repair parts and facilities to support the system; and performance of level of repair analysis. Maintenance planning constitutes a sustaining level of activity beginning with the development of the maintenance concept and continuing through the accomplishment of LSA during design and development, the procurement and acquisition of support items, and through the consumer use phase when an ongoing system support capability is required. It covers interim producer or contract support of the system during its early phases of consumer operations as well as the procedures for system upgrading and the installation of modification kits.

The maintenance plan should minimally include the maintenance concept; estimated corrective and preventive maintenance man hour requirements for each major assembly, and the identification of overhaul considerations. It prescribes maintenance actions, intervals, and locations (repair levels), together with the personnel numbers and skills, technical data, tools, equipment, facilities, and spares for each significant item of a system or equipment.

The maintenance plan is a document that describes the requirements and tasks to be accomplished for achieving, restoring, or maintaining the operational capability of materiel. The maintenance plan is published early in a materiel program and is updated periodically thereafter. It includes each maintenance significant item in either new or off-the-shelf materiel. The maintenance plan, based on the materiel maintenance concept, defines the maintenance resources required, and establishes their available-for-issue dates, in addition to allocating the tasks to the appropriate maintenance levels.

The maintenance plan provides guidance to all the elements involved in maintenance and support of the materiel. In the process of developing the maintenance requirements, allocations, and schedules, maintenance engineering uses experience gained with the same or similar materiel to its full advantage. Maintenance operations to be performed on any item are assigned to specific maintenance levels in accordance with the following:

1. The primary mission, character, and mobility of the materiel system involved.
2. The economical distribution of funds, skills, technical supervisors, tools, shop equipment, repair parts, materials, and so on.
3. The time available for performing the work.

These operations vary from simple preventive maintenance services performed by the personnel who are using the equipment, to complex repair and rebuild techniques practices at depot maintenance shops.

From the allocation of maintenance responsibilities, affected organizations develop progressive more detailed schedules for maintenance of each maintenance significant item in order to control the accomplishment of all known tasks in accordance with established priorities. These schedules can be relatively firm for preventive maintenance activities accomplished on a periodic basis, needing adjustment only for variations caused by operational requirements and immediate workloads. Corrective maintenance schedules must be developed on the basis of reliability data, actual or estimated, with anticipated failures prorated over a period of time on the basis of the best judgment of materiel maintenance specialists and previous experience.

12.3   MAINTENANCE ENGINEERING

The major maintenance engineering contributions to a program are to ensure that the system is designed for ease and economy of support, to define and develop an adequate and economic maintenance support subsystem that will be available when the system is distributed, and to monitor and improve the subsystem until the system is removed from the inventory. Design for ease and economy of support is obtained by determining optimum levels of materiel reliability, maintainability, human factors, safety, and transportability design features, and transmitting these features as requirements to design engineers. Requirements decisions result from a series of maintenance engineering analyses that tradeoff materiel operational requirements, acquisition costs, and support costs.

The support subsystem is comprised of support resources such as trained personnel, repair parts, and technical manuals required to support the system after it is distributed. Maintenance engineering develops the basis for the subsystem by a series of maintenance engineering analyses that identify and refine the requirements for each type of support resources. The requirements specify where, when, how, why, with what, and by whom the necessary actions will be taken to retain equipment in or restore it to a serviceable condition. After system distribution, these requirements are modified when analysis of available data shows that improvements in maintenance economy and efficiency are feasible.

To accomplish its mission, maintenance engineering conducts two closely related types of planning: planning the support of the system and planning for the acquisition of resources to provide the planned support. The first type of planning is accomplished almost solely by maintenance engineering, and is constrained by operational requirements and system design. The other planning is accomplished mainly by organizational support elements in consonance with the maintenance engineering analyses, planning, and resulting requirements.

Maintenance engineering is a dynamic function. The depth of analyses and consequently the depth of detail in the generated plans and requirements are limited by available design and support data. These data are quite gross at the start of most system programs, but become increasingly detailed as time progresses. As a result of iterative analyses, maintenance engineering plans and design and support requirements are progressively refined. An exception to the foregoing occurs when a system program involves the procurement and distribution of off-the-shelf equipment. In this case, design data and operational requirements are available, and the development and refinement of plans and support requirements can be accomplished with relatively few iterations of the maintenance analysis process.

It is important to note that maintenance engineering is responsible for generating design and support requirements, monitoring actions taken to satisfy the requirements, and judging the adequacy of the actions, but usually is not responsible for taking the actions. For example, maintenance engineering might impose a system design requirement for modular packaging and a support requirement for removal and replacement of modules. The latter requirement is then refined into detailed requirements specifying the personnel skills and quantities and repair parts required and describing how the maintenance action will be performed. The design requirements are submitted to design personnel, and the support requirements are submitted to personnel in the personnel training, repair parts, and technical publications support organizations. In both cases, maintenance engineering takes no action to actually satisfy the requirements, but is responsible for ensuring that the requirements are satisfied according to schedules that are compatible with deployment schedules.

An analogy can be drawn between system engineering and hardware development, and maintenance engineering and support subsystem development. The system engineer establishes the overall design concept, performance requirements, and interfaces among functional system elements. Detailed design (e.g., system electronics or hydraulics is performed by other disciplines. Similarly, the maintenance engineer establishes the overall support concept, performance requirements for the support resources, and interface requirements. Detailed design of the resources is accomplished by other disciplines. Thus, the maintenance engineer is the system engineer for the maintenance support subsystem.

Maintenance engineering is a technical analysis and planning function rather than a function that physically performs maintenance. The end products of the analysis and planning are mission-ready end-item weapons and equipment. The maintenance engineering effort, therefore, is oriented toward end items as systems, as contrasted with considering end items that are associated with more than one system as a homogeneous group.

Maintenance engineering participates throughout the life cycle of a system acquisition program, and all significant decisions and findings are based on maintenance engineering analyses. During the conceptual phase for a new system, historical maintenance data and support concepts are researched for use in developing materiel technical requirements. Maintenance analyses are then conducted to develop a broad general plan for logistic support that identifies anticipated critical issues of supportability, the anticipated materiel logistic environment, goals for life-cycle support costs, and recommended maintainability and reliability parameters. Although all program decisions are important, the initial support decisions are the particular significance since, barring program reorientation, all subsequent support decisions are refinements of the initial decisions.

During the next phase, demonstration and validation of the selected approach, studies are conducted, prototype hardware may be designed, and final reports, which include plans for system development, are prepared. Maintenance engineering participates in the support aspects of these activities. It provides support guidance, conducts support tradeoffs, provides information for reliability and maintainability studies, and updates and expands the maintenance analyses, which are still generalized, but of increasing depth, since functional design information is available to augment historical data. The results of maintenance engineering activity are a firm system maintenance concept, support plans, and maintenance related specifications.

Maintenance engineering activity starts to peak as the system materiel is designed, developed, and tested. Previously described activities are continued, and, as soon as preliminary engineering drawings are available, formal documentation of maintenance engineering analysis data are instituted. This analysis is continuously updated as the design evolves and becomes more detailed. All design changes are evaluated to determine the impact on support parameters, and, in turn, maintenance analysis reveals deficiencies that require design changes. Although many design changes are anticipated prior to production, early analysis is necessary in order to provide early planning data for long-lead support resources. The early maintenance analysis should be conducted in accordance with the same procedure as that used later in the program, and the data generated should be in the same format, to the extent possible, and limited only by the degree of design detail available. Requirements for the complete support subsystem are refined, and support resources are developed. Production configurations of operating materiel and support equipment and the resources of appropriate support elements are tested as a system to determine the adequacy of the planned support. These maintenance engineering activities will result in final system support plans and an operating maintenance analysis data system.

As production is accelerated, maintenance engineering activity declines from its peak. Design change impacts are analyzed; compatibility is maintained between the design changes, the data system, and the materiel support plans; and the acquisition of support resources is monitored. Additionally, a plan is prepared for modifying the materiel if modifications are required after it is deployed.

During distribution, maintenance engineering evaluates and analyzes the maintenance and operating experience of deployed materiel. The efficiency and effectiveness of support are determined, in large part, by comparing field data with the maintenance analysis data that were previously compiled. Problems are solved by in-depth maintenance analyses. In other cases, both hardware changes and support plan modifications will be required. As the materiel life cycle approaches its conclusion and sufficient data relating to the future force structure become available, maintenance engineering prescribes technical criteria regarding the final disposal of the system, and a plan is prepared for removing the system from the inventory. Preparation and implementation of the plan are not maintenance engineering responsibilities.

Maintenance engineering depends heavily on historical maintenance data. It is virtually the only type of data on which to base decisions during the early part of a materiel program. Subsequently, design data on which to base technical decisions become available, but historical data remain as a valuable source of ideas and a tool with which to test the validity of analytical determinations. Analysis of previous experience reveals which characteristics have or have not proved satisfactory on existing items. Such analysis discloses major downtime contributors, indicates high-failure-rate items, identifies design features that benefit support, identifies prime contributors to high cost, indicates maintenance man hour requirements, helps identify trouble spots, and provides parameters for analyses.

Without the benefit of the operational and support history of previous systems, the maintenance engineer cannot perform necessary functions efficiently and effectively on a new system. This highlights an easily overlooked fact: data acquired and analyzed after distribution of a system benefit not only the system itself but also future systems.

System support resource requirements vary with design changes, and are termed "support parameters" when considered in this light. Similarly, design features that impact support requirements are termed "maintenance parameters." One of the most important maintenance engineering functions is to influence the maintenance parameters in order to reduce the cost of the support parameters by an amount greater than any increase accruing to materiel acquisition costs as a result of design changes. A proper course of action cannot be selected without a detailed tradeoff.

12.4 MAINTENANCE ENGINEERING–MAINTENANCE INTERFACE

Maintenance engineering and maintenance are two distinct disciplines with well-defined interfaces. Maintenance engineering assists in the acquisition of resources required for maintenance and provides policies and plans for the utilization of the resources in accomplishing maintenance. Maintenance activities make use of the resources in physically performing those actions and tasks attendant on the equipment maintenance function for servicing, repair, test, overhaul, modification, calibration, modernization, and conversion.

Maintenance engineering activities begin in the conceptual phase of a system program and continue throughout its life cycle. Maintenance activities begin at the start of the distribution phase and continue until disposal. During the distribution phase, when the two activities are concurrent, maintenance personnel document maintenance experience. Maintenance engineering analyzes these data and may establish requirements for equipment modifications and modified maintenance policies or plans.

Maintenance engineering defines, integrates, and evaluates the total support subsystem. Any changes in system technical requirements or in the support plan that are subjected to a maintenance analysis invariably will impact on more than one support element. Maintenance personnel are not responsible for considering the total support subsystem, which would, of course, duplicate maintenance engineering functions.

Although maintenance engineering and maintenance have the same objective, that is, mission-ready equipment at lowest cost, the environments in which they function are significantly different. Maintenance engineering is an analytical function and, as such, is methodical and deliberate. On the other hand, maintenance is a function that must be performed under adverse circumstances and great stress.

12.5 MAINTENANCE ENGINEERING OBJECTIVES

The fundamental objectives of maintenance engineering are to ensure that the new system is designed for ease of maintenance and that an adequate economic support subsystem is provided in a timely manner. These objectives must be attained concurrently to reach an optimum balance between design and support. It is possible to provide an optimum support subsystem for a poorly conceived design, but this subsystem would represent failure in the achievement of the maintenance engineering design objective, and consequently would not be comparable economically to a support subsystem for well-designed materiel. The design and support objectives are inseparable.

The fundamental objectives may be attained through identification and attainment of a series of contributing objectives. Each objective is very important, but is termed "contributing" because its accomplishment merely contributes to accomplishment of the fundamental objectives rather than their complete accomplishment. The contributing objectives are to:

1. Reduce the amount and frequency of maintenance.
2. Improve maintenance operations.
3. Reduce the amount of supply support.
4. Establish optimum frequency and extent of preventive maintenance to be performed.
5. Minimize the effect of complexity.
6. Reduce the maintenance skills required.

7. Reduce the volume and improve the quality of maintenance publications.
8. Provide maintenance information and improve maintenance educational programs.
9. Improve the maintenance organization.
10. Improve and ensure maximum utilization of maintenance facilities.

The actions required for attainment of the contributing objectives, and hence the fundamental objectives, must start when the system is being conceived and must continue until it is removed from the inventory. The actions impact design and the structure and application of support resources. Some of the actions, by strict definition, do not fall within maintenance engineering functions, but since they impact adequacy and economy of support, maintenance engineering must provide leadership ensuring that they are accomplished. The contributing objectives and some of the more important supporting actions of each are:

1. Reduction in amount and frequency of maintenance:
   a. Establish a support concept and qualitative design requirements when the system is being conceived.
   b. Establish quantitative maintainability and reliability design features early enough to permit their incorporation into the system development program.
   c. When feasible, stress modular packaging, quick go–no-go diagnostics, prognostics, and accessibility.
   d. Make maximum use of test and maintenance data in establishing and evaluating support element resources.
   e. Accomplish a teardown of system prior to preparation of final maintenance allocation charts and initial provisioning.
   f. Obtain and analyze maintenance, performance, and failure data from the field, and correct discrepancies.
   g. Perform no unnecessary maintenance.
   h. Carefully establish inspection procedures and criteria by which to determine repair eligibility.
   i. Publish lists of materiel to be cannibalized or salvaged when it becomes unserviceable.
2. Improvement of maintenance operations:
   a. Define and apply the best of current management and maintenance techniques.
   b. Research industry practices, participate in symposis, and review trade publications.
   c. When possible, establish standard commercial type test, measurement, and diagnostic equipment, tools, and handling equipment for use in maintenance shops.

  d. Develop uniform criteria and procedures for computing maintenance workloads.
  e. Establish a file of reference data on all work operations, including time and overhaul standards, layouts, tool and equipment requirements, and related information, to expedite planning and accomplishment of recurring operations.
  f. Develop and apply simplified internal budgeting techniques to control costs in maintenance shops.
3. Reduction in the amount of supply support:
  a. Reduce the number of varieties of equipment, components, and repair parts by standardization, eliminating nonessential items, phasing out obsolete materiel, emphasizing geographic standardization, and, when feasible, using restrictive procurement to augment existing inventories with identical items.
  b. Screen repair parts lists and eliminate duplications.
  c. Use cannibalization as a source of low-mortality repair parts not type-classified as standard during the latter part of the system life cycle.
  d. Maintain current inventories of systems requiring repair parts support.
  e. Develop and publish data identifying where repair parts are used, according to make, model, and serial number, if necessary, of end items, assemblies, and components.
  f. Determine and publish data pertaining to repair parts interchangeability.
  g. Periodically review authorizations of expendable supplies and assure compatibility between current authorizations and requirements.
4. Reduce the frequency and extent of preventive maintenance to be performed:
  a. Establish design requirements such as self-adjusting assemblies, self-lubricating bearings, and corrosion-resistant finishes.
  b. Apply diagnostic equipment and techniques to eliminate teardown inspections for determining required maintenance.
  c. Establish realistic preventive maintenance intervals based initially on historical and design data, and adjust the intervals when field experience data become available.
  d. Ensure that current preventive maintenance checklists are in the hands of the user.
5. Minimize the effect of complexity:
  a. Design system for maximum practical reliability and maintainability.
  b. Design system to permit accomplishment of maintenance by easy removal and replacement of modules or assemblies.
  c. Design to provide with the maximum practical number of discard-at-failure modules.
  d. Provide for diagnostics by built-in test equipment (BITE) and automatic test equipment (ATE) that is easy to operate and interpret.

6. Reduce the maintenance skills required:
   a. Establish, during the conceptual phase, an optimum support concept and qualitative design requirements that are compatible with system mission requirements.
   b. Establish quantitative maintainability and reliability design features early enough so that they can be incorporated into the system development program.
   c. When practical, stress simple go–no-go diagnostics and discard-at-failure modules.
   d. Establish design features that eliminate or minimize the need for maintenance.

7. Reduce the volume and improve the quality of maintenance publications:
   a. Use the most advanced and proven educational techniques for presentation of material.
   b. When such presentation is effective, present information with combinations of microfilm and narrated tapes.
   c. Make maximum use of illustrated, charts, and tables.
   d. Periodically review maintenance publications to ensure currency.
   e. Critically review maintenance engineering analysis data provided to equipment publications personnel as the basis for manuals.
   f. Conduct careful validation and verification programs for maintenance publications.
   g. Stress adherence to standard definitions and symbols.
   h. Make maximum use of manufacturers' manuals.

8. Provide maintenance information and improve maintenance educational programs:
   a. Establish and maintain a program for dissemination of digested maintenance information of general value to maintenance personnel.
   b. Use available communication media.
   c. Ensure that key management personnel, both directly and indirectly associated with maintenance, are adequately indoctrinated with the objectives and importance of maintenance engineering and maintenance by attending appropriate schools as part of their career development program.
   d. Insure that agencies involved in the maintenance indoctrination of personnel use current material.
   e. Conduct on-the-job maintenance training to augment formal training courses.

9. Improve the maintenance organization:
   a. Place maintenance activities in a position in the organizational structure that provides for authority commensurate with the continuously increasing scope and magnitude of their responsibilities.
   b. Periodically evaluate the personnel and equipment resources assigned to maintenance organizations by determining workloads and resource utilization rates, and make appropriate changes.

   c. Develop and apply improved standards for determining the mainte-
      nance resources required to accomplish actual maintenance workloads
      and similar standards for accurately predicting maintenance workloads
      that will be generated by the new systems.
10. Improve and ensure maximum utilization of maintenance facilities:
   a. Identify and segregate excess cost resulting from underutilized capa-
      bilities of maintenance shops, and take appropriate action.
   b. Combine maintenance functions and allied trade shops, and use cross-
      servicing agreements.
   c. Reduce the variety of facilities, special tools, and test, measurement,
      and diagnostic equipment of standardization, eliminating obsolete and
      nonstandard supply items, establishing uniform maintenance proce-
      dures and shop layouts, and conducting effective maintenance engi-
      neering activities during the development of special tools and test,
      measurement, and diagnostic equipment.

12.6 MAINTENANCE ENGINEERING AND RELATED DISCIPLINES INTERFACES

Maintenance engineering is the interface between system design and system
support. It influences design by levying requirements on the design disciplines
of reliability, maintainability, human factors, safety, and transportability. It
controls the design of system support since it is the sole activity that establishes
resource requirements that must be satisfied by the support subsystem. The
requirements are further refined and the resources are developed and acquired
by organizational entities called "support elements," which are established
for support equipment, repair parts and support, equipment publications,
personnel and training, facilities, supply and maintenance technical assistance,
contract maintenance, and transportation and packaging.

Figure 12.1 is a flowchart showing how maintenance engineering interfaces
with these elements. The focal point of the flowchart contains the three
inseparable maintenance engineering elements: analysis, planning, and documentation. Analysis and planning constitute the systematic process by which
maintenance engineering considers all factors bearing on timely and economic
support and reaches a decision. Documentation is a systematic recording of
the analysis process and the decisions reached. Analysis, planning, and documentation are accomplished within the broad spectrum of formality. At one
extreme, analysis and planning can involve the solution of a current problem
by the simple application of historical data and judgment. The companion
documentation could be correspondence documenting the solution and giving
the rationale, and a milestone in a plan. The middle of the spectrum is
represented by analyses involving tradeoffs among various support alternatives, reaching a decision, and documenting the tradeoffs and the decision.
Typical of the other spectrum extreme is formal analysis, planning and doc-

umentation involving detailed examination of system materiel to determine support requirements and other maintenance relevant data, and recording the results in a prescribed format.

The activities (except provisioning) shown in Figure 12.1 apply to all system program phases. Maintenance engineering starts influencing design and defining support requirements in the conceptual phase and continues until disposal. At any point during the program phases, maintenance engineering is aware of the status of design and the status of the support subsystem. If design were frozen at any point, subsequent maintenance engineering effort would be devoted to refinement of the support subsystem to the degree permitted by the depth of design information. This situation rarely occurs. Design is constantly evolving, even during deployment, as is the support subsystem. As a result, maintenance engineering is continuously receiving design and support status information, performing and documenting analyses, updating plans, and issuing requirements to the design and support functional elements. Maintenance engineering design requirements may not always be completely satisfied because of conflicts with system constraints such as allotted time, performance, size, weight, and available funds.

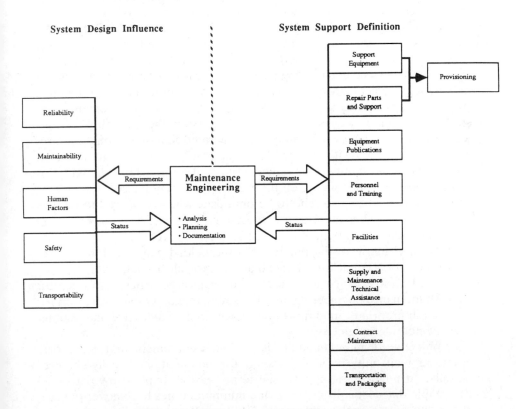

Figure 12.1 Maintenance engineering interfaces.

The interfaces among maintenance engineering and the related disciplines can best be described by briefly discussing the contribution of each discipline to support and the type of information that flows between maintenance engineering and the disciplines. The design disciplines will be discussed in the order in which they are shown in Figure 12.1.

Reliability is a characteristic of design that can be expressed briefly as the probability that equipment will perform without failure for a specified time under stated conditions. An analogous definition for maintainability is the probability that an item can be repaired in a specified time under stated conditions. These two design characteristics are very important. They combine to produce availability, which is the probability that materiel will be available for use, when required, under stated conditions. They are also the largest generator of support resource requirements, since failures resulting from unreliability generate the corrective maintenance workload, and the level of maintainability determines how economically the maintenance can be accomplished.

The leverage reliability exerts on the support elements can be appreciated by observing that an item that will function throughout its intended life cycle with no failures requires no maintenance corrective support other than the end items required to replace items lost in combat or otherwise destroyed. It would not be logical to plan for the repair of such equipment. Unfortunately, complex systems with 100 percent reliability are technically or economically impossible to produce, and therefore the maintenance support must be planned. Maintenance engineering participates in the establishment of initial reliability requirements. Major considerations are operational requirements, historical data, reliability state of the art, support resource requirements, and system acquisition costs. As the program progresses, the reliability requirements are refined whenever it can be demonstrated that operational requirements can be satisfied with reduced life-cycle costs. Reliability analyses continuously provide maintenance engineering with predicted reliability or observed reliability, depending on the materiel program phase.

Predicted reliability data tend to be optimistic when compared to failures that actually occur when materiel is in the hands of the user, because reliability engineers normally deal with inherent reliability—the reliability of the paper design—rather than with the reliability of the fielded materiel. Inherent reliability does not account for failures that might result from activities such as manufacturing, acceptance tests, user maintenance activities, and operator errors. Maintenance engineering ascertains how reliability data were derived and, when appropriate, modifies the data with field experience and maintenance engineering judgment.

The objective of maintainability is to design equipment that will satisfy operational availability requirements and can be maintained easily and economically. In relation to support, the term "easily" implies low personnel skills, simple diagnostic procedures, and minimum times to remove, replace, and test the failed, replaceable unit. The term "economically" implies accomplishment of the maintenance at lowest life-cycle cost. Maintainability and maintenance engineering objectives with regard to ease and economy of maintenance are the same. Maintenance engineering provides general requirements to maintainability by means of the maintenance concept, assists in the interpretation of the concept and in the conduct of design and support tradeoffs, and transmits specific requirements as they become available from analysis. Maintainability determines design features such as equipment packaging and diagnostics that economically satisfy both operational requirements and the maintenance concept and incorporates the features into system design. Design maintenance characteristics and predicted or observed repair times are transmitted to maintenance engineering.

Human factors and safety are disciplines closely related to each other and to maintainability. The objective of human factors is to design both operational and support equipment so that its use and maintenance are compatible with human capabilities. The objective of safety is to design the same equipment so that it can be operated and maintained safely. Maintenance engineering requirements for these disciplines are based on historical data, design analysis, and observation of activities involving the operation and maintenance of hardware. The disciplines transmit design information and safety procedures for maintenance engineering.

Transportability, in its broadest sense, is a design characteristic that establishes the transportation, handling, and packaging requirements for equipment. Some transportability features might be dictated by special operational requirements, such as a capability for equipment to be delivered by parachute. Others—such as compatibility with standard transportation and handling equipment, adequate tie-down and lift points, and compatibility with standard packaging and preservation techniques—are established by maintenance engineering. System design is monitored by maintenance engineering to ensure that transportability requirements are satisfied.

Maintenance engineering derives quantitative and qualitative resource requirements for each of the support elements by analyzing available data—including design information, historical data, and operational requirements—as they apply to the current maintenance concept. The requirements include delivery schedules that must be satisfied. The support functional elements feed back detailed plans for satisfying the requirements, and maintenance engineering develops a materiel support plan that defines how each type of resource will be used in logistics support and how it will be obtained. Typical products of each support element, other than a plan, and the nature of the requirements received by the element from maintenance engineering are discussed in the following paragraphs.

Support equipment includes test, measurement, and diagnostic equipment, handling equipment, tools, calibration equipment, and training equipment. Maintenance engineering transmits requirements to the support equipment element for both new and standard support equipment. The new equipment undergoes a design cycle identical to that of the operational equipment, and maintenance engineering influences the design as described previously. Use locations and quantities for all support equipment are refined, and requirements and supporting data for provisioning the equipment are transmitted to the support equipment element. Maintenance engineering plans the support of support equipment in the same manner that it plans the support of operational equipment.

Repair parts and support include repair parts and maintenance floats. Maintenance engineering identifies all requirements for repair parts and maintenance floats and generates other data required to provision the items. Requirements and documentation are transmitted to the repair parts and support functional element for satisfaction of the requirements.

After receiving maintenance engineering requirements, personnel from the support equipment and repair parts and support functions participate in provisioning activity. The provisioning activity has the objective of ensuring that support equipment and repair parts will be available in the proper locations when they are required. The full provisioning cycle involves documentation, selection, coding, determination of maintenance factors, cataloging, computation, procurement, production, and delivery. Maintenance engineering analysis provides the source data for the first four of these functions. Maintenance engineering generates documentation to support provisioning decisions, selects items, identifies the source of each item, establishes the lowest level of maintenance authorized to use the item, and provides guidance for the disposition of unserviceable items. Additionally, it provides maintenance factors showing the replacement rate requirement generated by distributed items. The two support functional elements and other agencies complete the provisioning process based on the maintenance engineering inputs. When the volume of data is large, as is the case in major material acquisitions, provisioning is accomplished with the assistance of automatic data processing equipment.

Equipment publications define the manner in which the operational equipment will be operated and maintained. In addition, publications include information pertaining to the maintenance and operation of the support equipment. Maintenance engineering inputs to the equipment publication element. Publication personnel augment the information by drawing and hardware analysis.

The objective of the personnel and training element is to train personnel in the numbers and skills required. Maintenance engineering specifies requirements for the numbers and skills and provides other information that assists in defining training requirements. The personnel and training element prepares courses of instruction, identifies requirements for training equipment, which are made a part of maintenance engineering requirements to the support equipment element, and accomplishes the instruction.

The facility element exists to satisfy all maintenance and storage facility requirements by either reprogramming the use of existing facilities or constructing new ones. Maintenance engineering describes the facility requirements in terms of utilization, plans or sketches, utility requirements, and other information required by the facility element to accomplish its function.

Supply and maintenance technical assistance is provided to the field to augment its organic supply and maintenance capability. Maintenance engineering establishes supply and maintenance technical assistance requirements based on cost-effectiveness considerations. Sometimes it is cost-effective to use the assistance during a limited period while organic capability is being established. However, the assistance frequently is provided throughout the operational phase of complex systems and for low-density systems with materiel quantities that do not justify the establishment of a normal full-range support program.

Transportation and packaging include the activities involved in moving equipment from the production line to the point of use and recycling it between the point of use and appropriate maintenance levels. Specifically, it involves preservation, packaging, packing, transportation, and handling. Maintenance engineering provides the transportation and packaging element with requirements that identify quantities, locations, schedules, and environmental constraints that impact preservation and packaging. The requirements emphasize nonstandard aspects of equipment that preclude normal transportation by military and commercial carriers, and aspects that necessitate special preservation and packaging techniques. Transportation and packaging may identify requirements for additional handling equipment, in which case the requirements will be transmitted to the support equipment element by maintenance engineering.

12.7 MAINTENANCE ENGINEERING ANALYSIS CONCEPTS

Maintenance engineering analysis is the dynamic catalyst in an integrated support program. During the early stages of a system acquisition cycle, the data identified by maintenance engineering analysis are general and parametric in nature. As the design progresses and a product baseline is identified, support requirements are defined in increasing detail. Interactions between maintenance engineering and design engineering activities must be many, varied, and continuing, particularly in the early phase. Logistic feasibility studies are made concurrently and are correlated closely with technical feasibility studies. A continual dialogue is maintained between design engineer and maintenance engineering as an inherent part of system development. This relationship maximizes possibilities for early identification of problems, thus forcing design versus support tradeoff decisions before the design if finalized. Maintenance engineering analysis efforts during the conceptual, validation, and early development phases are of special importance, having the potential for major impacts on design, system supportability, and life-cycle cost.

Maintenance engineering analysis provides for specific consideration of operator as well as maintenance requirements, and injects system support criteria into the design process at an early point in the acquisition cycle. Program essentials are analysis and definition of qualitative and quantitative support requirements, prediction of support costs in funds and other resources, and evaluations and tradeoffs.

12.8   MAINTENANCE ENGINEERING PROCESS

A systematic, comprehensive maintenance engineering analysis program that includes consideration of the projected system operational environment is conducted on an iterative basis throughout the acquisition cycle. This maintenance engineering analysis is the single analytical logistic effort within the system engineering process and is responsive to acquisition program schedules and milestones. Maintenance engineering analysis is a composite of systematic actions taken to identify, define, analyze, quantify, and process logistics support requirements. The analysis evolves as the development program progresses. The numbers and types of iterative analyses vary according to the program schedule and complexity. As maintenance engineering analysis evolves, records are maintained that provide the basis for logistic constraints, identification of design deficiencies, and identification and development of essential support resources.

Initially, maintenance engineering analysis develops qualitative and quantitative logistics support objectives. As the program progresses, these objectives are refined into design parameters for use in design–cost–operational availability–capability tradeoffs, risk analyses, and development of support capabilities. The initial effort evaluates the effects of alternative hardware designs on support costs and operational readiness. Known scarcities, constraints, or logistics risks are identified, and methods for overcoming or minimizing these problems are developed.

During design, analysis is oriented toward assisting the designer in incorporating logistics requirements into hardware design. The goal is to create an optimum system that meets the specification and is most cost-effective over its planned life cycle. Logistics deficiencies, identified as design evolves, become considerations in tradeoff studies and analyses.

Periodically, the design and the hardware are subjected to formal appraisals to verify supportability features, such as accessibility and compatibility of test equipment, as specified in the contract. As the program progresses, and designs become fixed, the maintenance engineering analysis process concentrates on providing timely, valid data for all areas of support.

Detailed logistics support requirements are identified as the design of the end item becomes firmly established. The range and depth of analyses vary, depending on the extent of system design definition and the goal of the analysis. Some analyses are highly iterative, while others are a one-time effort. Feedback and corrective action loops include controls to ensure that deficiencies are corrected and documented. Generally, detailed analysis of support requirements in concentrated on line replaceable units, modules, and major assemblies–subassemblies, plus necessary tools, test, and ground support equipment.

12.9 MAINTENANCE ENGINEERING TASK ANALYSIS

Support synthesis provides an organized basis on which to conduct support modeling evaluation of the proposed support subsystem and the framework for other analysis tasks. Synthesis is defined as the putting together of parts or elements so as to form a whole, or the assembly of various support approaches into conceptual support subsystems. Initiative and creativity are applied to influence equipment design for maintainability and logistics support. The analyst considers a wide variety of maintenance and support parameters within the restraints imposed by operational requirements and cost-effectiveness. Since a functional model or procedure with quantification is useful on all except the most minor acquisitions, synthesis data elements should be selected appropriate to the modeling technique used and the outputs required for the specific materiel procurement. Three basic areas should be considered in performing the synthesis: (1) variables representing the system–equipment must meet the purpose of the investigation, (2) the scope of the representation must be adequate, and (3) care must be taken in the manner of describing the synthesized support system. Characteristics of each approach should be defined and quantified.

A logistics design appraisal is an integrated part of program and design reviews held for the materiel. As a minimum, logistics design appraisals are conducted on completion of conceptual design, prior to the release of design drawings for full-scale development, and on completion of full-scale development. Informal support subsystem design appraisals are conducted at lower system indenture levels throughout full-scale development. The primary objective of the appraisal is to evaluate the projected design and, finally, the actual design on completion of the full-scale development phase. System design is reviewed for incorporation of support requirements from early in the conceptual phase through full-scale development. Specifically, the design appraisal considers the following:

1. Logistic support for the total system.
2. Physical configuration, including structural arrangement, installation, controls, displays, mounting, accessibility of subcomponents, and transportability.
3. Maintainability considerations, such as standard versus special test equipment, on-line versus off-line test equipment, component interchangeability, modularization, accessibility, criticality, standardization, and human-factors engineering.
4. Component reliability or malfunction rate and mode of subassemblies.

Subsequent to the support system design appraisal, a systematic follow-up is performed to insure incorporation of changes defined for logistics considerations.

Tradeoffs between support alternatives and equipment design parameters are made to provide an economical support subsystem that best satisfies the system operational requirements. The rationale and results of all tradeoffs made are provided as specified by the procuring activity.

Time factors are identified and determined for equipment operation, transportation, maintenance, and supply as an intrinsic part of all tasks of maintenance engineering analysis. These time factors are used to determine system downtime as a measure of system availability–effectiveness, maintenance man-hour requirements, maintenance time standards, and supply response requirements. Time factor determination is essential for quantitative maintainability prediction that involves the statistical combination of time-to-accomplish estimates. Time factor determinations begin in the conceptual stage on the gross maintenance functions and continue through full-scale development when hardware design has progressed to the point that specific features are known. Time factors normally are determined earlier and in more detail for those functions or functional sequences in which time is critical to mission success, safety, use of resources, minimization of downtime, and/or increasing availability. Examples of data outputs are task time in man-hours; task time elapsed; time line of critical tasks; maintenance man-hour figures per operating hour; maintenance manhour figures per year; maintenance manhour figures per maintenance action; mean time between maintenance actions; mean time between overhauls; mean time to repair; and other time data associated with operation, transportation, supply, and the maintenance cycle.

The functional requirements identification task identifies the support functional requirements as the frame of reference for developing support approaches. This task must be accomplished in time to provide a basis for concurrent consideration of support requirements with critical design decisions. Functional requirements identification progresses from gross functional levels, possibly with no mention of hardware in the conceptual stage, to a more formalized identification during full-scale development when the design has developed to the point that engineering drawings and hardware are defined in detail.

Failure mode and effects analysis (FMEA) is performed, unless otherwise specified in the contract, to identify predicted system failures and effects of the failures. The analysis provides timely identification of deficiencies in the total system. Deficiencies are corrected through design changes or by proper logistic support adjustments to the extent mandated by functional mission requirements and safety considerations. To the extent possible, inherent catastrophic or critical failures are alleviated. Failure mode and effects analysis is performed at the beginning, in the early stages of system definition and design. The analyst first uses system functional level breakdown and flow block diagrams, as developed by the design, reliability, and maintenance engineering activities. As design progresses, the FMEA extends down to the lowest functional level.

A repair level study is conducted to arrive at the optimum level of component discard and level of repair. The replacement unit size and the maintenance level are determined to define the various replace–repair action alternatives. Emphasis is placed on cost, operational availability, and operational effectiveness. Tradeoffs among these three factors and any overriding restraints, such as distribution requirements and supply line reliability, form the basis for replacement unit and capacity decisions. The support resource requirements generated by the various alternatives—including personnel and training, technical data, support equipment, facilities, and replacement–inventory parts—are evaluated to determine the optimum level decisions.

Maintenance and repair level decisions are made on the basis of a detailed review of the operational requirements of the system, the technical characteristics of system design, and the economics of support. It is an iterative process conducted throughout the conceptual, development, and production phases. A tentative maintenance allocation is necessary early in the program life cycle to analyze adequately the impact of preliminary design decisions. Constraints imposed by operational requirements may dictate the repair level decisions for certain items on the basis of mobility requirements, availability of resources, and other factors. Other decisions are made on the basis of optimum logistics cost-effectiveness and operational capability. The general decision process must allow rapid identification of those tasks that can be allocated immediately and those that require more detailed analysis. A systematic screening process must be established to eliminate the obvious discard-at-failure items first, and then analyze the remaining items at increasing levels of detail until each item maintenance and repair level is allocated.

Since a materiel support plan should be completely compatible with the logistic system, the considerations that follow should govern the assignment of repair levels.

Wherever the repair responsibility is allocated, care must be exercised to ensure that properly skilled personnel and necessary test equipment are available within the units assigned that responsibility.

In determining the level at which repair will be made, the program manager considers such factors as (1) technical feasibility of repair, (2) maintenance capability and economy of repair at the various levels, and (3) physical characteristics of the item. Other program manager responsibilities include the preparation or approval of repair specifications, the designation of repair facilities, and ensuring that required repair capability and capacity are established.

A maintainability prediction is conducted. The maintainability prediction is quantitative during the full-scale development phase. Prior to that time, quantification may be limited by uncertainty of design and scarcity of data;

however, best estimates must be used in conjunction with other analysis activities that determine repair levels, establish logistic resources, and optimize support characteristics. Output data from maintainability prediction are system maintainability values associated with hardware indenture levels. These values include maintainability allocations, mean time to repair, mean downtime, mean time between maintenance actions, and man-hours per operational increment. The data are used as inputs to the analytical determination of logistic support requirements.

Maintenance engineering task analysis is a detailed investigation of the maintenance–operational functions in which all tasks or actions required to accomplish them are identified and become the baseline data for the following:

1. The organization of specific maintenance procedures that must be conducted to sustain or to return the equipment to operating condition. These procedures form the basis of the equipment maintenance manuals.
2. Task time as vital for predicting maintenance time parameters.
3. Skill requirements and quantities of personnel necessary to perform the maintenance and operational tasks.
4. Tools, support equipment, expendable items, and spares and repair parts required to perform maintenance and operational tasks.
5. Minimizing the hazards associated with operating and maintaining the item.
6. Human-factors engineering studies.
7. Facility and space needs for performing tasks.

As in many of the other areas of maintenance engineering analysis, task analysis is evolutionary. Maintenance times and personnel requirements are estimated in the conceptual stage and iterated on a continuing basis as the design progresses through full-scale development. The FMEA is the primary source for corrective maintenance task identification. Particular attention is given to fault isolation, servicing, and corrective and preventive maintenance. Data resulting from the task analysis result in a complete description of the maintenance function and include such elements as task description, task number, sequential actions comprising a task, task frequency, man-hours per task, task elapsed time, personnel requirements per task, replacement parts per task, and support and test equipment per task. The task analysis must depict clearly the relationship between tasks and functions in performing complete jobs. Task analysis is performed in greater detail as the design is defined. When an initial design has been established (at the end of the validation phase), tasks are defined to the line replaceable unit level for use in determining manning requirements and level of repair. When detailed design data are available, tasks are broken down into step-by-step procedures and are used as the basis for technical data preparation.

In addition, an analysis is performed to optimize the safety characteristics of the system within the constraints imposed by the operational requirements. The analysis identifies hazards and specifies measures to minimize the danger to personnel, as well as the unique support requirements. The FMEA is used for identification of safety hazards. Responsibility for the system safety analysis should be independent of the system design function and should have recourse directly to top management.

Standardization reviews are conducted to achieve the maximum use of existing components, tools, support equipment, test, measurement, and diagnostic equipment and personnel skills without significantly inhibiting design improvement. New items introduced require justification that the items already in the system do not meet the approved characteristics or safety requirements, the state of the art and technology advances require the introduction of a new item, or the new item materially will increase the overall effectiveness and modernization of the equipment under development. Design improvement tradeoffs with the advantages of standardization may be cited as a reason for use of a nonstandard item. The key factor required in reducing the life-cycle costs and enhancing effectiveness of logistics support is to standardize for both physical and functional interchangeability. Standardization in this regard also requires:

1. Identicality of the end articles produced under contract, including identicality of internal parts, during the span of multiyear procurement, and across lead and follow-on contracts, when applicable.
2. Intra-end-article (intra-weapon-system–intra-aircraft–intraship) standardization to ensure the use of the minimum different components, equipment, and/or items within the end article wherever the closest tolerance or highest output could become the standard when horizontal standardization is not practical.
3. Intra-departmental standardization (the design reuse of reliable components and/or equipment already supported in the specific department).

12.10 MAINTENANCE

Basic Concepts

Maintenance is any action taken to retain a system in a serviceable condition or to restore it to serviceability. It includes inspection, testing, servicing, classification for serviceability, reclamation, repair, overhaul, rebuild, modification, retrofit, calibration, and refurbishment. Thus, the scope of maintenance tasks ranges from simple preventive maintenance services performed by the operator of equipment to complex maintenance operations performed in fixed shop facilities.

Each item is maintained in accordance with a maintenance concept that is established initially during the concept phase of a system program. Essentially, the concept establishes what, when, how, and where corrective and preventive maintenance is to be performed. The maintenance concept undergoes revision throughout the materiel life cycle. Prior to deployment, it is refined to reflect design changes, test results, and other new information. Subsequently, it may be revised as a result of field experience.

Maintenance concepts are based on tradeoffs and analyses of combinations of materiel design, maintenance actions, and maintenance locations that will satisfy operational requirements at lowest life-cycle cost. Maintenance concept decisions are the responsibility of maintenance engineering. Such decisions are extremely important, in that, for a given design, they establish the support resources required at each maintenance location and, consequently, establish life-cycle support costs.

Maintenance Levels

Currently, Department of Defense Directive 4151.16 defines three levels of maintenance. At the time of writing (present volume), delineation between maintenance levels is not yet clear. Organizational maintenance is the responsibility of and is performed by a using organization on its assigned equipment. Its phases normally consist of inspecting, servicing, lubricating, adjusting, replacing parts, and performing minor assemblies and subassemblies. Organizational maintenance is primarily preventive in nature. When it is properly performed, fewer breakdowns occur and fewer demands are made on supporting supply and maintenance activities. System repair at this level is limited to malfunctions that can be corrected through the removal and replacement of readily accessible assemblies or components. The objective of maintenance at this level is to quickly restore the system to operational service. Organizational maintenance personnel seldom repair the defective unit.

Intermediate maintenance is the responsibility of and is performed by designated maintenance activities for support of using organizations. Its phases normally consist of calibration, and repair or replacement of damaged or unserviceable parts, components, or subassemblies; the manufacture of critical nonavailable parts; and providing technical assistance to using organizations. Intermediate maintenance is normally accomplished in fixed or mobile shops, tenders, or shore-based repair facilities, or by mobile teams. Intermediate maintenance includes the repair of defective units and assemblies that have been removed at the organizational level. The defective item is repaired through the identification, isolation, and replacement of major assemblies and piece parts. Intermediate personnel are generally more skilled and better equipped than those at the organizational maintenance level.

Depot maintenance is performed by designated maintenance activities on segmented stocks of serviceable materiel. Maintenance at this level constitutes the highest type of maintenance support. It is expressly dedicated to those tasks that are above and beyond the capabilities of the lower levels. Its phases normally consist of inspection, test, repair, modification, alteration, modernization, and overhaul of weapons systems and parts. Depot maintenance is normally accomplished in fixed shops, shipyards, and other shore-based facilities, or by depot field teams. Depot maintenance is the responsibility of national level materiel managers. This category of maintenance overhauls economically reparable materiel and reconditions degraded stocks, which extends the service life for equipment and thereby delays procurement of replacement items. Depot maintenance workload requirements are forecast by national level item managers for a period of 5 fiscal years. These requirements consist of unserviceable assets on hand and those forecast to be returned, if needed, to fulfill the authorized acquisition objectives for serviceable major items.

12.11   MAINTENANCE FACTORS

The maintenance factor of a repair part is defined as the expected number of failures that will occur per year in a group of 100 end items containing the part. Failures that must be considered are those resulting from a deficiency in the inherent reliability of the repair part, as well as those resulting from the application of various K-factors; K-factors are applied in calculating maintenance factors, but the basic principles regarding K-factors versus inherent reliability remain unchanged.

The maintenance factor, because it is an indication of the expected number of failures for a repair part, plays a leading role in many areas of support planning. Its primary use is for provisioning. However, it is also a measure of the anticipated number of corrective maintenance actions that will be performed and, therefore, impacts requirements for all support resources in addition to those that are provisioned. Remembering that materiel life-cycle support costs represent the greater part of the materiel life-cycle costs, it is apparent that maintenance factors in most cases are the single greatest basic determinant of life-cycle costs.

Estimating a maintenance factor is an important task, and procedures have been established with which to accomplish the estimates. However, none of the procedures yet devised can be used without a thorough understanding of the relationship of maintenance factors to other materiel parameters and the significance and source of K-factors. The maintenance factor of a repair part directly affects the availability and reliability of the part and the availability, reliability, and maintainability of the parent end item. The predicted maintenance factor is therefore a very important system parameter and worthy of considerable attention.

Relationship to Failure Rate

Failures of a repair part may be categorized as burn-in, random, and wear-out. A repair part usually is classified as a random failure part or a wearout part. A random failure part is one that experiences a constant or near constant failure rate during its operational life, and a wear-out part is one that experiences an increasing failure rate. Some repair parts experience a decreasing failure rate early in their operational lives, but few sustain such a pattern. Generally, parts usually are not classified as burn-in failure parts.

At any point in the operational life of a repair part, the majority of failures of a repair part will be due to one of the three types of failure. The failure rate usually is considered to be constant over a finite period of time representing a portion of the operational life of the part, and reflects the cumulative effect of all types of failure. The estimate of the failure rate will reflect the effects of the three types of failure in light of the usage that was anticipated in the estimation time. The failure rate that the repair part actually experiences in operation in a particular deployment situation may depart from this estimate significantly as a result of factors that were not considered during the prediction process.

A maintenance factor is nothing more than a failure rate defined on the basis of 100 end items for a time period of 1 year. Normally, a failure rate is defined on a per-item basis for a period of time. However, the failure rate of an item usually is intended to represent the inherent failure rate and does not take into account any contributing factors such as usage, human error, and environmental effects. In other words, anticipated failures due to burn-in, randomness, and/or wear-out are considered the primary sources of failure. The maintenance factor, therefore, is seen to depart from the classical failure rate definition in this way. Namely, the maintenance factor is estimated to represent the failure rate and additional failures brought about by various contributing elements encountered in actual use.

If the maintenance factor is to be of any value for its primary function of estimating the range and depth of repair parts during provisioning, it must be influenced by appropriate parameters. As in the case of reliability where the generic failure rate of an item is factored by the influence of quality, test, usage, and so forth, to arrive at an operational failure rate, the basic maintenance factor must be influenced by the fact that there is not necessarily a one-to-one relationship between failures and replacements of repair parts. Thus, for the establishment of initial provisioning requirements for repairable materiel, the repairable concept and its ultimate impact on repair parts requirements must be considered in addition to the maintenance factor.

Relationship to Availability

The maintenance factor representing the failure rate of a repair part is generally assumed to be constant over a finite period of time. For practical purposes, this means that a maintenance factor represents the expected rate of failure for a period of 1 year, allowing for adjustments for future periods of time. The availability of a repair part is directly related to the mean time between failure, which is the reciprocal of a constant failure rate. Therefore, the estimated maintenance factor directly affects the predicted availability of the repair part. This, in turn, contributes to the availability of the end item, which is of primary importance.

Relationship to Reliability and Maintainability

The unreliability of a repair part represents the probability of failure of that part in a time increment. Therefore, reliability is a key parameter in maintenance factor determinations. Reliability improvement of a repair part necessitates a reestimation of the maintenance factor.

The relationship of the maintenance factor to the maintainability of a repair part is not as direct, and depends on the particular measure of maintainability used. If the mean downtime, representative total downtime is used, the maintenance factor influences the resupply process by controlling the demand rate for replacement parts and maintenance actions. This, in turn, influences the rate at which the repair part is repaired and restored to a usable condition, assuming the repair facility has the capability for repair. If only repair time is used as a measure of maintainability of a repair part, the maintenance factor of that part does not affect the inherent maintainability of the part. The maintainability of the end item–system (inherent or otherwise) is directly affected by the maintenance factors of the member parts.

12.12 FAILURE RATE

Usually, a failure rate is estimated on the basis of reliability test results, previous experience, and engineering judgment. The failure rate value, however, is representative of failures that can be expected for a given item when that item experiences usage that was anticipated in the design. The effects when the item is used in various assemblies, along with other items performing a common function, are not represented, nor are the effects that might be produced by maintenance or operation. As such, the estimated failure rate probably will be determined, in its purest form, by the manufacturer of the item.

Care must be taken to ensure that the manufacturer has accounted for stray factors such as geographic phenomena, in estimation of the failure rate. This would be acceptable if the phenomena would always be encountered regardless of the distribution area, but would considerably complicate matters if such conditions applied to only one distribution area. In other words, the contractor's estimated failure rate should include the effects of all conditions met in usage that theoretically would be met universally.

The failure rate exerts so much influence on the maintenance factor that the importance of thoroughly understanding its basis cannot be overemphasized. Generally speaking, failures in a part will increase as either operational time or stress increases, and will increase even more rapidly if both parameters are concurrently increased. There is a direct relationship between all of the other contributing elements and either operating time or stress. The usage rate is a direct measurement of operating time. All the other elements contribute to stress. Clearly, the modification of a failure rate that already reflects the effects of one or more of the contributing elements, by those elements, can result in significant maintenance factor errors.

If there is a single, most important step in estimating maintenance factors, probably it is precisely defining the parameters that are used in calculating the failure rate. The rest of the estimation process is based on the assumption that this is accomplished.

12.13   INTEGRATION OF OPERATIONS AND MAINTENANCE

Since a specified operational readiness level must be maintained and periodically demonstrated, and the purpose of a maintenance system is to meet operational requirements with minimum loss of time due to repairs, scheduling of operations and maintenance activities must be integrated. This necessitates a scheduling and workload control function within the field unit that will provide for management of all the unit's resources in performing those activities required to satisfy the unit's mission requirements. This management is accomplished through centralized scheduling of all maintenance and maintenance-related activities that directly impact the unit's tactical capabilities.

Conditions that result in the unit being in a status below full tactical capability require control to the fullest extent. In order to alleviate such conditions a number of controls should be instituted. For example, a scheduling and workload control element should be made totally responsible for scheduling necessary maintenance activities and integrating them with the unit's operational schedule; the schedule should then be approved by the unit commander; the approved schedule then becomes directive on all affected maintenance and support organizations; and any deviations from the schedule should be limited to only those required as a result of failures within critical equipment. The detailed scheduling, implementation, and execution of these maintenance activities are the responsibility of the affected maintenance organization. However, these schedules are established so as not to conflict with operational requirements and are coordinated with the scheduling and workload control element. Normally, the unit's scheduling and workload element schedule these maintenance activities to be accomplished within a given time span, with the maintenance area given the prerogative of establishing the specific date that permits the most effective use of its assigned resources. The complexity of the scheduling task can be illustrated best by a brief review of typical activities that require maintenance support and that must be considered in scheduling maintenance activities.

Corrective Maintenance

Corrective maintenance is performed to restore materiel to a satisfactory condition by correcting a malfunction that has caused materiel performance to fall below a specified level.

Preventive Maintenance

Preventive maintenance involves systematic inspection, detection, and correction of incipient failures in the system before they occur, or before they develop into major defects. Since the basic purpose of a preventive maintenance program is to reduce equipment failures, the scheduling and workload control element schedules, coordinates, monitors, and controls the program for all the materiel systems. Usually, the magnitude of the preventive maintenance activity, at both the system level and the item level is so great that it is imperative that preventive maintenance be closely controlled and monitored to verify that all requirements are executed in a timely manner. Additionally, preventive maintenance schedules first are established as a result of historical data and engineering judgment. A thorough and continuing analysis of preventive maintenance reports may reveal that periods between maintenance can be lengthened, or that maintenance can be eliminated.

Limited Life Component Replacement

Limited life component maintenance is governed by the basic interfaces, constraints, and controls applicable to preventive maintenance.

Training

New personnel training, cross training, and individual proficiency training may be required. Any training periods or activities requiring on-equipment instruction must be scheduled so as not to interfere with the activities.

12.14   EVALUATION OF TEST AND SUPPORT EQUIPMENT FOR OPERATIONAL AND POSTPRODUCTION SUPPORT

Test and support equipment should be specifically selected or designed to meet any unique requirements, the expected operating environment of the end item, and the capabilities of operating and maintenance personnel. Allocated test and support equipment may turn out to be more complex than the equipment being supported, thus requiring additional logistics support.

Logistics involvement in test and support equipment includes a determination of what is required, the quantity required, and the schedule of availability. The purpose of the support and test equipment program is to ensure that the required support and test equipment is available in a timely manner. The ability to perform the required scheduled and unscheduled maintenance depends on the adequacy of support and test equipment identified or developed with the prime system. Support and test equipment consists of tools, metrology, and calibration equipment; monitoring and checkout equipment; maintenance stands; and handling devices. It also includes production test or support equipment that is modified and delivered for field use.

The support and test equipment program encompasses all life-cycle phases. It requires the application of tailored ILS planning techniques. The identification of support and test equipment places additional requirements on ILS plans, maintenance plans, and provisioning plans; that is, all the logistics support requirements necessary for the operation and maintenance of the support and test equipment itself. The main consideration for obtaining support and test equipment must be cost, schedule, performance, and the ability to acquire adequate management capability for follow-on support.

The requirements for support and test equipment are derived through the LSA process, are published in the maintenance plan, and differ for each level of maintenance. Support and test equipment quantities are a function of the equipment being supported, product reliability and maintainability, the maintenance concept, and the number of maintenance locations.

Test Measurement and Diagnostic Equipment (TMDE)

Items of TMDE serve to extend the basic human senses for purposes of enabling the diagnosis and troubleshooting process. TMDE is defined as any system or device used to evaluate the operational condition of a system or equipment to identify and/or isolate any actual or potential malfunction. TMDE must give a measurement or indication of the operational condition of the system or unit under test. Practically speaking, TMDE provides the man–machine interface, enabling assessment of system performance parameters that the maintenance technician cannot directly see, hear, touch, or smell.

TMDE has a significant influence on the overall diagnosis and troubleshooting process. This influence is felt primarily in the form of time and in the form of skill and training requirements. TMDE also has a significant influence on operational availability and maintenance and logistics requirements.

The test equipment for materiel may range from manual test equipment to more elaborate and complex automatic test equipment. The complexity of the systems requiring detailed and extensive monitoring and checking to ensure performance virtually has eliminated the manual, step-by-step, probing type of testing. More stringent requirements related to operational availability of equipment, minimum downtime of equipment, and reaction time, combined with the decrease in skills available, dictate that the prime consideration in relation to test equipment for materiel be the person–machine interface.

In the early stages of equipment design the type of test equipment to be used for system monitoring and maintenance must be selected and repair policies and overall maintenance plans defined. The factors involved in the decision include the mission and operational characteristics of the equipment, personnel resources, operational environment, logistics support requirements, development quantities and time, and cost. Trade studies should be made before incorporation of automatic test equipment in new designs is specified since, as a general rule, automatic test equipment should be considered only when one or more of the following conditions prevail: turnaround time or downtime must be held to an absolute minimum, many repetitive measurements must be made, availability and readiness test requirements dictate its use, and maintenance loads warrant its use.

The operational availability of TMDE is a function of its reliability, maintainability, and the closely related parameter, durability. These parameters determine the frequency with which TMDE maintenance is performed and the resources required to perform the maintenance.

TMDE that is not built-in can impact system operational availability, even though the TMDE is operating properly. The impact can result from time lost during test setup and/or preparation and, if the TMDE normally is not stored in the immediate vicinity of the prime equipment, from time lost during transportation. Maintenance engineering should ensure that the prime equipment and the test equipment are designed to provide simple, foolproof methods for any required test setups. Also, careful consideration should be given to proposed TMDE storage locations and TMDE transportation times that will be required in an operational environment.

TMDE must be transported, either as an entity or as part of the prime equipment, from the point of manufacture to the point of use, and subsequently to support the prime system and to be maintained. Thus, it generates logistic requirements and costs. Some test equipment is considered inherently untransportable unless special handling and transportation factors are considered. Handling and transportation considerations include the requirements for special containerization and the mode of transportation, with its subsequent environmental consideration of shock, vibration, and temperature. Included in the transportation factor is consideration of the total mobility of test equipment, that is, whether it is an integral part of the prime equipment or a separate item of test equipment that the maintenance technician must obtain from some remote location in order to perform diagnosis and troubleshooting on the prime item.

Test equipment calibration requirements represent another important logistic consideration for maintenance engineering. Identifying these requirements and providing a calibration capability when the system is distributed is as important an ensuring that all the system support equipment is available when the system is distributed. Uncalibrated TMDE inherently is incapable of performing its function properly. Without a required TMDE calibration capability, system availability becomes zero the first time the prime equipment requires calibration.

TMDE calibration requirements are associated closely with tolerance requirements established for operation of the prime equipment. Normally, it is not feasible to eliminate TMDE calibration requirements. However, with proper planning, sometimes it is feasible to minimize the requirements. To ensure that a calibration capability exists when test equipment is distributed, initial calibration requirements are established during the system conceptual phase.

The design and/or selection of TMDE also influences the skill and knowledge levels required of the maintenance technician. TMDE can be made to perform selected portions of the malfunction analysis process automatically, thus reducing the amount of knowledge required of the technician. Computer-based systems use the computer to accomplish automatic malfunction analysis of complex electronic materiel. The use of the computer to perform automatic fault isolation can minimize the skill level requirement of the maintenance technician and reduce the required corrective maintenance time. The computer-based system normally encompasses features that include self-test, dynamic operational monitoring, automatic built-in fault isolation routines, automatic visual displays or instructions, and software routines for detailed diagnosis.

Typically, the computer-based system provides automatic fault isolation to a module, circuit board, or other replaceable part or assembly for a large percentage of the total failures. Dynamic monitoring of selected system functions during operational missions permits the computer to detect materiel malfunction immediately and to advance automatically into a preestablished fault isolation routine. The results of this routine may permit identification of a replaceable part or assembly, a group of replaceable items, or a major assembly. In the case of the latter, further isolation to a specific replaceable item normally is accomplished by the use of additional software routines. The identification of additional corrective action steps to be taken may be indicated by visual indicators. These indicators, combined with detailed instructions in technical manuals, provide the maintenance technician with the information needed to perform the corrective maintenance action.

Since a computer-based system uses the computer to control both operational and fault isolation processes, it is of prime importance to ensure satisfactory operation of the computer. This function normally is achieved through the use of built-in self-test features or through the use of these features in combination with detailed software routines. Computer-controlled TMDE normally is not used to troubleshoot nonelectronic materiel. However, the complexity of such materiel is increasing, and the maintenance technician must be provided with increasingly sophisticated TMDE in order to maintain skill and diagnostic time requirements at acceptable levels. Table 12.1 identifies some TMDE considerations, while Table 12.2 elaborates on the advantages and disadvantages of test equipment.

TMDE Design Principles

When analyzing for TMDE selection or design, the maintenance engineer should consider several basic criteria:

1. TMDE must be compatible with the modular maintenance concept. TMDE design, selection, acquisition, and allocation must support this concept to the extent feasible, as determined by technological, economic, and operational considerations.
2. Easy-to-use and interpret go–no-go built-in test equipment will be incorporated in the designs of all systems whenever technically and economically feasible.
3. Multipurpose automatic TMDE capable of fault identification–isolation, diagnosis, and failure prediction must be developed and procured for use at all levels of maintenance consistent with cost and efficiency considerations and on the basis of level of replacement–repair authorized in the maintenance allocation chart.
4. TMDE designs must provide for standard, foolproof quick-connect/ disconnect capability of TMDE to or from the end item or system under test without, when practical, the need for manual insertion of sensors–transducers into the unit under test.
5. TMDE configurations must be determined through economic analysis, consideration of force structure, qualitative and quantitative personnel and training requirements, related support equipment requirements, and mean time to repair requirements.

TABLE 12.1   TMDE Considerations

Categories of Test Equipment	Types of Test Equipment	Functional Test	Types of Test Indications
Special-purpose	Built-in	System	Go–no-go
General-purpose	Automatic–semi-automatic	Item	Quantitative
	Go–no-go	Open loop	Marginal
	Collating	Closed loop	
	Computer software	Static	
	(prime equipment)	Dynamic	
		Marginal	

TABLE 12.2 Test Equipment Advantages and Disadvantages

Test Equipment	Advantages	Disadvantages
Category		
Special-purpose	Accurate, simple for task, meets special need of material	High cost, short life, high risk, field impact, scheduling problems for availability, unique materiel
General-purpose	Inexpensive, readily available, long life, supportable, user familiarity, user confidence, versatility	Requires ingenuity for adaptability, time consuming in maintenance process
Type		
Built-in	Minimizes external support equipment, availability, minimizes downtime due to transport, no probing or manual connections in fault isolation, configuration status current with equipment, readily identifies performance degradation, no special transport or storage requirements	Prime equipment heavier, larger, more power demands, complex, higher cost, increase in maintenance, calibration integral to prime equipment and difficult owing to inseparability, self-checking for test feature to ensure performance required, inflexibility in test procedures, may be expended (e.g., missile system)
Automatic–semiautomatic	Rapid, increases test capability, controlled testing and consistency in test, eliminates human errors, reduces skill level and training for basic prime equipment task	Large, heavy, expensive, highly specialized, requires self-checking features, test point consideration in design for applicability, sensitive to design changes, complex, less reliable than manual, increases skill and training required for maintenance of test unit
Go–no-go	Simplifies decisions and maintenance tasks, information clear, concise, and decisive	Unique design circuitry, test unit costs high, scheduling problems, nonversatility for detailed circuit analysis
Collating	Reduces number of indicators, checking time, and error, simplifies troubleshooting	Similar to go–no-go and automatic, does not pinpoint specific signal malfunction

6. Sophisticated TMDE must be concentrated at the highest level of maintenance considered most productive and cost-effective.

Support Equipment

Support equipment includes all equipment required to perform the support function except that integral to mission equipment or required to perform mission operational functions. Support equipment should be interpreted as including tools; test equipment; automatic test equipment; organizational, intermediate, and depot support equipment; and related computer programs and software and maintenance repair kits. The acquisition of support equipment under the provisioning process begins with the LSA process and is directly dependent on the established maintenance concept. These factors are vital to establish repair and throwaway criteria, base–depot spares stockage, management skills needed for each level of repair, and the technical data required.

The stratification concept is used when it is known that support equipment must be developed on new items and that no other acquisition option is suitable. Five options are available—prototype development, early development, deferred development, normal development, or development of special test equipment. Each of these will be discussed in detail in the following paragraphs.

Prototype Development. The decision to develop prototype support equipment should be made during the demonstration–validation phase. Actual hardware and software design and development, including engineering testing, may begin at this time and continue into FSD. Prototype is done when the leadtime is long, the need date is early, and sensitivity to system design change is high. Prototype items are usually high-cost, special items (not common). The intent is to accelerate the design process concurrently with the system design provided enough engineering data are available on the system design. When developing support equipment, the project manager must make a conscious decision to accept design changes as the system changes. This method has the advantage of design concurrency, and cost competition, and removes some pressure on the prime contractor to perform high-priority work on the system.

Early Development. The decision to invoke early development is also made in the demonstration–validation phase, but design and development does not start until early in FSD and also requires an accelerated effort. Early development items are similar to prototype items except that sensitivity to system changes is low and the risk of follow-on changes is small. When enough engineering data are available, this method is especially appealing to competitive contracting and independent development by nonprime sources.

Deferred Development. The decision to defer development is made when the need date can be met by alternatives. This type of item is characterized by long to short leadtime and high sensitivity to system changes, may be special or common, and is a high-cost item. Owing to the characteristics of this type of item, the project manager may decide that deferring development has less risk in costs and is practical because of support alternatives. The prime contractor usually develops deferred items when the system baseline has stabilized and when higher priority system development work has been completed. This method also reduces the need for accelerated design work and ends the needs for expensive engineering data early in the FSED phase. Deferred development is decided in the D&V phase, but development does not start until the production phases.

Normal Development. The normal or common development decision is made during the D&V phase, but development does not start until well into the FSD phase and may run into the production phase. Normal development items may be special or common, are low in cost, have low sensitivity to system design changes, have short to medium leadtime, and have an early need data. Early effort should be made to baseline early development items by the end of FSD. This method is practical to use with the provisioning process.

Special Test Equipment (STE). The use of STE developed by the contractor for validation, test, development, or production program support is an acceptable alternative to developing new complex support equipment. STE is normally used for depot maintenance. Candidates should be evaluated to determine suitability to do depot diagnostics, fault isolation, and tests to ensure that operational life expectancy is consistent with life-cycle needs and to assess the availability and adequacy of logistics support for the extended period of use. Caution should be used in selecting STE for depot use as contractors are not required to maintain configuration control of STE.

12.15   EVALUATION OF MANPOWER AND PERSONNEL IN OPERATIONS AND POSTPRODUCTION SUPPORT

Manpower and personnel involves the identification and acquisition of operating and support personnel with the appropriate skills to operate and maintain the system over its lifetime. Manpower and personnel requirements define the training program for operating and maintenance personnel and the need for training devices to support training throughout the life cycle. Projections of manpower requirements should reflect current and forecast attrition rates as well as the capability of the raw personnel inputs. Manning determinations and their derivation from system design activities must be accomplished to ensure that manpower requirements meet equipment requirements. Manpower requirements must be based on related ILS elements as well as human-factors engineering that will ensure optimum person–machine interface. Manning and training requirements can be estimated from the initial operational and maintenance analysis in the LSA. Personnel requirements are a function of the tasks that must be performed and the time required for their performance. To determine the number of personnel, the work-hours available for task performance must be considered.

12.16   EVALUATION OF TRAINING AND TRAINING DEVICES FOR OPERATIONAL AND POSTPRODUCTION SUPPORT

The training–training devices category includes the process, procedures, techniques, and equipment used to train personnel to operate and maintain the system throughout its life cycle. Training requirements must be developed along with training curricula and must reflect the operations and maintenance concepts and the technical data plans.

The principle factors to be considered for training include schedules for training plans, conferences, institutions of training; determination of training equipment requirements and their required support; impact of training program leadtimes; and training throughout the life cycle of the system. Initial training estimates are based on experience. The initial operations and maintenance analyses from the LSA enable the estimation of training requirements. Types of training devices must be identified; this includes simulators and mock-ups. Classroom facilities and locations must also be identified. In addition, the types of training must be identified, that is, factory training, instructor and key personnel training, new equipment training teams, and resident training. Courses must be written and schedules developed. Training devices must be available for the training program.

Training in system operation depends on planned utilization and operating scenarios, whereas maintenance training relies heavily on reliability, availability, and maintainability analysis. These data are used by the training staff to evaluate the operations and maintenance tasks and divide them into related task groupings and skill level categories.

Successful implementation of the training course demands proper time-phasing with respect to the other elements of the program. Training requirements analysis cannot be completed until maintenance engineering has identified the tasks and times required for task performance. A primary impediment to scheduling training courses is the availability of training devices. The lack of equipment precludes hands-on training. Other factors include the availability of tools, test equipment, and technical publications.

12.17 POSTPRODUCTION SUPPORT

Providing the Plan

Each system will have support problems that are unique to that system and many of these will be unanticipated. The ILS manager should include postproduction support as a line item in the budget to accommodate the resultant changes.

Task 403, *Postproduction Support Analysis,* of MIL-STD-1388-1A, *Logistics Support Analysis,* should be performed during FSD. The "postproduction support plan" (PPSP) should be completed prior to Milestone II and updated with the "integrated logistics support plan" (ILSP). The PPSP should be maintained current as long as the system is in the active inventory and should focus on such issues as (1) system and subsystem readiness objectives in the postproduction time frame; (2) organizational structures and responsibilities in the postproduction time frame; (3) modifications to the ILSP to accommodate the needs of PPS planning; (4) resources and management actions required to meet PPS objectives; (5) assessment of the impact of technological change and obsolescence; (6) evaluation of alternative PPS strategies to accommodate production phase-out (second sourcing, preplanned product improvement, standardization with existing hardware, engineering level of effort contracts in the postproduction time frame, life-of-type buys, contract logistics support vs. organic support, etc.); (7) consideration of support to the materiel system if the life of the materiel system is extended past the original forecast date; (8) data collection efforts in the early deployment phase to provide the feedback necessary to update logistics and support concepts; (9) potential for foreign military sales (FMS) and its impact on the production run; and (10) provisions for utilization, disposition, and storage of government tools and contractor-developed factory test equipment, tools, and dies.

Planning Criteria

Postproduction support activities include those management and support activities necessary to ensure attainment of readiness and sustainability objectives within economic parameters after termination of the production phase. Planning should be currently maintained as long as the materiel system is in the active inventory. As a minimum, planning should include:

1. Identification and assessment of the impacts on both the major system and the support system as a result of expected production phase-out and technological change or obsolescence forecast.
2. System/subsystem readiness objectives in the postproduction time frame.
3. Resources and management actions and responsibilities required to satisfy postproduction support objectives.

4. Evaluation of alternative postproduction support strategies to accommodate obsolescence or production phase-out.
5. Support strategy of systems declared obsolete.
6. Actions needed to obtain cost effective competition of PPS requirements.

Establishing a Competitive Environment

Relying on a single industrial source for critical support entails risks in the areas of cost and availability of needed spares and repair parts during the operational phase and particularly after termination of end-item production.

Postproduction Support Decision Meeting

The program manager should conduct a PPS decision meeting prior to the final production order to avoid major nonrecurring charges if follow-on production is required later. This meeting should consider the advisability of purchasing major items from the manufacturer, such as (1) major manufacturing structures, (2) forgings and castings, (3) insurance items to cover crash–battle damage or fatigue, (4) proprietary data, and (5) raw material and updating the PPSP on the basis of the latest available data.

Other Remedies

When faced with the imminent loss of production sources for unique spares and repair parts, two basic options are available to logistic managers: (1)

SPARE AND REPAIR PARTS ACTIONS	
INCREASE SUPPLY	**DECREASE DEMAND**
• Develop a reprocurement technical data pacakge and alternate production sources • Withdraw from disposal • Procure Life-of-Type Buy • Seek substitute (interchangeable) parts • Redesign system to accept standard component if not interchangeable • Purchase plant equipment; establish an organic depot capability • Subsidize continuing manufacture • Draw (cannibalize) from marginal, low priority systems.	• Restrict the issue to critical applications in support of combat essential items • Phase out less essential systems employing the same parts • Restrict issue to system applications where no substitute is available • Accelerate replacement of the system

NOTE: For additional actions, see DODD 4005.16, Diminishing Manufacturing Sources and Material Shortages Program.

Figure 12.2 Logistic actions to reduce impact of loss of parts production sources.

increase the supply or (2) decrease the demand. A combination of actions listed in Figure 12.2 is often the most practical approach. These remedies are generally less effective and more costly than actions taken earlier in the production cycle.

Funding of Engineering and Publications Support

There is generally a continuing need to correct hardware design, specifications, and software after the completion of system development. Changes to technical manuals and/or technical orders are also needed to reflect the system and software changes and to correct other deficiencies reported by operator and maintenance personnel. Once the system is no longer in production, major problems for engineering and publications support arise.

12.18 SUMMARY

Effective maintenance support ensures that the users of the equipment are not deprived of its use for any appreciable length of time, and extends the economic service life of the equipment. Maintenance alleviates supply problems by extending the useful life of equipment through preventive maintenance practices by overhaul and by repair for return to the user or for return to the supply stocks. Maintenance is dependent on supply for the assemblies, kits, repair parts, and bulk materiel needed for repair or overhaul. The necessary tools and test, measuring, and diagnostic equipment are acquisitioned by supply.

Planning for maintenance support of equipment responds first to readiness requirements and next to economics in the commitment of maintenance resources. Maintenance planning allocates maintenance tasks and resources to those maintenance categories and determines where the work can be most efficiently accomplished, while maintaining or improving readiness. To ensure cost-effectiveness, level of repair analyses are performed during the maintenance planning process. The resulting plan provides the necessary capability to using units, ensuring them a degree of self-sufficiency commensurate with operational needs.

The ability to perform the required unscheduled and scheduled maintenance depends on the adequacy of the support and test equipment identified or developed with the prime system and developed through the LSA process and published in the maintenance plan. Test and support equipment quantities are a function of the quantities of equipment being supported, product reliability and maintainability, the maintenance concept, and the number of maintenance locations at each level of maintenance. Test and support equipment is unique in that it does not fall within the category of spares and repair parts or any of the other support elements. Yet, it is required for it provides an inherent ability to monitor, control, test, measure, evaluate, repair, and calibrate the weapons system. In the past, test and support equipment has largely been ignored when performing logistics evaluations addressing support for the system. It must be remembered that test and support equipment requires the same degree of support as the system itself. Logistics requirements for test and support equipment must be planned for and implemented with the same consideration that is accorded the system supported.

The LSA is also a significant contributor to the identification of personnel requirements to include skills, grades, and numbers of personnel required for the operation and support of a materiel system.

Training differs from education in that it is task-oriented and places greater emphasis on the acquisition of skills, whereas education is concept-oriented and more general in nature. The analysis of training requirements is a function of the training course objective. Training in product or system operation depends on planned utilization and operating scenarios, whereas maintenance training relies heavily on RAM analysis.

Readiness and RAM experience during the operational phase is employed to adjust the logistic support resources programmed during the FSD and production phases. Performance and RAM deficiencies must be detected and corrected as early as possible in the operational phase. Thus, the objective of planning performed during system development is to ensure that readiness objectives are met and sustained through the operational phase, including the postproduction period. Planning deferred until problems are encountered will be limited in effectiveness.

APPENDIX 1
DEFINITIONS OF TERMS

Acquisition. (1) A very broad term generally including quantity determination, procurement, and distribution to satisfy logistic needs. It can also include contract definition, development, design and test, evaluation, production, installation, purchasing, and contract administration. (2) The process consisting of planning, designing, production, and distributing a weapon system—equipment. (3) The process of acquiring supplies and equipment, facilities and services for use by/within the Department of Army, including life-cycle systems management of hardware and software, formulation of requirements, research, development, testing, procurement, production, fielding, operation, support, and disposal of army materiel. (4) The purchasing, renting, leasing, or otherwise obtaining of personnel, services, supplies, and equipment from authorized sources as prescribed by the Defense Acquisition Regulation.

Acquisition Plan (AP). Derived from the acquisition strategy; summarizes acquisition background and need, objectives, conditions, strategy, and related functional planning (with emphasis on contractual aspects); it provides detailed planning for contracts and milestone charting.

Acquisition Strategy (AS). (1) The conceptual framework for conducting materiel acquisition, encompassing the broad concepts and objectives which direct and control the overall development, production, and deployment of a materiel system. It evolves in parallel with the system's maturation. Acquisition strategy must be stable enough to provide continuity, but dynamic enough to accommodate change. (2) Conceptual framework for conducting materiel acquisition, encompassing broad concepts and objectives that direct and control overall development, production, and deployment of a materiel system. Evolves in parallel with maturation of the system. Must be stable enough to provide continuity but dynamic enough to accommodate change.

Affordability. (1) The demonstration that a system can be procured, operated, and supported efficiently and effectively for the programmed and budgeted resources (DoDD 5000.1). (2) Function of cost, priority, and availability of fiscal and manpower resources.

Allocated Baseline. (1) Development specifications (Type B) that define the performance requirements for each configuration item of the system. (2) The initial approved configuration identification; this is the baseline to which systems and equipment are controlled. (3) An allocated configuration identification that is a baseline initially approved by the customer. See "Allocated Configuration Identification."

Allocated Configuration Identification (ACI). (1) Current, approved performance oriented specifications governing the development of configuration items that are part of a higher-level CI, in which each specification (a) defines the functional characteristics that are allocated from those of the higher-level CI, (b) establishes the tests required to demonstrate achievement of its allocated functional characteristics, (c) delineates necessary interface requirements with other associated configuration items, and (d) establishes design constraints, if any, such as component standardization, use of inventory items, integrated logistic support requirements. (2) Performance specifications guiding the development of configuration items that are a part of a higher-level CI. These specifications cover functional characteristics allocated from those of the higher-level CI, tests to demonstrate achievement of the functional characteristics, interface requirements, and design constraints.

Appropriation. An authorization to incur obligation for specified purposes and to make payments out of the treasury.

Associated Support Items of Equipment (ASIOE). An end item required for the operation, maintenance, and/or transportation of a basis of issue plan (BOIP) item. ASIOEs are listed on the BOIP of the item they support. ASIOE's have their own LIN and are separately documented into table of equipment (TOE).

Availability. (1) The probability that a system or equipment will, when used under specified conditions, operate satisfactorily and effectively. Also, the percentage of time or occurrences a product will operate properly when requested. Inherent availability (A_i) is "pure, as designed." It considers only corrective maintenance time. Achieved availability (A_a) includes preventive maintenance time but an ideal support environment. Operation availability (A_o) considers total downtime, which includes administrative and supply times. (2) A measure of the degree to which an item is in an operable and committable state at the start of a mission when the mission is called for at an unknown (random) time (MIL-STD-1388-1A). (3) Measure of the degree to which an item is in operable and committable state at the start of the mission, when the mission is called for at an unknown (random) point in time.

Availability (Achieved). The proportion of time a system is operating, considering operating time and total maintenance (scheduled and unscheduled) downtime. The formula is

$$A_a = \frac{OT}{OT + TCM + TPM}$$

where *OT* and *TCM* are as defined under "Availability (Inherent)" and *TPM* represents the total preventive maintenance downtime in clock hours during the stated *OT* period.

Availability (Inherent). The proportion of time a system is operating, considering operating time and unscheduled (corrective) maintenance downtime. The formula is

$$A_i = \frac{OT}{OT + TCM}$$

where *OT* is the operating time during a given calendar time period and *TCM* is the total corrective maintenance downtime in clock hours during the given period.

Availability (Operational). (1) A measure of the degree to which a system is either operating or is capable of operating in any given time period when used in its typical operational and support environment. (2) The proportion of time a system is either operating or is capable of operating, when used in a specific manner in a typical maintenance and supply environment. All calendar time is considered. The formula is

$$A_o = \frac{OT + ST}{OT + ST + TCM + TPM + TALDT}$$

$$= \frac{\text{total calendar time--total downtime}}{\text{total calendar time}}$$

where *OT* and *TCM* are as defined under "Availability (Inherent)," *ST* is standby time (not operating, but assumed operable) per given calendar time period, *TPM* is as defined under "Availability (Achieved)," and *TALDT* is total administrative and logistics downtime spent waiting for parts, maintenance personnel, or transportation per given calendar time period.

Baseline. (1) A configuration identification document or a set of such documents formally designated and fixed at a specific time during a CI life cycle. Baseline plus approved changes from baselines constitute the current configuration identification. For configuration management there are three baselines: functional, allocated, and product. (2) An approved reference point, at a specific time, for control of future changes to a product's performance, construction, and design.

Baseline Comparison System (BCS). A current operational system, or a composite of current operational subsystems, which most closely represents the design, operational, and support characteristics of the new system under development (MIL-STD-1388-1A).

Baseline Cost Estimate (BCE). (1) A document prepared by the materiel developer; detailed estimate of acquisition and ownership normally required for high-level decision; provides the basis for subsequent tracking and auditing. (2) Detailed estimate of acquisition and ownership normally required for high-level decisions; provides the basis for subsequent tracking and auditing. (3) A document prepared by the materiel developer; a detailed estimate of acquisition and ownership costs; normally required for high-level decisions. The BCE is the basis for all alter tracking auditing.

Block Design. Development of system improvements in blocks while the basic system is still being developed and continue into production. This replaces or supplements the heel-to-toe development procedures previously inherent in U.S. Army policy. It greatly improves responsiveness to battlefield threat.

Break-even Analysis. (1) Analysis of proposed procurement and facilitization to compare potential cost of establishing a second source with potential savings due to competitive pressure from the second source. (2) Analysis of proposed procurement and facilitization to compare potential cost of establishing a second source (facilities, educational buy, TDP, and rights costs) with potential savings due to competitive pressure from the second source.

Budget. A planned program for a fiscal period in terms of estimated costs, obligations, expenditures, source of funds for financing, reimbursements anticipated, and other resources to be applied.

"Build-to" Specifications. Specifications developed during detail design and prototype fabrication; contain the information necessary to fabricate, assemble, test, and produce equipment and facility items. In MIL-STD-490 these are identified as "Product Specifications."

Calibrate. Verify the accuracy of test equipment and ensure performance within tolerance, usually compared to a reference standard that can be traced to a primary standard.

Certification. A process, sometimes incremental, by which evidence that a product meets contract or other requirements is provided.

Change Control Board. The same as configuration control board.

Change Documentation. Specification change notice, engineering order, engineering change proposal, design change notice, notice of revision, and so on.

Change Identification Number. A number assigned to a data package designing an equipment engineering change; used to control, sequence, and account for production, implementation, and actions related to change.

Change Impact Analysis. The logic and reasoning process that permits quantification of the results of a change before the change is attempted, in order to predict the outcome.

Change Request (Engineering Change Request). A document that is used by the project staff to request a change to the approved product configuration.

Class I Change. A change affecting the contract specification, price, weight, delivery schedule, reliability, performance, interchangeability, interface with other products, or safety.

Class II Change. Any change not falling within the Class I change definition given above.

Class I Drawing. A drawing for which the government retains responsibility for preparation and maintenance.

Class II Drawing. A drawing for which the company retains responsibility for preparation and maintenance.

Commercial Components, Products, or Items. Products or items in regular production which are sold in substantial quantities to the general public and industry at established market or catalog prices. The term also includes products developed by other government agencies, U.S. Military Services, and other countries.

Commercial Item. An item regularly used for other than government purposes and sold during normal business operations.

Commonality. Materiel or system that are interchangeable. Each can be used or operated and maintained by personnel trained on the other system without more specialized training. Also, repair parts and components can be interchanged and applied to consumable items without adjustment.

Comparability Analysis. An examination of two or more systems and their relationships to discover resemblances or differences.

Compatibility ECP. An ECP priority used for changes required during system installation and checkout that are necessary to make the system work (design deficiency correction). Also used to process changes to system requirements after the design requirements baseline is established (not a design deficiency).

Component. (1) A part, subassembly, assembly, or combination of these items joined together to perform a function. (2) An item identified, authorized, cataloged, and issued as part of another item.

Computer Firmware. An assembly composed of a hardware unit and a computer program integrated to form a functional entity whose configuration can not be readily altered during normal operation; the computer program is stored in the hardware unit as an integrated circuit with a fixed logic configuration that will satisfy a specific application or operational requirement.

Computer Program. A series of instructions or statements in a form acceptable to computer equipment, designed to cause the execution of an operation or series of operations. It is a configuration item when it satisfies an end-use function and is designated for configuration management; may vary widely in complexity, size, and type from a special-purpose diagnostic program to a large command and control program, and will represent a requirement or set of requirements allocated from the functional or allocated baseline(s). Also, computer software and other software.

Computer Resource Management Plan (CRMP). The primary program management document that describes the development, acquisition, test, and support plans for computer resources.

Computer Resource Working Group (CRWG). Advisory board formally chartered by the program manager with the coordination of the operating, supporting, and participating commands for each system in which MCCR are likely to be used. Members will actively participate in all aspects of the program involving computer resources.

Computer Resources Support. The facilities, hardware, software, documentation, manpower, and personnel needed to operate and support embedded computer systems, one of the principal ILS elements.

Computer Software Component. A functional or logically distinct part of a computer software configuration item. May be top-level or lower-level.

Computer Software Quality. Degree to which the attributes of software enable it to perform its specified end item use. Also, software quality.

Concept Demonstration and Validation Phase. Normally the second phase in the acquisition process. Consists of steps necessary to resolve or minimize logistics problems identified during concept exploration, verify preliminary design and engineering, accomplish necessary planning, fully analyze tradeoff proposals, and prepare contract required for FSD.

Concept Exploration–Definition Phase. Initial phase of the materiel acquisition process. During this phase, the AS is developed, system alternatives are proposed and examined, and the materiel requirements document is refined to support subsequent phases.

Concept Exploration (CE) Phase. The identification and exploration of alternative solutions or solution concepts to satisfy a validated need (MIL-STD-1388-1A).

Concept Formulation. The effort made before a decision to conduct engineering development. Includes system studies and experimental hardware tests.

Conceptual Phase. The period preceding the definition phase. It begins with determination of broad project objectives and ends with the start of definition phase; it includes concept formulation, general equipment design approach, feasibility evaluation, block diagram, and equipment layout.

Configuration. (1) The physical and functional characteristics of systems, equipment, and related items of hardware or software, and the relative arrangement and contours of these: The shape of a thing at a given time. The specific parts used to construct a machine. (2) The functional and/or physical characteristics of hardware–software as set forth in technical documentation and achieved in a product. (3) The complete technical description required to build, test, accept, operate, maintain, and logistically support equipment. Also, the physical and functional characteristics of the equipment.

Configuration Accounting. The reporting and recording of changes made to the approved configuration.

Configuration Audit Review. A technical review comparing each CI documentation description with the prototype to ensure the documentation's accuracy and adequacy for manufacture and its conformance to the CI description prepared during the development effort.

Configuration Control. The evaluation, coordination, and approval or disapproval of all changes to the equipment configuration defined by the baseline.

Configuration Control Board. A group of technical and administrative project personnel who are responsible for reviewing and assessing engineering changes to the CI after the baseline has been approved.

Configuration Element. An item subject to configuration management.

Configuration Identification. (1) The current approved or conditionally approved technical documentation for a configuration item as set forth in specifications, drawings, and associated lists, and documents referenced therein. (2) The currently approved technical data describing the approved configuration of the product or the process for identifying these data, the product, and changes made to them.

Configuration Item (CI). (1) An aggregation of hardware–computer programs or any of its discrete portions which satisfies an end-item use function and is designated by the government for configuration. (2) An aggregation of hardware–software, or any of its discrete portions, that satisfies an end use function and is designated by the government for configuration management. CIs may vary widely in complexity, size, and type, from an aircraft, electronic, or ship system to a test meter or round of ammunition. (3) A collection of hardware or software, or any of its parts, that satisfies an end use and is designated by the government or customer for configuration management. Also, computer program configuration item or computer software configuration item.

Configuration Item Development Record. Provides status of progress for a particular configuration item.

Configuration Item Identification Number. A seven-digit alphanumeric permanent number assigned to identify a configuration item. It is a unique identifier for the particular item, but also as an indicator of a common series, lot, or hierarchical dependence.

Configuration Item Specification Addendum. A new specification (addendum) directly referenced to an existing specification.

Configuration Management (CM). (1) The process that identifies functional and physical characteristics of an item during its life cycle, controls changes to those characteristics, provides information on status of change actions, and audits the conformance of configuration items to approved configuration. (2) A discipline applying technical and administrative direction and surveillance to (a) identify and document the functional and physical characteristics of a configuration item, (b) control changes to those characteristics, and (c) record and report change processing and implementation status. (3) The discipline of providing systematic and uniform configuration identification, control, and accounting of an equipment and its parts.

Configuration Management Office. The organization within the system program office that is responsible for (a) formulating, issuing, and maintaining all configuration management documentation; (b) administration support to the CCB; (c) direction and supervision of the uniform specification program; and (d) the transferring of all configuration documentation to the customer.

Configuration Management Plan. (1) Defines government and bidder or contractor interaction and schedules procedures for conducting the configuration management program. (2) A definition of configuration management policies, methods, and procedures for a particular program or project.

Configuration Status. The actual configuration of an equipment at a given time in relation to an approved configuration or baseline.

Configuration Status Accounting. The recording and reporting of the information that is needed to manage configuration effectively, including a listing of the approved configuration identification, the status of proposed changes to the configuration, and the implementation status of approved changes.

Consumables. Materials that are used up during operation of a product, as are gasoline and oil in an automobile.

Contract Definition. (1) The phase of a system's life cycle during which preliminary design and engineering are verified or accomplished, and firm contract and management planning are performed, in connection with major industrial or government–industry joint projects. (2) Normally a competitive period or phase that involved the verification or completion of preliminary design of a CI. Includes firm contract and management planning.

Contract End Item. A deliverable item that is formally accepted by the customer. The same as CI.

Contract Maintenance. The procedure for changing a contract to incorporate approved changes and corrections to the text, including preparing documents for recording the authorization and history of changes and revisions.

Contract Work Breakdown Structure. Contract elements delineated for accounting purposes and workload dissemination.

Contractor. (1) Any individual, partnership, company, corporation, or association having a formal agreement or contract with a procuring activity to furnish things or services, at a specified price or rate. (2) The entity designing, developing, and building the equipment for the customer or government. Same as seller.

Corrective Maintenance. (See also "Unscheduled Maintenance" and "Repair.") (1) Involves unscheduled maintenance or repair actions performed as a result of failures or deficiencies, to restore items to a specified condition. (2) All actions performed, as a result of failure, to restore an item to a specified condition. Corrective maintenance can include any or all of the following steps: localization, isolation, disassembly, interchange, reassembly, alignment, and checkout.

Cost. Refers to the dollars paid by the government for an item or service.

Cost Analysis. An itemized list of labor, materials, test equipment, tooling, fixtures, and documentation required to make an engineering change to an equipment.

Cost and Operational Effectiveness Analysis (COEA). (1) A documented investigation of the comparative effectiveness of alternative means to meet a defined threat. The cost of developing, producing, distributing, and sustaining each alternative system in a military environment for a time preceding the combat application.

Cost Baseline. A validated and formally approved listing of aggregate program costs that reflects all program directive document (PDD)-delineated efforts. The cost baseline is a part of the program management control system (PMCS) documentation.

Cost-Effectiveness. (1) A term describing the relative value of a system; measures the costs of acquisition and utilization against system effectiveness. (2) The property of a design or administrative change that will (a) improve the capability of a product at minimum increased cost or (b) reduce product capability at minimum increased cost or (b) reduce product capability by an acceptable amount with a resulting major cost reduction. Also, the measure of an item's ability to fulfill a specific need at minimum cost.

Cost Estimating Relationship (CER). A statistically derived equation that relates life-cycle cost or some portions thereof directly to parameters that describe the performance, operating, or logistics environment or system.

Cost–Schedule Control Systems Criteria (C/SCSC). The set of standards (criteria) used to determine the adequacy of a contractor's cost–schedule control system and the manner in which it is used.

Cost Work Breakdown Structure. Similar to Program Work Breakdown Structure only cost terminology is used; primarily for accounting purposes as well as cost analysis exercises.

Critical. Category of items crucial to performance and more vital to operation than are noncritical items. The "significant few."

Critical Component. A component within a CI that requires an approved specification to establish technical or inventory control at the component level.

Critical Design Review (CDR). (1) Determines that the detail design satisfies the performance and engineering specialty requirements of the development program. The CDR is performed late in the prototype subphase when the design detail is essentially complete but prior to drawing release and fabrication of formal test articles. (2) A formal technical review of design in order to identify specific engineering documents for release to production and to establish a basis for spares provisioning, preparation of manuals, and other support activities depending on the detail definition of the equipment.

Critical Failure. A failure (or combination of failures) that prevents an item from performing a specified mission. (*Note*: Normally only one failure may be charged against one mission. "Critical failure" is related to evaluation of mission success.)

Criticality. The relative importance or priority of items within specific equipment with respect to successful operation of the equipment.

Debugging. The examining or testing of a procedure, program, or equipment for detecting or correcting errors.

Defense Guidance (DG). Provides Secretary of Defense guidance to the DoD "components" for the preparation of their "program objective memorandum."

Defense Resources Board (DRB). Established to supervise the OSD review of the DoD components' POM and budget submissions and manage the program and budget review process.

Demonstration–Validation Phase. The period when selected candidate solutions are refined through extensive study analyses; hardware development, if appropriate; test and evaluations (MIL-STD-1388-1A).

Deployment. (1) The process of planning, coordinating, and executing the deployment of a materiel system and its support. (2) Functions to be performed and system elements required to initially transport, receive, process, install, test, check out, train, operate, and as required emplace, house, store, or deploy the system into a state of full operational capability.

Design Constraints. Envelope dimensions, weight, shape, mounting configuration, RFI characteristics, component standardization, use of inventory items, and integrated logistics support policies to be followed.

Design Criteria. Specific standards or goals for design of the product; for example, lightweight, compactness, high reliability, simplicity, safety, low power consumption, high accuracy, long operating life, ease of maintenance, flexibility, and versatility.

Design Interface. The relationship of logistics-related design parameters, such as R&M, to readiness and support resource requirements. These logistics-related design parameters are expressed in operational terms rather than as inherent values and specifically relate to system readiness objectives and support costs of the materiel system one of the principal elements of ILS.

Design Parameters. Qualitative, quantitative, physical, and function value characteristics that are inputs to the design process, for use in design tradeoffs, risk analyses, and development of a system that is responsive to system requirements.

Design Reviews. (1) Determination of the technical adequacy of the system engineering and design efforts in meeting system requirements. (2) An examination of the equipment design, construction, data, and operation to ensure that it meets customer requirements.

Design to Cost (DTC). (1) An acquisition management technique to achieve

Department of Defense system designs that meet stated cost requirements. Cost is addressed on a continuing basis as part of a system's development and production process. The technique embodies early establishment of realistic but rigorous cost objectives, goals, and thresholds and a determined effort to achieve them (DoDD 4245.3). (2) A management concept wherein rigorous cost goals are set during development. The control of system costs (acquisition, operations, and support) to those goals is achieved by practical tradeoffs between operational capability, performance, costs, and schedules. Addressed on a continual basis as part of a system's development and production process. (3) Acquisition management cost control technique that requires engineering knowledge of the system to identify potential unit production and operations and support cost (O&S) savings. Rigorous but realistic objectives are established in "requirements–tech base" and "Proof-of-principle" (POP) phases (traditionally the "concept" and "demonstration–validation" phases). In the late POP and "development prove-out" phases [traditionally "full-scale development" (FSD) phase] firm DTC goals are established for both unit production and O&S. When appropriate, cash or other incentives are established to encourage the contractor to meet the proposed goals. Design to cost is also implemented on product improvement programs (PIPs) and nondevelopmental items (NDIs) if significant design change is involved.

Design-to-Cost Goal. A specific cost established as a goal for a specific configuration, established performance characteristics, and a specific number of systems at a defined production rate.

Design to Operations and Support Cost (DTOSC). The activity involved with the development of a unit cost goal based on the recurring costs associated with a learning–experience curve, specific quantity and a given schedule. A cost per unit that the government can expect to pay for a given quantity of item delivered on a specific date.

"Design to" Specifications. Specifications containing the performance, design, and verification (test) requirements for an item of equipment or facility. They are developed prior to detail design of the item and provide the basis for design. In MIL-STD-490 these are identified as the "system specification and development specifications."

Design-to-Unit Production Cost (DTUPC). (1) Contractual provision that is the anticipated unit production price to be paid by the government for recurring production costs, based on a stated production quantity, rate, and time frame. (2) The activity involved with the development of a unit cost goal based on the recurring costs associated with a learning–experience curve, specific quantity and a given schedule. A cost per unit that the government can expect to pay for a given quantity of item delivered on a specific date.

Development. (1) The process of working out and extending the theoretical, practical, and useful applications of a basic design, idea, or scientific discovery. The design, building, modification, or improvement of the prototype or original model of a vehicle, engine, instrument, or the like as determined by the basic idea or concept. (2) The effort involving design, testing, evaluation, and redesign of a product before acceptance for production.

Development Test (DT). (1) The engineering test to provide data on safety, the achievability of critical system technical characteristics, refinement and ruggedization of hardware configurations, and determination of technical risks. This testing is performed on components, subsystems, materiel improvement, nondevelopmental items (NDI), and hardware–software-integrated and related software. DT includes the testing of compatibility and interoperability with existing or planned equipment and systems and the system effects caused by natural and induced environmental conditions during the development phases of the materiel acquisition process. (2) Testing of materiel systems conducted by the materiel developer using the principle of a single, integrated development test cycle to demonstrate that the design risks have been minimized, that the engineering development process is complete, and that the system will meet specifications; and to estimate the system's military utility when it is introduced. DT is conducted in factory, laboratory, and proving ground environments.

Development Test and Evaluation (DTE). Test and evaluation conducted to assist the engineering design and development process and to verify attainment of technical performance specifications and objectives (DoDD 5000.3)

Developmental Baseline. The approved configuration identification during the development phase, established after design review.

Developmental Configuration. Software and associated technical documentation that defines the evolving configuration during the development phase.

Deviation. Written authorization to depart from a specification or other document requirement before the departure is made.

Direct Costs. Any expenses that can be associated with specific products, operations, or services.

Disposal. The act of getting rid of excess or surplus property under proper authorization. Disposal may be accomplished by, but is not limited to, transfer, donation, sale, abandonment, destruction, or recycling.

Disposition. The decision to rework, use as is, return to vendor, or discard an item that does not meet specifications or quality standards.

Distribution. (1) The functional phase of logistics that embraces the dispensing of materials, supplies, equipment, products, or services, according to need, requisition, orders, plans, and so forth; includes the authorized delivery of such things (2) The process of issuing the required copies of documents to project members and customer representatives.

Downtime. That portion of calendar time during which an item or equipment is not in condition to fully perform its intended function.

Durability Failure. A malfunction that precludes further operation of the item and is sufficient in cost, safety, or time to restore that the item must be replaced or rebuilt.

Early Comparability Analysis (ECA). Identifies those manpower personnel, and training "high-driver" tasks that can be limited or eliminated in the design of new or improved systems. ECA also helps develop preliminary manpower, personnel and training constraints, and/or guidelines.

Economic Order Quantity (EOQ). The amount of an item that should be ordered at one time in order to obtain the lowest combination of inventory carrying and order–production costs.

Economic Repair. Capability of being restored to sound condition at a cost less than the value of the estimated remaining useful life of the item concerned, based on life expectancy, acquisition, or replacement cost or other relevant factors.

ECP Number. A unique alphanumeric designator assigned to an engineering change proposal.

ECP Package. A compilation of data submitted to customer for approval of a change. It includes standard ECP form, change description and drawings, cost estimates, and schedule revisions.

Embedded Training. Training that is delivered by an equipment system in addition to the primary operational function. The training is made available by components of the equipment that take advantage of the overall system capabilities.

End Item. A combination of items that form a product that accomplishes a specific task or function.

End User. The individual or organization that employs an article or system to accomplish the purpose for which it was designed and intended. This is normally the terminal point of the logistics system intended to accomplish a task, except for the disposal phase.

Engineering. The profession in which a knowledge of the mathematical and natural sciences gained by study, experience, and practice is applied with judgment to develop ways to utilize economically, the materials and forces of nature for the benefit of humankind.

Engineering Change. (1) Any design change that will require revision to the contract specifications or engineering drawings, or documents referenced therein. (2) Any change in design or performance of an item after establishment of its configuration identification.

Engineering Change Proposal (ECP). (1) Proposal to change design or engineering features of materiel under development or production. Includes proposed engineering change and documentation by which the change is described and suggested. (2) A document that describes an engineering change.

Engineering Data. (1) Design related drawings, supporting indexes, specifications, referenced standards, and related technical documents and software used in the design, manufacture, fabrication, or erection of an item, or prepared by a design activity relating to design, performance, manufacture, test, or inspection. (2) Specifications, drawings, parts, and wire lists.

Engineering Development. (1) Refers to a CI that is still in the design and evaluation state and has not been approved for production or operation. (2) RDTE funding category that includes development programs being engineered for U.S. Military Service use, but not yet approved for procurement or operation.

Equipment. (1) All items of a durable nature that are capable of continuing or repetitive utilitarian use by an individual or organization (DoD 5000.8). (2) An item designed and built to perform a specific function as a self-contained unit or to perform a function in conjunction with other units. It is the same as a product.

Facilities. (1) Inability to perform the basic function; inability to perform within previously specified limits; malfunction. (2) The permanent or semi-permanent real property assets required to support the materiel system, including conducting studies to define types of facilities or facility improvements, locations, space needs, environmental requirements, and equipment (DoDD 5000.39), one of the principal elements of ILS.

Failure. (1) The inability of an item to perform within previously defined limits. (2) The event, or inoperable state, in which an item or part of an item does not, or would not, perform as previously specified (see MIL-STD-721).

Failure Analysis. The logical, systematic examination of an item, or its design, to identify and analyze the probability, causes, and consequences of real or potential malfunction.

Failure Mode. The consequence of the mechanism through which the failure occurs (for example, short, open, fatigue, fracture, or excessive wear). (See MIL-STD-721.)

Failure Mode and Effects Analysis (FMEA). Reliability analysis of what items are expected to fail and the resulting consequences of failure.

Failure Mode, Effects, and Criticality Analysis (FMECA). (1) An analysis to identify potential design weaknesses through systematic, documented consideration of the following: all likely ways in which a component or equipment can fail; causes for each mode; and the effects of each failure (which may be different for each mission phase) (MIL-STD-1388-1A). (2) Narrative description of probable effects of failure for each failure mode. Included is criticality of the failure; for example, completely inoperable in some modes, or operable at a degraded level of performance.

Failure Rate. The number of failures per unit measure of life (cycles, time, miles, events, etc.) as applicable for the item (MIL-STD-721B).

Fast-Track Program. An acquisition program in which time constraints require the design, development, production, testing, and support acquisition processes to be compressed or overlapped (MIL-STD-1388-1A).

Field Change. An engineering change made to a CI officially accepted by the customer.

Field Operation. The use of an equipment at the deployment site, at a facility, or in an aerospace vehicle to accomplish its intended goal or mission.

First Article Test (FAT). Production testing that is planned, conducted, and monitored by the materiel developer. FAT includes pre-production and initial production testing conducted to ensure that the contractor can furnish a product that meets the established technical criteria.

First in–First out (FIFO). Dictum meaning "use the oldest item in inventory next." Contrasts with LIFO. FIFO accounting values each item used at cost of the oldest item in inventory.

Five-Year Defense Program (FYDP). The publication that records, summarizes, and displays the decisions that have been approved by the SECDEF as constituting the DoD program.

Fixed Costs. Expenses such as office facilities and training that do not vary directly with activity rates.

Follow-on Operational Test and Evaluation (FOTE). (1) FOTE is that OTE conducted as necessary after the full production decision during production and deployment of the system. FOTE is conducted to assess system training and logistics, and to verify correction of deficiencies, if required, and to ensure that initial production items meet operational effectiveness and suitability thresholds. FOTE will be scheduled and programmed as a normal part of an acquisition program. The operational IE will make maximum use of both production and preproduction qualification tests and other data sources (e.g., sample data collection, field user surveys) to assess FOTE issues minimizing the requirement for follow-on operation testing. (2) Test and evaluation conducted subsequent to a Milestone III production decision to obtain information lacking from earlier initial operational test and evaluation. Normally, FOTE is conducted subsequent to the decision to proceed beyond low rate initial production.

Force Development. The integration of allocated and projected resources into a time phased program to develop a force that is properly organized, equipped, trained, and supported to carry out the U.S. Army missions and functions worldwide. This includes force planning, programming, analysis, structuring, combat, and training developments.

Force Development Test and Experimentation (FDTE). (1) FDTE is conducted early on to support the force development and materiel development processes by examining the effectiveness of existing or proposed concepts of training, logistics, doctrine, organization, and materiel. FDTE is conducted early and can be scheduled as needed during any phase of the materiel acquisition process. They may be related to, combined with, or used to supplement OT. During the requirements formulation effort, FDTE may be used to determine essential and desirable capabilities or characteristics of proposed systems. Prior to Milestone II, FDTE will be used to assist in refining concepts of employment, logistics, training, organization, and personnel, in lieu of OT when operational issues are adequately addressed. FDTE also includes filed experiments designed to gather data through instrumentation to address a training development program or to support simulations, models, war games, and other analytical studies. Requirements for FDTE may also be generated by the results of combat developments, training developments, or training effectiveness analysis training and studies. (2) Tests that range from a small, highly instrumented, and high-resolution field experiment to a large, less instrumented, low-resolution, controlled scenario field test. Test data are evaluated largely by using subjective rather than analytical techniques. Tests are conducted to evaluate new concepts of tactics, doctrine, organization, and new forms of materiel.

Form. The shape of an equipment.

Form, Fit, and Function. The physical and functional characteristics of a CI as an entity, but not covering characteristics of the elements making up the CI.

Formal Test. A witnessed test conducted in accordance with approved test plans and procedures.

Full-Scale Development (FSD) Phase. (1) The period when the system and the principal items necessary for its support are designed, fabricated, tested, and evaluated. (2) Normally, third phase in the materiel acquisition process during which a system, including all items necessary for its support, is fully developed, engineered, fabricated, tested, and initially type-classified.

Function Analysis. Determination of the functions and their sequence and interdependence required to accomplish a mission objective, and the relating of (basic) requirements to the functions on which they impact.

Functional Area. (1) An overview of the life-cycle stream of activities, functions, and elements for a primary function. (2) A distinct group of performance requirements that are part of next-lower-level breakdown of the overall highest level performance requirements of a system.

Functional Baseline. (1) The technical portion of the program requirements (Type A specifications); provides the basis for contracting and controlling system design. (2) The initial approved functional configuration identification. (3) The approved functional configuration identification.

Functional Characteristics. Performance, operating, and logistics parameters and their tolerances; for example, range, speed, safety, reliability, and maintainability.

Functional Configuration Audit (FCA). (1) Verifies that the actual item that represents the production configuration complies with the development specification. (2) The formal examination of functional characteristics' test data for a configuration item, prior to acceptance, to verify that the item has achieved the performance specified in its functional or allocated configuration identification.

Functional Configuration Identification (FCI). (1) The current approved technical documentation for a configuration item that prescribes all necessary functional characteristics. (2) Technical data give the functional characteristics, demonstration tests, interface characteristics, and design constraints of a product.

Functional Level. Breakdown of the physical hierarchy of a product. Typical levels of significance from smallest to largest are as follows: part, subassembly, assembly, subsystem, and system.

Functional Support Requirements (FSR). A function (transport, repair, re-supply, recover, calibrate, overhaul, etc.) that the support system must perform for the end item to be maintained in or restored to a satisfactory operational condition in its operational environment (MIL-STD-1388-1A).

Functional Support. Systematized methodologies and procedures, or a common set of standards applied to materiel acquisition programs that include, but are not limited to, personnel, technical requirements planning, security, automated data processing, controller support, cost analysis, training, MANPRINT, safety, audit, logistics, product assurance, reliability, equal employment opportunity, obligation planning and reporting, production, industrial preparedness, value engineering, test measurement and diagnostic equipment (TMDE), public affairs, legal matters, inspector general, mobilization, procurement contracting, international cooperation, and small business.

General-Purpose Warehouse Space. Warehouse area other than controlled humidity, flammable, or refrigerated warehouse area. Such warehouse area may be further classified as either heated or unheated warehouse space.

General Support Equipment. That which has maintenance application to more than a single model or type of system, subsystem, device, article, or equipment. (See "Support Equipment.")

Government-Furnished Material (GFM). Material provided by the government to a contractor or comparable government production facility to be incorporated in, attached to, used with, or in support of an end item to be delivered to the government or ordering activity, or that may be consumed or expended in the performance of a contract. It includes, but is not limited to, raw and processed materials, parts, components, assemblies, tools, and supplies. Material categorized as government-furnished equipment (GFE) and government-furnished aeronautical equipment (GFAE) are included (MIL-STD-1388-1A).

Ground Support Equipment (GSE). A test used to check out the operation of a product or equipment used for such things as installing, launching, adjusting, repairing, or controlling a product.

Hardware. (1) Physical object(s), as distinguished from its capability or function. A generic term dealing with physical items of equipment, tools, implements, instruments, devices, sets, fittings, assemblies, components, parts, raw materials, and so on, as opposed to funds, personnel, services programs, plans, and similar termed "software." (See "Software.") (2) Any item built by the company or vendor (except software and firmware).

Health Hazard. An existing or likely condition, inherent to the operation or use of materiel, that can cause death, injury, acute or chronic illness, disability, and/or reduced job performance of personnel by exposure to

(a) shock–recoil, (b) vibration, (c) noise (including steady state, impulse, and blast overpressure), (d) humidity, (e) toxic gases, (f) toxic chemicals, (g) ionizing or nonionizing radiation (including X-rays, γ-rays, magnetic fields, microwaves, radio waves, and high-intensity light, (h) lasers, (i) heat and cold, (j) oxygen deficiency, (k) blunt–sharp trauma, (l) pathogenic microorganisms.

Health Hazard Assessment (HHA). The application of biomedical and psychological knowledge and principles to identify, evaluate, and control the risks to the health and effectiveness of personnel who test, use, or service systems.

Human Engineering. (1) The application of knowledge of human capabilities and limitations to the planning, design, development, and testing of systems, equipment, and facilities to obtain the best mix of safety, comfort, and effectiveness compatible with established requirements. (2) A design discipline that assures a product can be used safely and properly by people.

Human Factors. (1) All scientific biomedical and psychosocial facts and considerations that constitute characteristics pertaining to the nature of humankind. These include, but are not limited to, principles and applications in the areas of human engineering, personnel selection, training, life support, job performance aids, and human performance evaluation. (2) The application of knowledge regarding human characteristics to design of items to achieve effective person–machine integration and use. Also includes personnel selection, training, job performance aids, and performance evaluation. Human engineering is a subclass of human factors.

Human-Factors Engineering Analysis (HFEA). (1) An analysis, performed in support of acquisition milestone reviews, to identify any problems in MANPRINT (human-factors engineering, manpower, personnel, training, system safety, and health hazards) that may be sufficiently critical to preclude the system's proceeding into the next phase of the acquisition process. A secondary objective is to identify MANPRINT concerns that, while not critical in terms of program decisions, are resolvable, and must be addressed during the subsequent phase of the acquisition cycle. (2) Analysis performed in support of milestone decision reviews, to identify human-factors engineering problems that may be sufficiently critical to prevent the system from proceeding into the next phase of the acquisition process. Secondary objective is to identify those concerns that, while not critical in terms of program decisions, are resolvable, and must be addressed during the subsequent phase of the acquisition cycle.

Indenture. A method of showing relationships to indicate dependence.

Independent Evaluation (IE). The process used by the independent evaluators to independently determine whether the system satisfies the approved requirements. It will render an assessment of data from all sources and an engineering or operational analysis to evaluate the adequacy and capability of the system.

Inherent RAM Value. (1) Any measure of reliability or maintainability that includes only the effects of item design and installation, and assumes an ideal operating and support environment (DoDD 5000.40). (2) Any measure of RAM that includes only the effects of an item design and its application, and assumes an ideal operating and support environment.

Initial Operational Capability (IOC). (1) The initial operational capability is the first attainment of the capability by a unit and its support elements to operate and maintain effectively a production item or system.

Initial Operational Test and Evaluation (IOTE). Operational test and evaluation accomplished prior to the Milestone III production decision.

In-Process Review (IPR). (1) Review of a project or program at critical points to evaluate status and make recommendations to the decision authority. Conducted by the MATDEV/PEO. (2) U.S. Army acquisition programs other than DoD major or designated acquisition programs. (3) A review of a project at critical points: (a) evaluate the status of the project; (b) accomplish effective coordination and make cooperative, proper, and timely decisions bearing on the future of the project.

Installation. A fixed or relatively fixed facility location together with its real estate, buildings, structures, utilities, equipment, and so on. Also, that period of initial setup and adjustment and performance of a product in the customer's environment.

Insurance Items. Stocked articles or material that may be required occasionally or intermittently, but not subject to periodic replacement or wearout, for which prudence requires that there be some stock on hand at certain strategic points, due to the essence of the time required in procurement of the items or material involved.

Integrate. To put or bring (parts) together into a whole; to unify.

Integrated Logistic Support (ILS). (1) A composite of the elements necessary to ensure the effective and economical sustaining of a system or equipment, at all levels of maintenance, throughout its programmed life cycle; characterized by the harmony and coherence obtained between each of its elements and levels of maintenance. (2) A disciplined, unified, and iterative approach to the management and technical activities necessary to (a) integrate support considerations into system and equipment design; (b) develop support requirements that are related consistently to readiness objectives, to design, and to each other; (c) acquire the required support; and (d) provide the required support during the operational phase at minimum cost. (3) A composite of all support considerations necessary to ensure the effective and economical support of a system at all levels of maintenance for its programmed life cycle. (4) A grouping of elements for assuring effective and economical support of an equipment at all maintenance levels for its planned life cycle. These elements include (a) facilities, (b) maintenance, (c) support equipment, (d) logistics data, (3) spares and repair parts, (f) logistic support personnel, and (g) contract maintenance. (5) Composite of elements necessary to ensure effective and economical support of a system or equipment at all levels of maintenance for its programmed life cycle. (6) A composite of all support considerations necessary to ensure the effective and economical support of a system at all levels of maintenance for its programmed life cycle. A unified and iterative approach to the management and technical activities needed to (a) influence operational and materiel requirements and design specifications, (b) define the support requirements best related to system design and to each other, (c) develop and acquire the required support, (d) provide required operational phase support at lowest cost, (e) seek readiness and LCC improvements in the materiel system and support systems during the operational life cycle, (f) repeatedly examine support requirements throughout the service life of the system.

Integrated Logistics Support Management Team (ILSMT). A team of government and industry functional and management personnel formed to advise and assist the ILS manager with planning, coordinating, monitoring schedules and contractor performance, ensuring accuracy and timeliness of government inputs, and compliance with applicable requirements, regulations, specifications, standards, and so on.

Integrated Logistics Support Plan (ILSP). (1) The formal planning document for logistics support; is kept current through the program life; sets forth the plan for operational support; provides a detailed ILS program to fit with the overall program; provides decisionmaking bodies with necessary ILS information to make sound decisions in system development and production and provides the basis for the ILS portion of procurement packages. (2) Provides a composite of all support considerations necessary to ensure the effective and economical support of a system for its life cycle and serves as the source document for summary and consolidated information required in other documents of the program management documentation.

Integrated Logistic Support Planning. During acquisition requires management of the following selected activities or elements: (1) maintainability and reliability, (2) maintenance planning, (3) support and test equipment, (4) supply support, (5) transportation and handling, (6) technical data, (7) facilities, (8) personnel and training, (9) funding, (10) management data.

Integrated Support Plan (ISP). A comprehensive plan to demonstrate how a contractor intends to manage and execute the ILS program.

Integrated System Support. Considerations of logistics support aspects for a system in the context of the system's role in the force structure. Emphasizes interactive relationships such as standardization, interoperability, and resource implications (e.g., manpower, petroleum, oils, lubricants, storage, training site, ammunition) of fielding the new system.

Integration. The combining of different equipments into a subsystem or system so that they can work together harmoniously.

Interchangeability. Standardization of the functional and physical characteristics of items to enhance the substitution of one item for another.

Interface. (1) A common boundary between two or more items, characteristics, systems, functions, activities, departments, objectives, and so forth. That portion of anything that impinges on or directly affects something else. (2) A boundary or point common to two or more command and control systems, subsystems, or other entities against which or at which necessary information flow takes place. (3) A common boundary between two or more items. This boundary may be electrical, mechanical, functional, or contractual.

Interoperability. (1) The ability of systems, units, or forces to provide services to and accept services from other systems, units, or forces and to use the services so exchanged to enable them to operate effectively together. (2) The ability of systems, units, or forces to provide services to and accept services from, other systems, units, or forces and to use these services to enable them to operate effectively together. (3) Standardization of the functional characteristics of items to enhance their ability to work together in the same operational environment.

Inventory Management. That phase or function of logistics that controls the input, availability, and disposal of items within the total owned by any organization. Similar to but slightly broader than inventory control.

Iterative Methodology. Sequential and repetitive top–down development of a topic by identifying those actions (functions) required to accomplish the objective, allocating the (basic input) requirements to the appropriate functions (functional allocation), translating the requirements into solutions (synthesis or conceptual design) through system–design engineering studies, portraying the interdependence among the solution elements, researching and evaluating the alternate solutions and determining the most feasible solution, and analyzing the selected solutions to assess the impact on the requirements–design and other solution elements.

Justification of Major System New Starts (JMSNS). (1) The military component's submission on which the mission need determination is accomplished. The JMSNS is submitted with the program objectives memorandum (POM) in which funds for the budget year of the POM are requested. The Secretary of Defense will provide appropriate program guidance in the program decision memorandum. This action provides official sanction for a new program start and authorizes the U.S. Military Service, when funds are available, to initiate the "concept exploration" (CE) phase (adapted from DoDD 5000.1). (2) Defines a deficiency or opportunity such that there is a reasonable probability of satisfying a need by the acquisition of a single system.

Key Functional Characteristics. Critical characteristics that affect the satisfactory fulfillment of an equipment's operational requirements such as range of measurement of an instrument, payload capability of a rocket, or altitude capability of an aircraft.

Kits. Kits, as referred to in this regulation, are an assemblage of items identified for a specific application that requires repeated installations and removal to meet specified operational requirements for an item of equipment.

Last in–First out (LIFO). Dictum meaning "Use newest inventory next" (contrasts with FIFO); LIFO accounting values each item used at the cost of the last item added to inventory.

Lead Time. The allowance made for that amount of time required to accomplish a specific task, or to reach a specific objective.

Level of Supply. The quantity of an item authorized or directed to be kept on hand at a storage or distribution point, to meet predictable and anticipated future demands.

Life Cycle. (1) The series of phases or events that constitute the total existence of anything. The entire "womb-to-tomb" scenario of a product from the time concept planning is started until it is finally discarded. (2) The period covering the design, development, manufacture, operation, maintenance, logistics support, and repair of equipment. (3) All phases through which an item passes from conception through disposition.

Life-Cycle Cost (LCC). (1) All costs associated with the system life cycle, including research and development, production, operation and maintenance, and termination. (2) The total cost to the government of acquisition and ownership of the system over its full life. It includes the cost of development, acquisition, support, and where applicable, disposal. (3) Approach to costing that considers all costs incurred during the projected life of the system, subsystem, or component being evaluated. Includes cost to develop procure, operate, and maintain the system over its useful life.

Life-Cycle Phases. Streamlined acquisition process (SAP) consists of three phases; proof-of principle, development prove-out, and production–deployment. The DoD traditional life cycle consists of four phases: concept exploration, demonstration–validation, full-scale development, and production–deployment.

Limited Production. The initial, low-rate production of a system in limited quantity to be used in operational test and evaluation for verification of production engineering and design maturity and to establish a production base prior to a decision to proceed with production.

Line Item. An item called out in a separate line as deliverable to the customer in the work statement or contract.

Line Item Number (LIN). (1) Six-character alphanumeric identification of the generic nomenclature assigned to identify nonexpendable and type-classified expendable and durable items of equipment during their life-cycle authorization and supply management. (2) Six-character alphanumeric or numericalpha identification of the generic nomenclature assigned to identify nonexpendable and type-classified expendable and durable items of equipment during their life-cycle authorization and supply management.

Line Replaceable Unit (LRU). An LRU is an essential support item that is removed and replaced at field level to restore the end item to an operationally ready condition.

Logistic Support Analysis (LSA). (1) The selective application of scientific and engineering efforts undertaken during the acquisition process, as part of the systems engineering process, to assist in (a) causing support considerations to influence design, (b) defining support requirements that are related optimally to design and to each other, (c) acquiring the required support, (d) providing the required support during the operational phase at minimum cost. (2) An analytical technique used by integrated logistic support management to provide a continuous dialog between designers and logisticians. LSA provides a system to identify, define, analyze, quantify, and process logistics support requirements for materiel acquisition programs. Provide a system to identify, define, analyze, quantify, and process logistic support requirements for materiel acquisition programs.

Logistic Support Analysis Record (LSAR). (1) That portion of LSA documentation consisting of detailed data pertaining to the identification of logistic support resource requirements of a system–equipment. (2) File of logistic support information in standardized format, on acquisition programs for specific new or modified systems and equipments. Serves acquisition process using logistic data derived during all phases of the process to support logistic support analysis processes.

Logistic Supportability. The degree to which the planned logistics (including test equipment, spares and repair parts, technical data, support facilities, and training) and manpower meet system availability and wartime usage requirements.

Logistic Time. All replacement procurement time, except that time when the maintenance technician is engaged in the procurement activity.

Logistics. (1) The art and science of management, engineering, and technical activities concerned with requirements, design, and supply and maintaining resources to support objectives, plans, and operations. (2) The same as product support.

Logistics Management. The process by which human efforts are systematically coordinated to create economic and effective support throughout the planned life cycle of equipment, systems, projects, and operations.

Logistics over-the-Shore Operations (LOTS). The loading and unloading of ships without the benefit of fixed-port facilities in friendly or nondefended territory and, in time of war, during phases of theater development in which there is no opposition by the enemy.

Logistics R&D. Technology programs funded *outside* the weapon system development programs that may result in improved subsystem R&M, improved support elements needed in the operation and maintenance of weapon systems, and improved logistics infrastructure elements (DoDD 5000.39).

Logistics Support. Support necessary to ensure the availability of all resources required to sustain and maintain the full effectiveness of individuals, organizations, operations, projects, equipments, or systems. In the broad sense, it may involve acquisition, maintenance, disposal, and any or all of the many included functions. Achievement of logistics support is, therefore, the objective of most modern logistic action or endeavor. At times the term "logistics support" or "logistic support" is used to cover all logistic action, collectively.

Low-Rate Initial Production (LRIP). (1) Optional materiel acquisition phase used when circumstances preclude full-rate production decision. LRIP describes the low rate of output at the beginning of production to reduce the government's exposure to large retrofit programs and costs while still providing adequate numbers of hard-tooled production items for final test prior to full-rate production decisions. (2) Required phase (Milestone IIIA) in the acquisition process for major programs intended for a production decision greater than LRIP.

Maintainability (M). (1) The inherent characteristic of a design or installation that determines the ease, economy, safety, and accuracy with which maintenance actions can be performed. Also the ability to restore a product to service or to perform preventive maintenance within required limits. (2) The measure of the ability of an item to be retained in or restored to specified condition when maintenance is performed by personnel having specified skill levels, using prescribed procedures and resources, at each prescribed level of maintenance and repair. (3) The quality of equipment design and installation that simplifies inspection, test, servicing, and repair with a minimum of time, skill, and resources.

Maintainability Engineering. (1) The application of applied scientific knowledge and methods, and management skills, to the development of equipment, systems, projects, or operations with the inherent ability to be effectively and efficiently maintained (i.e., with favorable maintenance characteristics). It involves coordination with other systems engineering facets to provide the necessary effectiveness, considering all costs over the entire planned life cycle involved. (2) The application of scientific knowledge and engineering skills to the development of items of U.S. Army equipment aimed at providing an inherent ability to be maintained (i.e., favorable maintenance characteristics).

Maintenance. (1) The function of keeping items or equipment in, or restoring them to serviceable condition. It includes servicing, test, inspection, adjustment–alignment, removal, replacement, reinstallation, troubleshooting, calibration, condition determination, repair, modification, overhaul, rebuilding, reclamation, and the initial provisioning of support items. Maintenance includes both corrective and preventive activities. (2) Functions to be performed and system elements required to obtain and sustain logistical support for the system (i.e., continuing normal system operational readiness). (3) The operations involved in keeping equipment in working order or restoring it to normal operation after a malfunction occurs. (4) All actions necessary for retaining an item in, or restoring it to, a specified condition.

Maintenance Concept. (1) Statements and illustrations that define the theoretical means of maintaining equipment. It relates tasks, techniques, tools, and people. (2) A narrative description identifying the broad, planned approach to be employed in sustaining the system–equipment at a defined level of readiness or in a specified condition in support of the operational requirement. Provides the basis for the maintenance plan.

Maintenance Engineering. Developing concepts, criteria, and technical requirements for maintenance during the conceptual and acquisition phases of a project; providing policy guidance for maintenance activities and exercising technical and management direction and review of maintenance programs.

Maintenance Planning. The process conducted to evolve and establish maintenance concepts and requirements for the lifetime of a materiel system, one of the principal elements of ILS.

Major Systems Acquisition. A system acquisition program designated by the Secretary of Defense to be of such importance and priority as to require special management attention.

Major Waiver. A contractual departure from documentation involving health, performance, interchangeability, reliability, maintainability, repair effective operation, or weight (after the fact).

Manpower. (1) The total demand, expressed in terms of the number of individuals, associated with a system. Manpower is indexed by manpower requirements, which consist of quantified lists of jobs, slots, or billets that are characterized by the descriptions of the required number of individuals who fill the job, slots, or billets. (2) The personnel strength as expressed in terms of the number of men and women available to, or required by, the U.S. Military Service. (3) The personnel strength (military and civilian) as expressed in terms of the number of men and women available to the service. Consideration of the net effect of systems and items on overall human resource requirements and authorizations (spaces, to ensure that each system is affordable from the standpoint of manpower). It includes analysis of the number of people needed to operate, maintain, and support each new system being considered or acquired, including maintenance and supply personnel and personnel to support and conduct training. It requires a determination of the manpower changes generated by the system, comparing the new manpower needs with those of the old system(s) being replaced, and an assessment of the impact of the changes on the total manpower limits.

Manpower and Personnel. The identification and acquisition of military and civilian personnel with the skills and grades required to operate and support a materiel system over its lifetime at peacetime and wartime rates one of the principal elements of ILS.

Manpower and Personnel Integration (MANPRINT). (1) A comprehensive technical effort to support system effectiveness by integrating into the materiel development and acquisition process all relevant information concerning human-factors engineering, manpower, personnel, training, system safety, and health hazards. (2) The entire process of integrating the full range of human-factors engineering, manpower, personnel, training, health hazard assessment, and system safety throughout the materiel development and acquisition process.

Manpower Requirements Criteria (MARC). The number of direct workers required to effectively perform a specified work activity. A principal computational component of MARC is the estimate of annual maintenance man hours (AMMH) and its variations (AAMMH, IPAMMH, and DPAMMH), each of which represents different contributing factors to the overall maintenance manpower and personnel determination. AAMMH, IPAMMH, AMMH, and DPAMMH are MARC components of a system from the perspective of the factors each represents. These MARC components are defined below:

> *Annual Available Maintenance Man Hours (AAMMH).* The number of annual man-hours each repairer is expected to be available under sustained operating conditions (e.g., wartime).
>
> *Annual Maintenance Man Hours (AMMH).* The sum of the direct and indirect productive time required to repair an item.
>
> *Direct Productive Annual Maintenance Man Hours (DPAMMH).* The estimated wrench-turning time required to repair a component or assembly:

$$DPAMMH = \frac{\text{equipment usage rate}}{\text{mean time between repair}} \times \text{mean time to repair}$$

MANPRINT (Manpower and Personnel Integration). "MANPRINT" refers to the comprehensive technical effort to identify and integrate into materiel development and acquisition (to ensure system effectiveness) all relevant information and considerations concerning (a) human-factors engineering, (b) manpower, (c) personnel, (d) training, (e) system safety, and (f) health hazards. This comprehensive effort occurs prior to, during, and after the materiel acquisition process (MAP).

MANPRINT Joint Working Group (MJWG). The MJWG provides oversight and manages MANPRINT issues during the materiel acquisition process.

Market Investigation. (1) Conducted in response to the O&O plan to identify systems in existence in order that the acquisition strategy can be developed. (2) Process of gathering information before making acquisition decisions. Conducted initially during the requirements–technology base activities phase and, in greater depth, during the proof-of-principle phase.

Materiel. Consists of all items or things used or needed in any business, industry, undertaking, or operation as distinguished from personnel. As a military term it covers all items necessary for the equipment, maintenance, operations, and support of military activities, whether for administrative or combat use, excluding ships and aircraft. In general use it applies to all but personnel.

Materiel Acquisition Decision Process Reviews (MADP). Major management decision reviews conducted prior to entry into each successive phase of the materiel acquisition process. The purpose of the reviews is to evaluate the development and surface critical issues prior to approval for entry into the subsequent phase. There are three levels of reviews: (a) the Defense System Acquisition Review Council (DSARC) reviews for major systems requiring the Secretary of Defense approval of program decisions—after a weapons program progresses beyond DSARC II, the Service Secretaries assume responsibility for surveillance as directed by the Deputy Secretary of Defense; (b) the Army Systems Acquisition Review Council (ASARC) reviews for major systems requiring the Secretary of the Army approval of program decisions, including those requiring subsequent approval by the SECDEF; (c) in-process reviews (IPR) for nonmajor system.

Materiel Acquisition Process (MAP). The sequence of acquisition activities starting with the identification of an unmet mission need and extending through the introduction of a system into operational use.

Materiel Fielding Plan (MFP). (1) The plan to ensure smooth transition of the system from the developer to the user.

Materiel Requirements Document. (1) States concisely the minimum essential operational, technical, logistical, and cost information necessary to initiate development of procurement of a materiel system. (2) Document concisely stating the minimum essential operational, technical, logistical, and cost information necessary to initiate development or procurement of a materiel system.

Materiel System. (1) A final combination of subsystems, components, parts, and materiels that make up an entity for use in combat or in support thereof, either offensively or defensively, to destroy, injure, defeat, or threaten the enemy; includes the basic materiel items and all related equipment, supporting facilities, and services required for operating and maintaining the system. (2) An item, system, or all systems or materiel; includes all required system support elements.

Mean Down Time (MDT). Average time a system cannot perform its mission, including response time, active maintenance, supply time, and administrative time.

Mean Time between Failure (MTBF). (1) The average time–distance–events a product delivers between breakdowns. (2) For a particular interval, the total functional life of a population of an item divided by the total number of failures within the population. The definition holds for time, rounds, miles, events, or other measures of life units, a basic technical measure of reliability.

Mean Time between Maintenance (MTBM). MTBF plus the interval between scheduled preventive maintenance.

Mean Time to Repair (MTTR). (1) The average time it takes to fix a failed item. (2) The total elapsed time (clock hours) for corrective maintenance divided by the total number of corrective maintenance actions during a given period of time, a basic technical measure of maintainability.

Measures of Effectiveness. A particular value or set of values of system–subsystem effectiveness pertinent to one or more mission objectives.

Methodology. A term used in estimating to describe the methods used to develop an estimate (k.e., parametric, analogy, grass roots).

Military Specification (MIL SPEC). Documents intended primarily for use in procurement; contain descriptions of the technical requirements for items, materials, or services, including the procedures for determining whether the requirements have been met. Specifications for items and materials also contain preservation, packaging, and marking requirements.

Military Standards (MIL STD). Specifications and requirements documented and approved by DoD for the use in the procurement of aerospace weapons, equipment, components, raw materials, and services.

Minor Deviation. A departure from approved documentation that does not involve safety, health performance, interchangeability, reliability, maintainability, effective operation, repair, or weight.

Minor Waiver. A departure from approved documentation that does not affect safety, health, performance, interchangeability, reliability, maintainability, effective operation, repair, or weight.

Mission. The intended goals or functions of an equipment, system, product, or spacecraft.

Mission Area Analysis (MAA). (1) Continuing analyses of assigned mission area by DoD components, OSD, and OJCS to identify deficiencies or to determine more effective means of performing assigned tasks. From these mission analyses, a deficiency or opportunity may be identified that could lead to initiation of a major system acquisition program. (2) An assessment of the capability of a force to perform within a particular battlefield or functional area. The analysis designed to discover deficiencies in doctrine, organizations, training, and materiel, and to identify means of correcting these deficiencies; stressing first doctrinal solutions, then training solutions, then organizational solutions, and finally, materiel solutions.

Mission-Critical Computer Resources (MCCR). Consists of one or more of the following characteristics: (a) physically a part of, dedicated to, or essential in real-time performance of the mission of the system; (b) used for system engineering, integration, specialized training, diagnostic testing and maintenance, simulation, or calibration of the systems—including system and non-system training devices; (c) used for research and development of the system.

Mission Element. A segment of mission area critical to the accomplishment of the mission area objectives and corresponding to a recommendation for a major system capability as determined by a DoD component.

Mission Essential Functions. The minimum operational tasks that the system must be capable of performing to accomplish its mission profiles.

Mission Profile (MP). A time-phased description of the operational events and environments an item experiences from beginning to end of a specific mission (including the criteria for mission success or critical failures). This profile is used as the basis for mission reliability assessment. The MP will not include unscheduled downtime. MP types are (a) multifunctional (an item performing several tasks, such as a tank shooting, moving, and communicating), (b) single-function continuous (an item continuously performing one task, such as a surveillance radar), (c) single-function cyclic (an item performing the same task repeatedly), and (d) single-function one-time (an item performing only a one-time task).

Mission Reliability. The ability of an item to perform its required functions for the duration of a specified mission profile.

Mission Support Costs. An area of other government costs that includes the miscellaneous administrative costs incurred in the day-to-day operations of a program office.

Modification. (1) A change in configuration. (2) A change to an equipment and spares allowed only after the contract has been revised.

Modification Kit. A kit used for changing a production equipment accepted by the customer or released for installation at the launch site or deployment area.

Modification Work Order. A document that provides instructions for modification of an equipment.

Monte Carlo Methods. A catch-all label referring to methods of simulated sampling. When taking a physical sample is either impossible or too expensive, simulated sampling may be employed by replacing the actual universe of items with a universe described by some assumed probability distribution, and then sampling from this theoretical population by means of random number table.

Multiservice T&E. T&E conducted by two or more DoD components for systems to be acquired by more than one DoD component, or for a DoD component's systems that have interfaces with equipment of another DoD component.

Multiyear Appropriation. An appropriation that is available for incurring obligations for a definite period in excess of one fiscal year (i.e., ≥ 2 years).

Next Higher Assembly. The assembly into which another assembly goes.

Nondevelopmental Item (NDI). (1) A generic term covering materiel available from a variety of sources with little or no development effort by the Service. NDIs are normally selected from commercial sources, materiel developed and in use by other U.S. Military sources, government agencies, or other countries. (2) Those items available for procurement to satisfy an approved materiel requirement from existing sources (such as commercial items and items developed by other government agencies, U.S. Military Service, or other countries) requiring little or no additional development. (3) Those items determined by a materiel acquisition decision process (MADP) review. The item may be a commercial product or an item that has been developed and used by another U.S. Military Service, county, or government agency.

Operating and Support (O&S) Costs. (1) The cost of operation, maintenance, and follow-on logistics support of the end item and its associated support systems. This term and "ownership cost" are synonymous. (2) The sum of all costs related to operating and support equipment inventory over a specified time for a specified number of items. (3) The added or variable costs of personnel, materials, facilities, and other items needed for the peacetime operation, maintenance and support of a system during activation, steady-state operation, and disposal.

Operating and Support (O&S) Phase. The period in the system life cycle that starts with the delivery of the first unit to the customer and terminates with disposition of the system from the inventory.

Operational. Applies to actual use of a product.

Operational and Organizational Plan (O&O Plan). (1) The program initiation document in the materiel acquisition process; prepared prior to the ROC or JSOR to support acquisition of all new materiel systems. (2) An operational, organizational, training, and logistical plan for the employment of specific hardware systems within U.S. Army organizations. O&O plans are based on operational concepts and are developed in conjunction with those concepts. Each O&O plan should be able to trace its lineage through one or more functional concepts to the basic (umbrella) concept.

Operational Availability (A_o). The probability that, when used under stated conditions, a system will operate satisfactorily at any time; includes standby time and administrative and logistic delay time.

Operational Effectiveness. The overall degree of mission accomplishment of a system when used by representative personnel in the environment planned or expected for operational employment of the system considering organization, doctrine, tactics, survivability, vulnerability, and threat (including countermeasures, nuclear, and chemical and/or biological threats).

Operational Mission Failure. Any incident or malfunction of the system that causes (or could cause) the inability to perform one or more designated mission-essential functions.

Operational-Mode Summary (OMS). A description of the anticipated mix of ways equipment will be used in carrying out its operational role; includes expected percentage of use in each role and percentage of time it will be exposed to each type of environmental condition during the system life. The OMS will not include unscheduled downtime.

Operational Needs Statement (ONS). An ONS states a user's operational need for a materiel solution to correct a deficiency or to improve a capability that impacts on mission accomplishment. The ONS provides an opportunity outside of the combat–materiel developer community to initiate the combat development process.

Operational RAM Value. Any measure of reliability, availability, or maintainability (RAM) that includes the combined effects of item design, quality, installation, environment, operation, maintenance, and repair. (This measure encompasses hardware, embedded software, crew, maintenance personnel, equipment publications, tools, TMDE, and the natural, operating, and support environments.)

Operational Suitability. The degree to which a system can be satisfactorily placed in field use, with consideration given to availability, compatibility, transportability, interoperability, reliability, wartime usage rates, maintainability, safety, human factors, manpower supportability, logistic supportability, and training requirements.

Operational Test (OT). (1) The field test, under realistic combat conditions, of the system for use in combat by representative military users. OT provides data to assess operating instructions, training programs, publications, and handbooks. It uses personnel with the same military occupational specialty as those who will operate, maintain, and support the system when deployed. (2) Testing and evaluation of materiel systems accomplished with typical user operators, crews, or units in as realistic an operational environment as possible to provide data for estimating: (a) The military utility, operational effectiveness, and operational suitability [including compatibility, interoperatibility, reliability, availability, maintainability, supportability, operational person (soldier)–machine interface, and training requirements] of new systems; (b) from the user viewpoint, the system's desirability considering systems already available and the operational benefits and/or burdens associated with the new system; (c) the need for modification of the system; (d) the adequacy of doctrine, organization, operating techniques, tactics, and training for employment of the system and, when appropriate, its performance in a countermeasures environment.

Operational Test I (OTI). An operational test of a hardware configuration of a system, or components thereof, to provide an indication of utility and worth to the user. Testing should refine identified critical issues, report areas that should be addressed in the future OT and identify new ones for subsequent testing. OTI is accomplished during the validation phase on brassboard configuration, experimental prototypes to provide data leading to the decision to enter full-scale development.

Operational Test II (OTII). The test of engineering development prototype equipment prior to the initial production decision. The goal is to estimate an item's utility, operational effectiveness, and operational suitability in as realistic an operational environment as possible.

Operational Test and Evaluation (OTE). (1) Test and evaluation conducted to estimate a system's operational effectiveness and suitability, identify needed modifications, and provide information on tactics, doctrine, organization, and personnel requirements. (2) Test and evaluation conducted to estimate the system's utility; can be monitored during test and evaluation as well as operation. Some of these elements are currently required by service regulations and MIL SPECS and others will be established by the project manager, subject to review for adequacy, to influence the design and to control O&S cost.

Outline Test Plan (OTP). The formal document included in the Five Year Test Plan (FYTP) containing administrative information; and the test purpose, objective, scope, tactical context, resource requirements, and cost estimates.

Packaging. (1) The use of protective wrappings, cushioning, inside containers, and complete identification marking, up to but *not* including the exterior shipping container. (2) The total mechanical design of an equipment or item; includes parts layout, materials used, protective finish, thermal control design, and fabrication techniques.

Packaging, Handling, Storage, and Transportation (PHST). The resources, processes, procedures, design considerations, and methods to ensure that all system, equipment, and support items are preserved, packaged, handled, and transported properly, including environmental considerations, equipment preservation requirements for short- and long-term storage, and transportability, one of the principal elements of ILS.

Packing. The application or use of exterior shipping containers or other shipping media (such as pallets), and assembling of items or packages thereof, together with necessary blocking, bracing, or cushioning, weatherproofing, exterior strapping, and marking the shipping container or device.

Parametric Estimating. An estimating technique that employs one or more cost-estimating relationships. It involves collecting relevant historical data at an aggregated level of detail and relating it to the area to be estimated through the use of mathematical techniques.

Part Number. A number used to uniquely identify a part. It is usually the same as the drawing number, minus revision letter, or includes the drawing number. Its purpose is to control assembly and replacement of items on the basis of interchangeability.

Parts List. A list of all materials and parts that make up a module, subassembly, assembly, or equipment. A parts list may be a separate list or may be a part of the drawing.

Performance Specification. A specification that describes what is to be accomplished by specific equipment but does not describe how the equipment is to be designed.

Personnel. (1) The people employed by an organization. Also, the name of the organization concerned with people. (2) The supply of individuals, identified by specialty or classification, skill, skill level, and rate or rank required to satisfy the manpower demand associated with a system. This supply includes both those individuals who support the system directly (i.e., operate and maintain the system,) and those individuals who support the system indirectly by performing those functions necessary to produce and maintain the personnel required to support the system directly. Indirect support functions include recruitment, training, retention, and development (MIL-STD-1388-1A). (3) Military and civilian persons of the skill level and grades required to operate and support a system, in peacetime and war.

Physical Configuration Audit (PCA). (1) A technical examination of a designated configuration item to verify that the item "as-built" conforms to the technical documentation that defines the item. (2) Formal examination of the "as-built" configuration of a unit of a configuration item against its technical documentation in order to establish the item's initial production configuration identification. (3) A formal audit to verify that an "as-built" configuration item conforms to the defining technical documentation.

Postproduction Support (PPS). Systems management and support activities necessary to ensure continued attainment of system readiness objectives with economical logistic support after cessation of production of the end item.

Preliminary Design Review (PDR). (1) Conducted on each configuration item to evaluate the progress, technical adequacy, and risk resolution of the selected design approach, determine its compatibility with performance and engineering specialty requirements of the development specification, and establish the existence and compatibility of the physical and functional interfaces among the item and other items of equipment, facilities, computer programs and personnel (DSMC). (2) A formal review of a preliminary design of an equipment to establish compatibility of design, identify engineering documentation required, and define physical and functional interface relationships between different types of equipment. (3) Technical reviews held early in a program to view the acceptability of the concept and initial design information.

Preplanned Product Improvement (P³I). (1) Planned future evolutionary development of incremental improvements to system capability. (2) Planned future evolutionary improvement of developmental systems for which design considerations are effected during development to enhance future application of projected technology. Includes improvements planned for ongoing systems that go beyond the current performance envelope to current performance envelope to achieve a needed operational capability. (3) Planned future evolutionary improvement of developmental systems for which design considerations are affected during development to enhance future application of projected technology. Include improvements planned for ongoing systems that go beyond the current performance envelope to achieve a needed operational capability.

Preventive Maintenance. All actions performed in an attempt to retain an item in specified condition by providing systematic inspection, detection, and prevention of incipient failures.

Prime Contractor. An individual, company, firm, or corporation that has entered into a written formal contract to furnish another individual or organization products, equipment, or services is the prime contractor to the receiving individual or organization.

Process Specification. A document that defines the requirements and procedures for performing a process, such as soldering, welding, encapsulating, heat-treating, coating, and plating.

Procurement. (1) The process of obtaining personnel, service, supplies, materials, and equipment or facilities. It may include the functions of design, standards determination, specification development, selection of suppliers, financing, contract administration, and other related functions. (2) The purchase of parts or materials for the equipment. (3) The act of obtaining raw material, purchased parts and equipment, subcontract, and other production items. The obtaining of equipment, resources, property, or services by purchasing, renting, leasing, or other means.

Procuring Agency. (1) The government organization responsible for issuing a contract for work to be done or for equipment to be delivered. (2) Any DoD agency authorized to procure materiel.

Producibility. (1) The relative ease of producing an item or system that is governed by the characteristics and features of a design that enable economical fabrication, assembly, inspection, and testing using available production technology. (2) Refers to the ease of manufacture and assembly of an item, including access to its parts, tooling requirements, and realistic tolerances.

Producibility Engineering and Planning (PEP). (1) Applies to those RDTE funded planning and system production engineering tasks undertaken by the materiel developer on major or nonmajor end items or components to ensure a smooth transition from development into production. PEP, a system engineering approach, ensures that an item can be produced in the required quantities and in the specified timeframe, efficiently and economically, and will meet necessary performance objectives within its design and specification constraints. As an essential part of all engineering design, it is intended to identify potential manufacturing problems and suggest design and production changes or schedule tradeoffs that would facilitate the production process. (2) System engineering approach ensuring that an end item can be procured in the required quantities and in the specified time frame, efficiently and economically, and will meet necessary performance objectives within its design and specification constraints. As an essential part of all engineering design, it is intended to identify potential manufacturing problems and suggest design and production changes or schedule tradeoffs that would facilitate the production process.

Product Assurance Plan. Implements a product assurance program including reliability, availability, and maintainability (RAM); quality hardware and software and system assessment to ensure user satisfaction; mission and operational effectiveness; and performance to specified requirements.

Product Baseline. (1) Specifications (Type C) that establish the detailed design documentation for each configuration item. Normally also includes process baseline (Type D) and material baseline (Type E). (2) Initial approved or conditionally approved product configuration identification.

Product Configuration Identification. The customer-approved technical data that define equipment configuration during production, operation, maintenance, and logistics support phases of the life cycle of that equipment. The PCI usually prescribes (a) required physical characteristics of the CI, (b) selected functional characteristics specified for acceptance testing, and (c) production acceptance tests to be conducted.

Product Definition. The definition of the product (or system) at each stage in the system life cycle. For example, Engineering must know what to design, Test and Evaluation must know what to test, Manufacturing must know what to produce, and Logistic Support must know what to operate and support at each stage of the system life cycle. Product definition includes the generation of operational requirements, technical requirements, specifications, configurations, etc.

Product Improvement. Effort to incorporate a configuration change involving engineering and testing effort on end items and depot repairable components, or changes on other than development items to increase system effectiveness or extend the useful life.

Product Improvement Program–Proposal (PIP). Proposed configuration change involving substantial engineering and testing efforts on major end items and depot repairable components, or changes on other than development items to increase system effectiveness or extend the useful life.

Production and Deployment Phase. (1) The period from production approval until the last system is delivered and accepted. (2) The last phase in a system's acquisition life cycle during which the system, including support and training equipment, data, facilities, and spares, will be produced and deployed for operational use.

Production Baseline. A three-part description of a specific major program entering production: (a) Part I—requirements—a set of minimum system performance requirements that must be met by the system in production in order to satisfy the system operational requirements; (b) Part II—schedule— the production delivery schedule, including options for the program; (c) Part III—total program cost—an estimate of the total cost of the system through the end of the production phase and the average unit production cost for the production program.

Production Cost. Considers the procurement appropriation costs, both contractor and government, associated with the fabrication, assembly, and delivery of a system in the quantities required to support DoD objectives. It includes the usable end item, support equipment, training, data, modifications, and spares.

Production Qualification Test (PQT). Tests ensuring the effectiveness of the manufacturing process, equipment, and procedures. These tests are conducted on a number of samples taken at random from the first production lot, and are repeated if the process or design is changed significantly, and when a second or alternate source is brought on line. These lines are also conducted against contractual requirements.

Production Rate. The maximum number of end items produced in a given time period such as a month or a year.

Production Readiness Review (PRR). (1) Formal examination of a program to determine whether the design is ready for production, production engineering problems have been resolved, and the producer has accomplished adequate planning for the production phase. (2) A formal review to ascertain the satisfactory completion of specific actions prior to production start.

Program. A related series of efforts to attain a broad scientific or technical goal. Also, used in a specialized sense, as follows: reliability program, quality assurance program, test program, and computer program.

Program Acquisition Cost. The sum of development and production costs. Initial spares are also included. The terms "program acquisition cost" and "program cost" are often used interchangeably.

Program Decision Memorandum (PDM). A document that provides decisions of the Secretary of Defense on program objective memoranda and the joint program assessment memorandum.

Program Directive (PD). Provides a clear definition of the approved program that is consistent with the approved acquisition strategy, funded program requirements, and the budget. The PD is a document of the program management control system.

Program Element (PE). A combination of manpower, equipment, and facilities related to a mission capability or activity. The PE is the basic building block of the DoD's Five-Year Defense Program (FYDP). A program element is identified by a specific six-digit alphanumeric code that indicates the mission and/or activity. Programs normally maintain the same PE code or series of codes (different ones at various phases) throughout the program life cycle and is the program identifier in all program financial submittals (e.g., POMs).

Program Executive Officer. Individual responsible for administering a defined number of major and/or nonmajor acquisition programs.

Program Management Plan (PMP). A document that records program decisions, contains the user's requirements, provides appropriate analysis of technical options, and contains the life-cycle plans for development, testing, production, training, and logistic support of materiel items.

Program Objective Memorandum (POM). (1) A document submitted to the SECDEF by the heads of the DoD components that recommends the total resource requirements within the parameters of the SECDEF fiscal guidance.

Program Specification. Establishes requirements for a related series of projects designed to accomplish a broad scientific or technical goal.

Program Work-Breakdown Structure (PWBS). The total work-breakdown structure for a program containing all the effort needed for a total weapon system. The contract work-breakdown structure (CWBS) is a subset of the PWBS. See "Contract Work-Breakdown Structure."

Project–Program Management. A concept for the business and technical management of specified projects or programs based on a designated, centralized management authority, or project–program manager, who is responsible for planning, directing, and controlling the definition, research, development, acquisition (in the broad sense), and initial logistics support of the user to the extent necessary to provide a balanced project or program that will accomplish the objectives effectively. It involves support of the project–program manager either (1) by the regular organization of functional activities, assigned specific tasks within the project–program, and responsible to the project–program manager for the execution of such tasks only or (2) by temporarily assigning portions of the regular functional organization to the project–program.

Project Summary Work-Breakdown Structure (PSWBS). A summary WBS tailored to a specific material item by selecting applicable elements from one or more summary WBSs or by adding equivalent elements unique to the project.

Proof of Principle. Technical documentation and troop experimentation conducted with brassboard configurations, subsystems, or surrogate systems, using troops in a realistic field environment. The demonstration includes an engineering evaluation, thereby replacing initial DT/OT. The process examines the organization and operational concept, provides data to improve requirements and evaluation criteria, and provides data on which to base the decision to enter the development prove-out phase.

Provisioning. (1) The process of determining and selecting the varieties and quantities of repair parts, spares, special tools, test and support equipment that should be procured and stocked to sustain and maintain equipments and systems for specified periods of time. It includes identification of items of supply, establishing data for catalogs, technical manuals, allowance lists, and instructions and schedules for delivery of provisioned items. (2) The process of determining and acquiring the range and quantity (depth) of spares and repair parts, and support and test equipment required to operate and maintain an end item of materiel for an initial period of service. (3) The act of providing spare parts for an equipment.

Qualification Testing (QT). Verifies the design and the manufacturing process and provides a baseline for subsequent acceptance tests. The completion of preproduction qualification test and evaluation before Milestone III decisions is essential and will be a critical factor in assessing the system's readiness for production. Production qualification TE shall be conducted on production items.

Quality Assurance. (1) Function of management ensuring that newly procured materiel conforms to the stated quality, performance, safety, and reliability standards of the TDP and contract performance specifications. (2) The system, procedures, and activities for ensuring that an item will perform satisfactorily in actual operation by verifying that materials, construction, and testing meet the requirements of project and procurement specifications. (3) Function of management ensuring that newly procured materiel conforms to the stated quality, performance, safety, and reliability standards of the TDP and contract performance specifications.

RAM Engineering. The design, development, and manufacturing tasks by which RAM is achieved.

RAM Parameter. A measure of reliability, availability, and maintainability (RAM) in which the units of measurement are directly related to operational readiness, mission success, maintenance manpower cost, or logistic support cost.

Rationalization, Standardization and Interoperability (RSI). Action that increases effectiveness of allied forces through more efficient or effective use of committed Department of Defense resources.

Readiness Drivers. Those system characteristics that have the greatest effect on a system's readiness values. These may be design (hardware or software), support, or operational characteristics.

Rebuild–Recondition. Total tear-down and reassembly of a product to the latest configuration.

Redundancy. Two or more parts, components, or systems joined functionally so that if one fails, any or all of the remaining components are capable of continuing with the function accomplishment. Failsafe, backup.

Reliability. (1) The probability that a system, equipment, device, machine, part, or any other item will perform its intended function adequately, without failure for a specified time period under specified conditions. (2) The duration or probability of failure-free performance under stated conditions. (3) The probability that an item can perform its intended function for a specified interval under stated conditions (for nonredundant items, this is equivalent to definition 2. For redundant items, this is equivalent to mission reliability. (4) The probability that an item will perform its function for a specified period in its intended environment. (5) A fundamental characteristic of materiel expressed as the probability that an item will perform its intended function for a specified interval under stated conditions. Durability is a special case of reliability.

Reliability, Availability, and Maintainability (RAM). (1) RAM requirements are those imposed on materiel systems to ensure that they are operationally ready for use when needed, will successfully perform assigned functions, and can be economically operated and maintained within the scope of logistics concepts and policies. RAM programs are applicable to materiel systems, test measurement and diagnostic equipment (TMDE), and training devices and facilities developed, produced, maintained, procured, or modified for military use. "Reliability" is the duration of probability of failure-free performance under stated conditions. "Availability" is a measure of the degree to which an item is in operable and committable state at the start of the mission. "Maintainability" is the ability of an item to be retained in or restored to specified condition when maintenance is performed by personnel having specified skill levels, using prescribed procedures and resources, at each prescribed level of maintenance and repair. (2) Those requirements imposed on materiel systems to insure that they are operationally ready for use when needed, will successfully perform assigned functions and can be economically operated and maintained within the scope of logistics concepts and policies. RAM programs are applicable to materiel systems, test measurement and diagnostic equipment (TMDE), Training devices and facilities developed, produced, maintained, procured, or modified for U.S. Army use. Reliability is the duration or probability of failure-free performance under stated conditions. Availability is a measure of the degree to which an item is in operable or committable state at the start of the mission. Maintainability is the ability of an item to be retained in or restored to a specified condition when maintenance is performed by personnel having specified skill levels, using prescribed procedures and resources, at each prescribed level of maintenance and repair.

Reliability Centered Maintenance (RCM). (1) A systematic approach for identifying preventive maintenance tasks for an end item in accordance with a specified set of procedures and for establishing intervals between maintenance tasks. (2) Precept that uses an analytical methodology or logic for influencing design maintainability and reliability establishing specific maintenance tasks for materiel systems or equipment. (3) A means for developing an integrated maintenance program, from designer and producer down to the ultimate user, that will result in safe, reliable, maintainable, and supportable equipment–commodities in the U.S. Military Services, capable of performing in support of required mission at least cost. RCM is a program that uses logic developed to ensure that the inherent design reliability and safety of an item is achieved while performing the least amount of maintenance, considering cost of the total life cycle of the material.

Reliability Engineering. The application of scientific and management knowledge, methods, and skills to the development of things with the inherent or built-in ability to perform intended functions adequately, without failure for a specified period of time (or equivalent measure of operation), under specified conditions, and the coordination of such quality with all other systems engineering elements.

Reliability Improvement Warranty (RIW). (1) A contractual commitment, included in an acquisition or overhaul contract, that provides the contractor with a financial inducement to improve a system to reduce repair or replacement costs and thus enhance field operational reliability. (For example, a common form of RIW provides a firm fixed price to the contractor to repair or replace, within a specified turnaround time, all equipment that fails (subject to specified exclusions) during a specified period of long duration (such as 5 years). In this form of RIW, the contractor may increase profits by introducing engineering changes (at no added cost to the government) that are cost-effective in reducing repair or replacement costs. An RIW may also include a guarantee of a specified reliability level, with the contractor obliged to upgrade all existing units at the contractor's expense if reliability falls below the specified level. (2) Contractual provision to motivate the contractor to improve system reliability.

Repair. The restoration or replacement of parts or components of facilities or equipment as necessitated by wear, tear, damage, failure of parts, and similar to return the facility or equipment to efficient operating condition. It may be accomplished by overhaul, reprocessing, or replacement of parts or materials, or by any combination of these actions. It may include preparation, fault location or isolation, procurement of items, correction, adjustment, calibration, and final test.

Repair Level Analysis (RLA). The repair level analysis limits the depth of maintenance task analysis in the LSA process by distinguishing between repairable and nonrepairable components and by selected the most cost-effective repair level. An RLA is normally conducted on all line replaceable units.

Repair Parts. (1) Individual parts or assemblies required for the maintenance or repair of equipment, systems, or spares. Such repair parts may also be repairable or nonrepairable assemblies, or one-piece items. Consumable supplies used in maintenance or repair, such as wiping rags, solvents, lubricants, and similar are not considered repair parts. (2) Those support items that are an integral part of the end item or system that are coded as nonrepairable. (3) Piece parts or nonrepairable assemblies for repairing spares or major end items.

Repairable (or Repairable Item). An item of durable nature that has been determined, by application of engineering, economic, and other factors, to be feasible of restoration to serviceable condition through regular repair procedures within an organization (not requiring return to the original manufacturer for repair).

Repair Time. The time spent replacing, repairing, or adjusting all items suspected to have been the cause of the malfunction, except those subsequently shown by interim test of the system not to have been the cause.

Replaceable Item. An item that is functionally interchangeable with another item, but that differs physically from the original item, necessitating special operations (filling, drilling, reaming, etc.) before it can be installed.

Replacement (or Replacement Item). An item that is functionally interchangeable with another item but differs physically from the original part to the extent that installation of the replacement requires such operations as drilling, reaming, cutting, filing, and shimming, in addition to normal attachment or installation operations.

Replenishment (Recurring) Spare Parts. Those parts procured on other than production contracts. These requirements cover support provided after the initial spare parts procurements and extending throughout the program life of the system or end item of equipment.

Request for Proposal (RFP). (1) Request for the manufacturer to submit a proposal supported by cost breakdown; provides a description of the item to be procured. An RFP may include specifications, quantities, time and place of delivery, method of shipment, packaging and instruction manual requirements, materiel to be furnished, and data requirements, both support and administrative. (2) A solicitation document used in negotiated procurements. It usually contains a description of the items or services to be procured, the terms and conditions, type of contract, schedules, work statement, specifications, listing of the items to be delivered, funding, data requirements, and instruction for the preparation of technical management and cost proposals.

Required Operational Capability (ROC). (1) A document that concisely states the minimum essential operational, technical, logistic, and cost information necessary to initiate full-scale development or procurement of a materiel system. (2) A document that states concisely (usually in four pages or less) the minimum essential operational, technical, logistical, and cost information necessary to initiate full-scale development or procurement of a materiel system.

Research and Development (R&D) Costs. The cost of developing a new or improved capability to the point where it is ready for serial production. This category includes those expenses that occur within the validation and full-scale development phases of the acquisition cycle. Included are all costs for the testing that are associated with the R&D effort directed to the elements of the work-breakdown structure: major system equipment, training, peculiar support equipment, systems test and evaluation, system–project management, data, operational–site activation, and common support equipment.

Research, Development, Test, and Evaluation (RDTE) Cost. The sum of all costs (contractor and government) resulting from applied research, engineering design, analysis, development, test, evaluation, and managing development efforts related to a specific system. See "Life Cycle Cost."

Retrofits. (1) The science of material and information flow. This concept includes the flow of material from the raw material stage through the production stage, to the consumer. (2) The incorporation of an engineering change in equipment accepted by the customer or in service. (3) The application of measures or controls to correct deficiencies in fielded systems.

Retrofit Kit. See "Modification Kit."

Revision. A modification of a drawing or design after it has been officially released by engineering.

Risk Analysis. The evaluation of the situation, environment, or set of conditions to determine the technical, financial, or business risks inherent in the venture or mission. Can be computed using complex models, expert opinions, or intuitive judgment.

Routine ECP. The priority given to ECPs that cannot be assigned a compatibility, emergency, or urgent priority.

Safety. Elimination of hazardous conditions that could cause injury. Protection against failure, breakage, and accident.

Safety and Health Hazard Assessment and Analyses. The documented quantitative determination of system safety and health hazards. This includes the evaluation of hazard severity, hazard probability, operational constraints, and the identification of required precautions, protective devices, and training requirements and/or restrictions.

Safety Assessment Report. A formal summary of the safety data collected during the design and development of the system. In it, the materiel developer summarizes the hazard potential of the item, provides a risk assessment, and recommends procedures or other corrective actions to reduce these hazards to an acceptable level.

Safety Stock. The quantity of an item, in addition to the normal level of supply, required to be on hand to permit continuing operation with a specific level of confidence if resupply is interrupted, there are unforseen minor interruptions to replenishment, or demand varies in an unpredictable manner.

Sample Data Collection (SDC). A method for obtaining information on the performance and maintainability of an item of equipment. Data are obtained directly from observations made in the field. An effort is made to see that the sample from which the feedback is obtained in representative of the total population.

Scheduled Maintenance. (Sometimes called "preventive maintenance.") (1) Consists of preplanned actions, according to an established time or use table, performed in an attempt to keep an item in a specified operating condition by means of systematic inspection, detection, and prevention of incipient failure. (2) Preventive maintenance performed at prescribed points in the item's life.

Selected Acquisition Report (SAR). (1) A document prepared for the SECDEF by a DoD component that summarizes current estimates of technical, schedule, and cost performance in comparison with the original plans and current program. (2) Standard, comprehensive, summary status report on DoD acquisition programs for management with DoD. (3) A report required to be provided for management in DoD at least annually by designated (selected) major programs. It summarizes and provides a status of cost and programmatic data relative to a development estimate.

Sensitivity Analysis. Repetition of an analysis with different quantitative values for selected parameters or assumptions for the purpose of comparison with the results of the basic analysis. If a small change in the value of the variable results in a large change in the results, the results are said to be sensitive to that parameter or assumption.

Serviceability. A characteristic of an item, piece of equipment, system, product, or similar that makes it easy to perform routine operating checks, fueling, lubrication, and the like (i.e., routine operator maintenance), after it is put into operation.

Shelf Life. The period of time during which an item can remain unused in proper storage without significant deterioration.

Shop Replaceable Unit (SRU). A unit installed in an end item of equipment or system that is replaceable only in a designated repair facility (shop environment).

Simulation. (1) A model of a set of conditions or an environment of inter-related elements exercised in a manner to gain knowledge of conditions that may develop under various circumstances. (2) A model in which means other than or in addition to sets of mathematical equations are used to represent the basic events and activities in the process being described. Simulation is often employed where the process cannot be described by traceable mathematical expressions. In simulation, detailed events of the process are often followed in sequence, and actions or decisions at each point are based on predetermined rules in the form of program logic or decision tables. Monte Carlo sampling may be employed to determine the value of probabilistic variables or the occurrence of probabilistic events.

Single Source. Characterized as one source among others in a competitive marketplace that, for justifiable reason, (e.g. immediate or previous experience, or current contractual involvement), is found to be most advantageous for the purpose of contract award. (Sometimes used interchangeably with the term "sole source.")

Site Activation Costs. The costs incurred to bring a site to operational readiness, including facility construction, the installation and checkout of all system and supporting equipment, and acceptance of the site by the operating command.

Software. (1) Efforts, plans, or paperwork to sustain or support projects, operations, equipment, assemblies, items, and so on including such things as engineering and design, technical data, plans, schedules, and computer programs. Software excludes physical parts, materials, equipment, and tools. (2) Also, computer program and computer software. (3) Concerning computer programs and instructions. In a general sense reports, drawings, sketches, computer programs or tapes, photos, and so on, as opposed to hardware.

Soldier–Machine Interface (SMI). Consideration through system analysis and psychophysiology of equipment design and operational concepts to ensure that they are compatible with the capabilities and limitations of operators and maintenance personnel. Also referred to as "soldier–materiel interface" and "soldier–machine interaction." SMI is included in MANPRINT.

Source Selection. The formal procurement process used within DoD or a company to (a) call for proposals, (b) evaluate proposals, (c) pass recommendations to higher authority, or (d) final awarding of a contract by the selection authority.

Source Selection Evaluation (SSE)–Source Selection Process. The process wherein the requirements, facts, recommendations, and government policy relevant to an award decision in a competitive procurement of a system–project are examined and the decision made.

Spares and Repair Parts. Spares are components or assemblies used for maintenance replacement purposes in major end items of equipment. Repair parts are those "bit and pieces," such as individual parts or nonrepairable assemblies, required for the repair of spares or major end items.

Spares (or Spare Parts). (1) Components, assemblies, and equipment that are completely interchangeable with like items installed or in use that are or can be used to replace items removed during maintenance and overhaul. (2) Those support items that are an integral part of the end-item or system that are coded as repairable.

Special Test Equipment (STE). All electrical, electronic, hydraulic, pneumatic, mechanical, or other items or assemblies of equipment that are of such a specialized nature that, without modification or alteration, the use of such items or assemblies is limited to testing in the development or production of particular supplies or parts or in performance of particular services.

Special Tools. Tools that are required for use with only one end item.

Specification Change Notice. A document that describes changes to an approved specification. The SCN is made a part of the specification after customer approval.

Specification Tree. A drawing showing the indentured relationships among specifications independent of the assembly or installation relationships of the items specified. The tree shows the dependency of specifications on other specifications.

Specifications. Documents that clearly and accurately describe the essential technical requirements for materials, items, equipments, systems, or services, including the procedures by which it will be determined that the requirements have been met. Such documents may include performance, support, preservation, packaging, and marking requirements.

Standard Configuration Identifiers. The complete set of numbers used to identify the configuration of the equipment and its spares. This set consists of six numbers: (a) specification identification number, (b) CI number, (c) serial number, (d) item identification and part number, (e) Change identification number, and (f) code identification number.

Standard Deviation. A measure of average dispersion (departure from the mean) of numbers, computed as the square root of the average of the squares of the difference between number and their arithmetic mean.

Standard Item. A part, component, material, subassembly, or assembly or equipment that is identified or described accurately by a company, industry, federal, or military standard document or drawing.

Standardization. The process of establishing, by common agreement, the engineering criteria, terms, principles, practices, materials, items, processes, parts, equipment, subassemblies, and assemblies to achieve the greatest practicable uniformity of items recurrently used, bought, stocked, or distributed, and of engineering practices, and to effect optimum interchangeability of equipment, parts, and components.

Standardization and Interoperability. Any action that increases the effectiveness of alliance forces through more efficient or effective use of committed defense resources.

Stock Control. The process of maintaining inventory data on the quantity, location, and condition of items due in, on-hand, and due out, to determine quantities available, required, or both, and to facilitate distribution and management of stock.

Stock Due-in. The quantity of items or materials expected to be received under outstanding procuring and requisitioning documents, and the quantity expected from other sources such as transfer, reclamation, and recovery.

Stock Due-out. The quantity of items or materials, requisitioned by ordering or using activities that is not immediately available for issue, but is recorded as a commitment for future issue.

Stock Number. (1) A consumer or user number assigned by the stocking organization or its parent organization to each group of articles or materials treated as if identical within the using supply system, and repetitively procured, stocked, and distributed. (2) A federal identification number for an equipment specified in a contract by government agency for inventory control.

Subassembly. Two or more parts that form a portion of an assembly replaceable as a whole, but having a part (or parts) that is (are) individually replaceable.

Subcontractor. (1) Any supplier, distributor, vendor, or firm that enters into a formal contract to furnish items, services, parts, or supplies to a prime contractor (see "Prime Contractor"). (2) One who performs a subtask for the company that has the equipment contract.

Subsystem. A major functional subassembly or group of items that is essential to operational completeness of a system.

Summary Work-Breakdown Structure (Summary WBS). Consists of upper three levels of a WBS prescribed by MIL-STD-881A and having uniform element terminology, definition, and placement in the family-tree structure. The upper three levels of a summary WBS have been organized with the following categories of defense materiel items: aircraft systems, electronic systems, missile systems, ordinance systems, ship systems, space systems, and surface vehicle systems.

Supply. The procurement, distribution, maintenance in storage, and salvage of items that are consumed in use or become part of other items thus losing their identity; and of items, assemblies, equipment, systems, and machinery. (Excludes procurement of land.)

Supply Management. Continuing actions of planning, organizing, directing, controlling, and reporting the use of personnel, money, materials, and facilities to provide supplies and equipment to users or consumers. It includes requirement computations, procurement, inventory control, distribution, maintenance-in-storage, issue, and salvage or disposal.

Supply Support. (1) Consists of all the spare parts, repair parts, consumables, and related materials and documents necessary for scheduled and unscheduled maintenance, taking into consideration location, transportation, time, and overall availability to ensure maximum continuity and effectiveness of operations. (2) All management actions, procedures, and techniques used to determine requirements to acquire, catalog, receive, store, transfer, issue, and dispose of secondary items. This includes provisioning for initial support as well as replenishment supply support, one to the principal elements of ILS.

Supply System. The organizations, offices, facilities, methods, techniques, and trained personnel utilized to provide supplies and equipment to users or consumers and to take care of requirements computation, planning, procurement, inventory control, distribution, maintenance-in-storage, issue, and salvage or disposal of items and materials.

Support Concept. A complete system level description of a support system, consisting of an integrated set of ILS element concepts, which meets the functional support requirements and is in harmony with the design and operational concepts.

Support Costs. Direct and indirect costs not directly attributable to the actual, physical fabrication and assembly of an end item.

Support Equipment. (1) Items required to maintain equipment and systems in effective operating condition under various environments. Support equipment includes general and special purpose vehicles, power units, stands, test equipment, tools, and test benches needed to facilitate or sustain maintenance actions, detect or diagnose malfunctions, or monitor the operational status of equipment and systems. (2) All equipment (mobile or fixed) required to support the operation and maintenance of a materiel system. This includes associated multiuse end items, ground-handling and maintenance equipment, tools, metrology and calibration equipment, test equipment, and automatic test equipment; also includes the acquisition of logistics support for the support and test equipment itself, one of the principal elements of ILS. (3) Equipment required to make the CI operational in its intended environment, for example, ground equipment or computer programs. (4) Includes all equipment required to perform the support function except that which is an integral part of the mission equipment. It does not include any of the equipment required to perform mission operation functions. Support equipment includes handling equipment, test equipment, automatic test equipment (when the automatic test equipment is accomplishing a support function), organizational field and depot support equipment, tools, and related computer programs and software. Further, it consists of peculiar support equipment (PSE) that is unique to a system and common support equipment that is in the customer inventory.

Support Resources. The materiel and personnel elements required to operate and maintain a system to meet readiness and sustainability requirements. New support resources are those that require development. Critical support resources are those that are not new but require special management attention as a result of schedule requirements, cost implications, known scarcities, or foreign markets.

Support System. Collectively, those tangible logistic support resources required to maintain a materiel system in an operationally ready condition. It is developed with the materiel system and merged with the ongoing logistic systems on production and deployment. The following elements of integrated logistics support (ILS) constitute the support system: support and test equipment, supply support, transportation and handling, technical data, facilities, and trained personnel. The other elements of ILS are the means by which the support system is developed and implemented.

Supportability. (1) The degree to which system design characteristics and planned logistics resources, including manpower, meet system peacetime readiness and wartime utilization requirements. (2) That characteristic of materiel indicative of its ability to be sustained at a required readiness level when supported in accordance with specified concepts and procedures. (3) The degree to which system design characteristics and planned logistics resources, including manpower, meet system readiness and utilization requirements.

Supportability Assessment. An evaluation of how well the composite of support considerations necessary to achieve the effective and economical support of a system for its life cycle meets stated quantitative and qualitative requirements. This includes integrated logistic support and logistic support resource related O&S cost considerations.

Supportability-Related Design Factors. Those supportability factors that include only the effects of an item's design. Examples include inherent reliability and maintainability values, testability values, and transportability characteristics

System. (1) A combination of assemblage of correlated hardware, software, facts, principles, doctrines, ideas, methods, procedures, and people or any combinations of these, all arranged or ordered toward a common objective, taking into consideration the important interrelationships involved. (2) A grouping of subassemblies, assemblies, and equipment that provide a specific function. A system may also include personnel and facilities. (3) A composite, at any level of complexity, of personnel, procedures, materials, tools, equipment, facilities, and software. The elements of this composite entity are used together in the intended operational or support environment to perform a given task or achieve a specific production, support, or mission requirement. (4) An item or group of items of materiel, including hardware, software, and personnel elements. (5) A group of subassemblies or devices or individual units of hardware (e.g., subsystems) that collectively meet or serve the total performance requirements of one or more defined functions (e.g., aircraft system, ship system, land vehicle system). (6) A combination of equipment end items, assemblies, components, modules, and/or parts assembled as a single functional unit to perform a specific task or mission. For the purpose of this policy, "system" is not restricted solely to weapon and/or reportable systems.

System Acquisition Process. A sequence of specified decision events and phases of activity directed to achievement of established program objectives in the acquisition of Department of Defense system and extending from approval of a mission need through successful deployment of the DoD system termination of the program.

System Concept Paper. (1) Supports the DoD Milestone I decision, documents the results of the concept exploration phase and provides the acquisition strategy for the program. (2) The system concept paper is the decision management documentation prepared for a MDR I decision. The format and content is contained in AR 70-1. (3) The SCP will summarize mission requirements and identify program alternatives based on initial studies and analyses of design concepts; alternative acquisition strategies; expected operational and associated capabilities; industrial base capacity; readiness and support requirements; manpower, personnel, training, and safety requirements; and cost estimates. The test and evaluation master plan (TEMP), as described in DoD Directive 5000.3, will outline the test and evaluation program. The SCP format is described in DoD instruction 5000.2 and becomes the program baseline for the demonstration–validation phase of the program. (*Source*: DoDD 5000.1.) (4) The SCP is a top-level document used to summarize the results of the concept definition phase; to describe the DoD component's acquisition strategy, including identification of the best concepts to be carried into the demonstration–validation phase for further development, and reasons for elimination of alternative concepts; and to establish broad program cost, schedule, and operational effectiveness and suitability goals and objectives to be met and reviewed at the next milestone. The purpose and content of the TEMP is described in DoD Directive 5000.3.

System Description. A narrative characterization of the purpose, function, size, mobility, maintenance, and performance of a system. This word representation is intended to convey a mental image of design aspects that enable the system to counter the postulated threat.

System Design Concept. An idea expressed in terms of general performance, capabilities, and characteristics of hardware and software oriented either to operate or to be operated as an integral whole in meeting a mission.

System Downtime. The time interval between the commencement of work on a system malfunction and the time when the system has been repaired and/or checked by the maintenance technician, and no further maintenance activity is executed.

System Effectiveness. The probability that a system can successfully meet an overall operational demand within a given time, when operated under specified conditions—or the ability of a system to do the job for which it was intended—or a measure of the degree to which a system can be expected to achieve its objective or purpose.

System MANPRINT Management Plan (SMMP). The SMMP is a living document that will be updated as needed throughout the materiel acquisition process. It serves as a planning–management guide (an audit trail), to identify the tasks, analyses, tradeoffs, and decisions that must be, or have been, made to address MANPRINT issues during the materiel acquisition process.

System Milestone Schedule. A schedule of events marking significant points in time in the course of development, production, and fielding of a system. Some of these typically are points where funds are available, production decisions are made, contracts are signed, deliveries are begun, and the first hardware is fielded.

System Readiness Objective. (1) A criterion for assessing the ability of a system to undertake and sustain a specified set of missions at planned peacetime and wartime utilization rates. System readiness measures take explicit account of the effects of system design R&M, the characteristics and performance of the support system, and the quantity and location of support resources. Examples of system readiness measure are combat sortie rate over time, peacetime mission capable rate, operational availability, and asset ready rate. (2) Measures relating to the effectiveness of an operational unit to meet peacetime and wartime mission requirements considering the unit set of equipage and the potential logistic support assets and resources available to influence the unit's operational readiness and sustainability. Peacetime and wartime SRO will differ according to usage rate, operational modes, and mission profiles and operational environments. SRO must relate quantitatively to system design parameters (e.g., RAM) and to support resource requirements.

System Safety. The application of engineering and management principles, criteria, and techniques to optimize safety within the constraints of operational effectiveness, time, and cost throughout all phases of the system of facility life cycle.

System Specification. A general specification containing technical and mission requirements for the system as a whole and apportioning these requirements to subsystems or equipment for meeting mission goals. It also defines interfaces between the different items.

System Test Support Package. An assemblage of support elements provided prior to and used during development and operational tests to validate the organizational, direct support, and general support maintenance requirements and capability. The maintenance test support package includes all required draft equipment publications (operator through general support maintenance equipment manuals); parts and accessories; special and common tools; test, support, calibration, and maintenance shop facilities; and personnel skill requirements.

Systems Engineering. (1) The application of scientific and engineering methods to the study, planning, design, construction, direction, and evaluation of person–machine systems and system components, whereby the relationships and utilization of the various parts of the systems are fully planned before firming up the design of the hardware in order to achieve the best balance among operational, economic, and logistic support factors. (2) System engineering is the application of scientific and engineering efforts to (a) transform an operational need into a description of system performance parameters and a system configuration through the use of an iterative process of definition, synthesis, analysis, design, test, and evaluation; (b) integrate related technical parameters and ensure compatibility of all physical, functional, and program interfaces in a manner that optimize the total system definition and design; (c) integrate reliability, maintainability, safety, survivability, and human and other such factors into the total engineering effort to meet cost, schedule, and technical performance objectives (MIL-STD-499).

Tailoring. The process by which the individual requirements (sections, paragraphs, or sentences) of the selected specifications and standards are evaluated to determine the extent to which each requirement is most suitable for a specific materiel acquisition and the modification of these requirements, where necessary, to ensure that each tailored document invoked states only the minimum needs of the Government (MIL-STD-1388-1A).

Task Analysis. A process of reviewing actual job content and context to classify information into units of work within a job. The process provides a procedure for isolating each unique unit of work, a procedure for describing each unit accomplished, and descriptive information to assist in the design and testing of training products.

Technical Data. (1) Recorded information regardless of form or character (such as manuals and drawings) of a scientific or technical nature. Computer programs and related software are not technical data; documentation of computer programs and related software are. Also excluded are financial data or other information related to contract administration, one of the principal elements of ILS. (2) Recorded information, regardless of form, used to define, produce, test, evaluate, modify, deliver, support, maintain, or operate a configuration item; technical data may be recorded as graphic or pictorial delineations in media such as drawings or photographs, text in specifications or related performance or design-type documents or in machine form such as punched-card magnetic tape, disk, computer memory printouts, or computer memory. (3) Technical writing, reproductions, drawings, or other graphic representations and works of a technical nature that are specified to be delivered pursuant to a contract. Excludes financial reports, cost analyses, management, and other information incidental to contract administration.

Technical Data Package (TDP). (1) A generic term applicable to various types of technical data when used for procurement and other specified purposes. A TDP is a composite of drawings, specifications, plans, standards, models, and such other data as may be necessary to describe existing materiel so that they may be procured by the method contemplated, maintained, installed, packaged, transported, used, or developed. (2) Generic term applicable to types of technical data when used for procurement purposes. Composite of specifications, plans, drawings, standards, and such other data as may be necessary to describe existing materiel so they may be procured by the method contemplated. (3) A generic term applicable to types of technical data when used for procurement purposes. It is a composite of specifications, plans, drawings, standards, and such other data as may be necessary to describe existing materiel so they may be procured by the method contemplated.

Technical Evaluation (TE). Addresses the system's technical issues and criteria and the acquisition and fielding of an effective, supportable, and safe system by assisting in the engineering design and development and verifying attainment of technical performance specifications, objectives, producibility, adequacy of the technical data package, and supportability; and determining safety, health hazards, human factors, and MANPRINT aspects. Technical evaluation encompasses the use of models, simulations, and test beds, as well as prototypes or full-scale development models of the system.

Technical Manual. A publication containing instructions designed to meet the needs of personnel engaged in operating, maintaining, servicing, overhauling, installing, or inspecting the equipment.

Technical Test–Testing (TT). Generic term encompassing technical feasibility tests, development tests, qualification tests, joint development tests, and contractor and foreign tests.

Technology Integration Steering Committee (TISC). A decision body that considers high-payoff materiel concepts for transition to the proof-of-principle phase. The TISC also provides an early focus of high-payoff battlefield system concepts that exploit breakthrough technology.

Test and Evaluation Master Plan (TEMP). (1) A broad plan that relates test objectives to required system characteristics and critical issues, and integrates objectives, responsibilities, resources, and schedules for all TE to be accomplished (DoDD 5000.3). (2) It is prepared for all defense system acquisition programs. The TEMP is a broad plan that relates test objectives to required system characteristics and critical issues and integrates objectives, responsibilities, resources, and schedules for all TE to be accomplished. Replaces coordinated test plan. (3) A documented method or scheme of action, or a program for testing the performance of a system throughout its development and production cycles. Tests are performed to ensure that the system fulfills the design intent, will perform its mission to counter the threat, and is compatible with planned operating philosophies.

Test and Support Equipment. Consists of all special tools and checkout equipment, metrology and calibrations equipment, maintenance stands and handling equipment required for maintenance. It includes external and built-in test (BIT) equipment considered part of the supported system or equipment.

Test Design Plan (TDP). A formal document developed by the test organization that states the circumstances under which a test and/or evaluation will be executed, the data required from the test, and the methodology for analyzing test results.

Test, Measurement, and Diagnostic Equipment (TMDE). (1) Any system or device used to evaluate the operational condition of a system or equipment to identify or isolate any actual or potential malfunction. (2) Any system or device used to evaluate the operational condition of an end item or subsystem thereof, or to identify and/or isolate any actual or potential malfunction. TMDE includes diagnostic and prognostic equipment, semiautomatic and automatic test equipment (with issued software), and calibration test or measurement equipment.

Test Program Set (TPS). The package that enables an ATE operator to diagnose, using automatic test equipment, a line or shop replaceable unit, printed-circuit board, or similar item(s). The package includes appropriate interconnect devices, automated load module tape(s), equipment publications, and other necessary articles that allow the ATE operator to perform a diagnostics–screening quality assurance function.

Test Readiness Review. A formal review to determine completeness of software test procedures and preparation for testing.

Time-Phased Procurement. The programming and funding of certain nonrecurring elements of a production program in a fiscal year different from that in which the usable end item is funded.

Top–Down. Starting with the highest level of hierarchy and proceeding through progressively lower levels.

Total Life-Cycle Competition Strategy (TLCCS). Describes the technical and contracting methods for maximizing effective full and open competition (FOC) at the manufacturing source level throughout the system's life cycle. Addresses entire system to include end item(s), components, and spare parts in light of breakout, spares acquisition integrated with production (SAIP), and acquisition of technical data and data rights.

Total System Downtime. The time interval between the reporting of a system malfunction and the time when the system has been repaired and/or checked by the maintenance technician, and no further maintenance activity is executed.

Tradeoff. (1) Action or decision generally concerned with the evaluation of alternatives and with compromises to obtain the best mix of logistic support characteristics, system performance, and total or real cost. (2) An evaluation of a design change to determine its importance in regard to benefits versus disadvantages (high cost, delays, etc.).

Trainer. The agency that trains personnel to operate and maintain development items or systems.

Training. (1) The pragmatic approach to supplementing education with particularized knowledge and assistance in developing special skills. It is helping people to learn to practice an art, science, trade, profession, or related activities. Basically more specialized than education. (2) Consideration of the training necessary and time required to impart the requisite knowledge, skills, and abilities to quality U.S. Military Service personnel for use, operation, maintenance, and support of military systems or items. It involves (a) the formulation and selection of engineering design alternatives that are supportable from a training perspective, (b) the documentation of training strategies, and (c) the timely determination of resource requirements to enable the military training system to support system fielding. Human-factors engineering techniques are used to determine the tasks to be performed by system user, operator, maintenance and support personnel; the conditions under which they must be performed; and the performance standards to be met. Training is linked with personnel analyses and actions in that availability of qualified personnel is a direct function of the training process. As a minimum, the following must be considered:

- Training effort and costs versus system design
- Training times
- Training program development, considering aptitudes of available personnel
- Sustainment training, as distinguished from training associated with initial system fielding
- Developmental training, as distinguished from initial entry training
- Training devices—design, development, and use
- Training base resourcing—manpower and personnel implication
- New equipment training (NET)
- Formal training base instructions, versus on-the-job training (OJT) in units
- Unit training
- Operational testing of the adequacy of training programs and techniques

Training Developer. The agency that develops the training strategy and requirements for both institutional and unit training.

Training Device. (1) Items that simulate or demonstrate the function of equipment or systems such as three-dimensional models, mock-ups, or exhibits. They are designed, developed, or procured only for training support. (2) Item designed, developed, and procured solely to simulate or demonstrate the function of equipment or systems to meet training support requirements. (3) Any three-dimensional object developed, fabricated, or procured specifically for improving the learning process. Training devices may be either system devices or nonsystem devices.

a. System devices are designed for use with one system or item of equipment, including subassemblies and components.

b. Nonsystem devices are designed to support general military training and/or for use with more than one system or item of equipment, including subassemblies and components.

Training Device Need Statement (TDNS). An initial training requirement that identifies a need in response to a training deficiency. The TDNS is similar to the initial O&O plan because it provides the front-end agreement on the requirement to initiate the concept exploration phase. the TDNS provides specific guidance for follow-on actions to solve the training need.

Training Equipment. The maintenance and operating training aids and related software used for training operating and maintenance personnel.

Training and Training Support. The processes, procedures, techniques, training devices, and equipment used to train civilian and active duty and reserve military personnel to operate and support a materiel system. This includes individual and crew training; new equipment training; initial, formal, and on-the-job training; and logistic support planning for training equipment and training device acquisitions and installations, one of the principle elements of ILS.

Transportability. (1) Capability of efficiently and effectively transporting an end item of military equipment or component, over railways, highways, waterways, ocean, and airways by either carrier, towing, or self-propulsion. (2) A design characteristic related to the construction of a CI for transport to its destination. Considerations include weight, size, dangerous or hazardous features, and sensitivity to shock, vibration, temperature, and humidity. (3) The inherent capability of an item to be moved by towing, self-propulsion, or carrier, via railway, highway, waterway, pipeline, ocean, and airway, using existing equipment or equipment that is planned for the movement of the item. (4) The capability of materiel to be moved by towing, self-propulsion, or carrier through any means, such as railways, highways, waterways, pipelines, oceans, and airways. (Full consideration of available and projected transportation assets, mobility plans and schedules, and the impact of system equipment and support items on the strategic mobility of operating military forces is required to achieve this capability.)

Transportability Approval. A statement by the appropriate U.S. Military Service transportability agent that an item of materiel, in its shipping configuration, is transportable by the mode(s) of transportation specified in development guides or materiel requirements, or meets amended transportability characteristics approved by higher authority.

Transportability Engineering. The performance of those functions required to identify and measure the limiting constraints, characteristics, and environments of transportation systems; the integration of these data into design criteria to use operational and planned transportation capability effectively; and the development of technical transportability guidance.

Transportability Guidance. Published information concerning loading, securing, moving, and handling operations to ensure safe and effective logistics transportation of an item of military equipment, or component thereof, by railway, highway, waterway, ocean, airway, and off-road route either as cargo or by towing or self-propulsion. The information includes the technical and physical characteristics that affect transportability, such as loading, blocking, bracing, tiedown, and anchoring; validated dimensions with metric equivalents; significant considerations for movement by air, land, and water transportation; sectionalization to conform to the limits of the various modes of transportation; center of gravity and distribution of load; shipping cube for both operational and sectionalized configurations; and transportation regulations, special procedures, and special permits for movement.

Turnaround Time. The interval between the time a repairable item removed from use and the time it is again available in full serviceable condition.

Turnaround Time (TAT). The time required to return an item to use between missions.

Unscheduled Maintenance (UM) or Emergency Maintenance (EM), or Corrective Maintenance (CM). Restoration of a failed item to usable condition.

Unscheduled Maintenance. Corrective maintenance required by item conditions.

User Test (UT). Generic term that includes force development test and experimentation (FDTE), innovative test and concept evaluation program (CEP), initial operational test and evaluation (IOTE), follow-on operational test and evaluation (FOTE), on-site user test (OSUT), and joint operational test. (See AR 71-3.)

Validation. (1) Evaluation to ensure compliance with specific requirements. (2) In terms of cost models, a process used to determine whether the model selected for a particular estimate is a reliable predictor of costs for the type of system being estimated.

Warehousing. Those operations and storage activities concerned with the receipt, storage, care, preservation, packaging, packing, marking, issue of items, and documentation and record-keeping incidental to such operations.

Warranty. Guarantee that an item will perform as specified for at least the given time.

Weapon Support and Logistics R&D. Technology programs funded *outside* the weapon system development programs that may result in improved subsystem R&M, improved support elements needed in the operation and maintenance of weapon systems, and improved logistics infrastructure elements.

Weapon System. The sum total of prime mission equipment and all the peripheral elements that are necessary to operate and maintain the equipment as a mission-ready unit. Weapon system includes support equipment, spares, supplies, trainers, people, technical orders, and facilities. Often referred to collectively as the "system."

Work-Breakdown Structure (WBS). (1) A product-oriented family tree composed of hardware, software, services, and other tasks; the WBS results from project engineering effort during the development and production of an item and completely defines the project. A WBS displays and defines the product to be developed or produced and relates the elements of work to be accomplished to each other and to the end products. (2) A method of diagramming the way that work is to be accomplished by separating the work content into individual elements.

APPENDIX 2
GLOSSARY OF ACRONYMS AND ABBREVIATIONS

AAE	Army acquisition executive
ABDR	Aircraft battle damage repair
ACI	Allocated Configuration Identification
ACSN	Advance change–study notice
ADM	Acquisition decision memorandum (replaces SDDM & SADM)
ADP	Automated data processing
ADPE	Automated data processing equipment
AFTOMS	Air Force Technical Order Management System
AG	Army guidance or availability guidance
A_i	Inherent availability
ALCM	Air-launched cruise missile
ALO	Authorized Level of Organization
AMC	Army Materiel Command
AMMS	Acquisition managment milestone system
AMSAA	(AMC) U.S. Army materiel systems analysis activity
A_o	Operational availability
AP	Acquisition plan
AQL	Acceptable quality level
AR	Army regulation
AS	Acquisition strategy
ASAP	Army streamlined acquisition program (or process)
ASG ILCO	Assistant Secretary General for Infrastructure, Logistics, and Council Operations
ASIOE	Associated support item of equipment
ASPR	Armed Services procurement regulation

ATE	Automatic test equipment
ATOS	Automated Technical Order System
ATTD	Advanced technology transition demonstration
AURS	Automated unit reference sheet
BCE	Baseline cost estimate
BEG	Budget estimate guidance
BES	Budget estimate submission
BIT	Built-in test
BITE	Built-in test equipment
BOA	Basic ordering agreements
BOIP	Basis of Issue Plan
BSM	Basic sustainment materiel
C³I	Command, control, communications, and intelligence systems
CAD	Computer-aided design
CAE	Computer-aided engineering; component acquisition executive
CALS	Computer-aided acquisition and logistics support
CARC	Component acquisition review council
CCB	Configuration control board
CDA	Catalog data activity
CDR	Critical design review
CDRL	Contract data requirements list
CDS	Concept description sheet
CE	Concept exploration (phase)
CEP	Concept evaluation program
CETS	Contractor engineering and technical services
CFE	Contractor-furnished equipment
CFP	Concept formulation package
CFSP	Contract field services personnel
CI	Configuration item
CITIS	Contractor integrated technical information system
CLS	Contractor logistic support
CLSSA	Cooperative logistic supply support arrangement
CM	Configuration management
CMS	Contractor maintenance services
COB	Collocated operating bases or command operating budget
COEA	Cost and operational effectiveness analysis

COMMZ	Communications zone
CONUS	Continental United States
CPM	Contractor performance management
CRA	Continuing resolution authority
CRISD	Computer resources integrated support document
CRMP	Computer resource management plan
CRWG	Computer resource working group
CSC	Computer software component
CSCI	Computer software configuration item
CSOM	Computer system operator's manual
CSS	Combat service support
CWBS	Contract work-breakdown structure
D&V	Demonstration and validation (phase)
DA	Department of the Army
DAB	Defense Acquisition Board (formerly DSARC)
DAE	Defense acquisition executive
DAP	Designated acquisition program
DBDD	Database design document
DEP	Draft equipment publications
DFARS	Defense Federal Acquisition Regulation Supplement
DG	Defense guidance
DID	Data item description
DLA	Defense Logistics Agency
DMDC	Defense Management Data Center
DoD	Department of Defense
DoD-STD	Department of Defense Standard
DoDD	DoD Directive
DoDI	DoD Instruction
DoDISS	Department of Defense Index of Specifications and Standards
DOTE	Directorate of Operational Test and Evaluation
DPAMMH	Direct productive annual maintenance Man-Hours
DPSC	Defense Personnel Support Center
DRB	Defense Resources Board
DRP	Direct requisitioning procedure
DS	Direct support
DSAA	Defense Security Assistance Agency
DSARC	Defense Systems Acquisition Review Council
DSS	Direct support system

DSMC	Defense Systems Management College
DT	Development test
DTC	Design to cost
DTE	Developmental test and evaluation
DTS	Defense Transportation System
DTUPC	Design-to-unit production cost
DVAL	Demonstration and validation (same as D&V)
ECA	Early comparability analysis
ECP	Engineering change proposal
ECR	Engineering change request
ED	Engineering development
EIS	Environmental impact statement
EMP	Electromagnetic pulse
ETSS	Engineering and technical services specialists
EUTE	Early user test and experimentation
FAR	Federal Acquisition Regulation
FCA	Functional configuration audit
FD/SC	Failure definition and scoring criteria
FDTE	Force development test and experimentation
FEA	Front-end analysis
FFBD	Functional flow block diagram
FFW	Failure-free warranty
FGC	Functional group code
FMEA	Failure mode and effects analysis
FMECA	Failure mode, effects, and criticality analysis
FMS	Foreign Military Sales
FOC	Full and open competition
FOTE	Follow-on operational test and evaluation
FSD	Full-scale development
FSED	Full-scale engineering development
FY	Fiscal year
FYDP	(1) Five-Year Defense Plan; (2) Five-Year Defense Program
GFE	Government-furnished equipment
GFM	Government-furnished material
GFS	Government-furnished software
GITIS	Government integrated technical information system
GS	General Support

GSA	(1) General Supply Administration, (2) General Services Administration
HARDMAN	Hardware versus manpower
HEL	Human engineering laboratory
HFE	Human-factors engineeering
HFEA	Human-factors engineering analysis
HHA	Health hazard assessment
HHAR	Health hazard assessment report
HNS	Host nation support
HWCI	Hardware configuration item
ICD	Interface control document
ICE	Independent cost estimate
ICS	Interim contractor support
ICTP	Individual and collective training plan
ICWG	Interface control working group
IDD	Interface design document
IDSM	Intermediate direct support maintenance
IE	Independent evaluation
IEP	Independent evaluation plan
IER	Independent evaluation report
ILS	Integrated logistics support
ILSMT	Integrated logistics support management team
ILSP	Integrated logistics support plan
IMP	Information management plan
IOC	Initial operational capability
IOTE	Initial operational test and evaluation
IPF	Initial production facilities
IPR	In-process review
IPS	Integrated program summary
IPSS	Initial pre-planned supply support
IRA	Industrial resource analysis
IR&D	Independent research and development
IRN	Interface revision notice
IRS	Interface requirements specification
ISSA	Interservice support agreement
ISD	Instructional system development
ISDB	Integrated support database
IWSDB	Integrated weapon system database

JCS	Joint Chiefs of Staff
JMSNS	Justification for major system new start
LAO	Logistics Assistance Offices
LCA	Logistics control activity
LCC	Life-cycle cost
LCSMM	Life-Cycle system management model
LD	Logistic demonstration
LIN	Line item number
LLCSC	Lower-level computer software component
LLI	Long-lead items
LLT	Long lead time
LLTI	Long-leadtime items
LOA	Letter of agreement or letter of offer and acceptance
LOC	Lines of communication
LOGMAP	Logistics master plan
LOGMAPS	Logistics master planning system
LOGSACS	Logistics structure and composition system
LORA	Level of repair analysis
LOR	Level of repair
LP	Limited procurement
LPT	LP—test
LPU	LP—urgent
LR	Letter requirement
LRIP	Low-rate initial production
LRP	Low-rate production
LRRDAP	Long-range research, development, and acquisition plan
LRU	Line replaceable unit
LSA	Logistic support analysis
LSAR	Logistic support analysis record
LSC	Logistics support costs
LSCG	Logistics support cost guarantee
LTPD	Lot tolerance percent defective
MAA	Mission area analysis
MAC	Military Airlift Command
MACOM	Major Army Command
MADP	Materiel acquisition decision process; mission area deployment plan

MAMP	Mission area materiel plan; materiel acquisition management plan
MANPRINT	Manpower and personnel integration
MARB	Materiel acquisition review board
MaxTTR	Maximum time to repair
MCCR	Mission-critical computer resources
MCCS	Mission-critical computer system
MDR	Milestone decision review
MEMO	Mission essential maintenance operations–capabilities
MER	Manpower estimate report
MFP	Materiel fielding plan
MIL-STD	Military standard
MJWG	MANPRINT joint working group
MLRS	Multiple launch rocket system
MLSC	Measured logistics support cost
MND	Mission need document
MNS	Mission need statement (formerly JMSNS)
MOU	Memorandum of understanding
MP	Mission profiles
MPT	Manpower, personnel, and training
MRO	Material release order
MRP	Material requirements planning
MRSA	U.S. Army materiel readiness support activity
MTBD	Mean time between demand
MTBF	Mean time between failure
MTBFG	Mean time between failure guarantee
MTBMA	Mean time between maintenance actions
MTBUMA	Mean time betwen unscheduled maintenance actions
MTFHE	Mean time to first human error
MTBHE	Mean Time Between Human Error
MTMC	Military Traffic Management Command
MTMCTEA	MTMC Transportation Engineering Agency
MTTR	Mean time to repair
NATO	North Atlantic Treaty Organization
NDI	Nondevelopment item
NETP	New equipment training plan
NOR	Notice of revision

NSN	National stock number
O&M	Operations and maintenance
O&O	Operational and organizational
O&S	Operation and support
OBCE	Operational baseline cost estimate
OBS	Obsolete
OMB	Office of Management and Budget
OMS	Operational-mode summary
ORF	Operational readiness float
ORLA	Optimum-repair level analysis
OSD	Office of the Secretary of Defense
OST	Order ship time
OT	Operational test–testing
OTE	Operational test and evaluation
OTEA	U.S. Army Operational Test and Evaluation Agency
P&A	Price and availability
P³I	Preplanned product improvement
PARR	Program analysis and resource review
PATE	Production acceptance test and evaluation
PBD	Program budget decision
PBG	Program and budget guidance
PCA	Physical configuration audit
PD	Program directive
PDES	Product data exchange specification
PDIP	Program development increment packages
PDM	Program decision memorandum
PDR	Physical or preliminary design review
PDSS	Postdeployment software support
PEO	Program executive officer
PEP	Producibility engineering and planning
PHST	Packaging, handling, storage, and transportation
PI	Product improvement
PIF	Provision of industrial facilities
PIP	(1) Product improvement program; (2) product improvement proposal
PIRN	Preliminary interface revision notice
PL	Public Law
PM	Program, project, or product manager

PM TRADE	Project manager, training devices
PMCS	Program management control system
PMD	Program management document
PMO	Program Management Office
POI	Program(s) of instruction
POL	Petroleum, oils, and lubricants
POM	Program objective memorandum
POQ	Period–order–quantity
PPBERS	Program performance and budget execution review systems
PPBES	Planning, programming, budgeting, and execution system
PPBS	Planning, programming, and budgeting system
PPQT	Preproduction qualification test
PPS	Postproduction support
PPSP	Postproduction support plan
PQT	Production qualification test
PRIMIR	Product improvement management information report
PRR	Production readiness review
PSWBS	Project summary work-breakdown structure
PTD	Provisioning technical documentation
PWBS	Project or program work-breakdown structure
QA	Quality assurance
QT	Qualification testing
R&D	Research and development
R&M	Reliability and maintainability
RAM	Reliability, availability, and maintainability
RCM	Reliability-centered maintenance
RDA	Research, development, and acquisition
RDD	Required delivery date
RDTE	Research, development, test, and evaluation
RFD	Request for deviation
RFP	Request for proposal
RFW	Request for waiver
RIW	Reliability improvement warranty
ROC	Required operation capability
RPSTL	Repair parts and special tools list
RRR	RAM rationale report

RSC	Reinforcement support category
RSI	Rationalization, standardization, and interoperability
SADM	Secretary of the Army decision memorandum
SAE	Service acquisition executive
SAIP	Spares acquisition integrated with production
SAP	Streamlined acquisition process
SAR	(1) Selected acquisition report; (2) safety assessment report
SCCB	Software configuration control board
SCMP	Software configuration management plan
SCN	Specification change notice
SDC	Software development cycle
SDDD	Software detailed design document
SDDM	Secretary of Defense decision memorandum
SDF	Software development file
SDL	Software development library
SDP	Software development plan
SE	Systems engineering
SECDEF	Secretary of Defense
SECDEF/OSD	Secretary of Defense/Office of the SECDEF
SEMP	Systems engineering management plan
SISMS	Standard integrated support management system
SKO	Sets, kits, and outfits
SMMP	System MANPRINT management plan
SOP	Standing operating procedure
SOW	Statement of work
SPE	System performance estimation
SPM	Software programmer's manual
SPS	Softwear product specification
SQEP	Software quality evaluation plan
SRS	Software requirements specification
SSA	Software support agency; system safety assessment; source selection authority
SSAC	Source selection advisory council
SSEB	Source selection evaluation board
SSP	Source selection plan
SSPM	Software standards and procedures manual
SSR	Software specification review

SSS	System–segment specification
SSWG	System safety working group
STD	Standard
STE	Special test equipment
STLDD	Software top-level design document
STP	Software test plan
STPR	Software test proceudre
STR	Software test report
SUM	Software user's manual
SWBM	Summary work-breakdown structure
TAT	Turnaround time
TC	Type classification
TC LPT	TC limited procurement test
TC LPU	TC limited procurement urgent
TDNS	Training device needs statement
TDP	Technical data package or technical design plan
TDR	Training device requirement
TE	Test and evaluation
TEA	Training effectiveness analysis; transportability engineering analysis
TEMP	Test and evaluation master plan
TISC	Technology integration steering committee
TIWG	Test integration working group
TLCCS	Total life-cycle competition strategy
TLCSG	Top-level computer software component
TLSC	Target logistics support cost
TMDE	Test, measurement, and diagnostic equipment
TPM	Technical performance management
TPRA	Task performance requirements analysis
TRADOC	Training and Doctrine Command
TRR	Test readiness review
TT/UT	Technical test–user test
WBS	Work-breakdown structure

APPENDIX 3
REFERENCES

AIR FORCE DOCUMENTS

AFR 65-2	*Provisioning of End Items of Materiel*
AFR 65-3	*Configuration Management*
AFR 66-1	*Maintenance Management*
AFR 71-1	*Packaging Management*
AFR 80-5	*Air Force Reliability and Maintainability Program*
AFR 80-14	*Test and Evaluation*
AFR 800-3	*Engineering for Defense Systems*
AFR 800-8	*Integrated Logistics Support Program*
AFR 800-12	*Acquisition of Support Equipment*
AFR 800-14	*Computer Resources Integrated Support Plan*
AFR 800-21	*Interim Contractor Support for Systems Equipment*
AFR 800-24	*Parts Control Program*
AFR 800-26	*Spares Acquisition Integrated with Production (SAIP)*
AFLCR 65-5	*Air Force Provisioning Policies and Procedures*
AFLCR 65-14	*Policy and Procedural Guidance for Interservice of Depot Level Maintenance*
AFLCR 66-17	*Depot Maintenance Support Policy*
AFLCM/AFSCM 800-4	*Optimum Repair Level Analysis (ORLA)*
AFLC/AFSC Pamphlet 800-34	*Acquisition Logistics Management*
AFM 26-1	*Manpower Policies and Procedures*

ARMY DOCUMENTS

AR 1-1	*Planning, Programming, and Budgeting within the Department of the Army*
AR 11-14	*Logistics Readiness*

AR 11-16	*The Cost Analysis Program*
AR 11-18	*The Cost Analysis Program*
AR 11-28	*Economic Analysis and Program Evaluation for Resource Management*
AR 18-7	*Data Processing Installation Management Procedures and Standards*
AR 32-15	*Classification and Inspection*
AR 34-3	*Battlefield Automated System Interoperability Management*
AR 70-1	*Army Research, Development, and Acquisition*
AR 70-10	*Test and Evaluation during Development and Acquisition of Materiel*
AR 70-15	*Product Improvement of Materiel*
AR 70-32	*WBS for Defense Materiel Items*
AR 70-37	*DoD Configuration Management*
AR 70-44	*DoD Engineering for Transportability*
AR 70-47	*Engineering for Transportability*
AR 70-61	*Type Classification of Army Materiel*
AR 70-64	*Design to Cost*
AR 71-9	*Materiel Objectives and Requirements*
AR 350-35	*New Equipment Training*
AR 350-38	*Training Device Policies and Procedures*
AR 381-143	*Logistics Policies and Procedures*
AR 385-16	*System Safety Engineering and Management*
AR 570-4	*Manpower Management*
AR 601-1	*Human-Factors Engineering Program*
AR 602-2	*Manpower and Personnel Integration (MANPRINT in Materiel Acquisition Process)*
AR 700-9	*Principles and Policies of the Army Logistics Systems*
AR 700-10	*Acquisition Management Milestone System*
AR 700-47	*Defense Standardization Program*
AR 700-60	*Department of Defense Parts Control Program*
AR 700-120	*Materiel Distribution Management for Major Items*
AR 700-127	*Integrated Logistic Support*
AR 700-139	*Army Warranty Program*
AR 702-3	*Army Materiel Reliability, Availability, and Maintainability (RAM)*
AR 702-13	*The Army Warranty Program*
AR 710-1	*Centralized Inventory Management of the Army Supply System*
AR 710-10	*Modification of Materiel and Issuing Safety of Use Messages*
AR 715-22	*High Dollar Spare Parts Breakout Program*
AR 725-50	*Requisitioning, Receipt, and Issue System*
AR 740-1	*Storage and Supply Activity Operations*

AR 750-1	*Army Materiel Maintenance Concepts and Policies*
AR 750-10	*Modification of Materiel*
AR 750-43	*Test Measurement and Diagnostic Equipment, Including Prognostic Equipment and Calibration Test/Measurement*
AR 1000-1	*Basic Policies for System Acquisition*

ARMY PAMPHLETS

DA PAM 700-50   *ILS Developmental Supportability Test and Evaluation*

BOOKS

Blanchard, Benjamin S., *Logistics Engineering and Managment,* 3rd ed., Prentice-Hall, Englewood Cliffs, NJ, 1986.

Blanchard, Benjamin S., and Wolter J. Fabrycky, *Systems Engineering and Analysis,* Prentice-Hall, Englewood Cliffs, NJ, 1981.

Daniels, M.A., *Principles of Configuration Management,* Advanced Application Consultants, Rockville, MD, 1985.

Hutchinson, Norman E., *An Integrated Approach to Logistics Management,* Prentice-Hall, Englewood Cliffs, NJ, 1987.

Jones, James V. *Engineering Design Reliability Maintainability and Testability,* TAB Books, Blue Ridge Summit, PA, 1988.

Jones, James V., *Logistics Support Analysis Handbook,* TAB Books, Blue Ridge Summit, PA, 1989.

Michael, Jack V., and William P. Wood, *Design to Cost,* John Wiley and Sons, New York, 1989.

Mosher, John, F., *Integrated Logistics Systems Handbook,* Vols. 1–6, Advanced Applications Consultants, Rockville, MD, 1983.

O'Conner, Patrick D.T., *Practical Reliability Engineering,* John Wiley and Sons, New York, 1983.

Patton, Joseph D., Jr., *Logistics Technology and Management the New Approach,* The Solomon Press, New York, 1986.

DEPARTMENT OF DEFENSE DIRECTIVES AND INSTRUCTIONS

DoDD 3235.1-H	*Test and Evaluation of System Reliability Availability and Maintainability, A Primer*
DoDD 4151.12-1	*Policies Governing Maintenance Engineering within the Department of Defense*
DoDD 4245.7	*Transition from Development to Production and Companion Manual,* DoD 4245.7-M

DoDD 5000.1	*Major System Acquisition*
DoDD 5000.3	*Test and Evaluation*
DoDD 5000.39	*Acquisition and Management of Integrated Logistics Support for Systems and Equipment*
DoDD 2000.9	*International Co-Production Projects and Agreements between the U.S. and Other Countries or International Organizations*
DoDD 2010.4	*U.S. Participation in Certain NATO Groups Related to Research, Development, Production and Logistics Support of Military Equipment*
DoDD 2010.6	*Standardization and Interoperability of Weapon Systems and Equipment within the North Atlantic Treaty Organization*
DoDD 2010.8	*Department of Defense Policy for NATO Logistics*
DoDD 2010.9	*Mutual Logistic Support between the U.S. and Other NATO Forces*
DoDD 2010.10	*Mutual Logistic Support between the U.S. and Other NATO Forces Financial Policy*
DoDD 2040.2	*International Transfers of Technology, Goods, Services, and Munitions*
DoDD 4245.3	*Design to Cost*
DoDD 4245.7	*Transition from Development to Production*
DoDD 5000.4	*OSD Cost Analysis Improvement Group*
DoDD 5000.36	*System Safety Engineering and Management*
DoDD 5000.40	*Reliability and Maintainability* (July 8, 1980)
DoDD 5000.43	*Acquisition Streamlining*
DoDD 7045.7	*Implementation of the Planning, Programming, and Budgeting System*
DoDD 7045.14	*The Planning, Programming, and Budgeting System* (PPBS)
DoDD 7110.1	*DoD Budget Guidance*

GOVERNMENT DOCUMENTS

An Appraisal of Models Used in Life-Cycle Estimation for United States Air Force Aircraft Systems, DTIC Technical Report, Defense Logistics Agency, October 1978.

Balahan, Harold S., Kenneth B. Tom, and George T. Harrison, Jr., *Warranty Handbook*; prepared for the Defense Systems Management College, Fort Belvoir, VA, June 1982.

Freihofer, James T., and Daniel S. Beach. "The Warranty Guaranty Clause: An Analysis of Its Use on the Spruance Class (DD-963) Shipbuilding Contract and Identification of Lessons Learned," thesis, Naval Postgraduate School, Monterey, CA, 1983.

Gondara, Arturo, and Michael D. Rich, *Reliability Improvement Warranties for Military Procurement,* Rand Corporation Report R-2264-AF, December 1977.

Hernandez, Richard J., and Leo E. Daney. "System Level Warranty Laws: Their Implications for Major U.S.A.F. Weapon System Acquisitions," thesis, Air Force Institute of Technology, Wright Patterson Air Force Base, Ohio, 1985.

Jacobson, Mark C., and Reagan L. Scaggs, "Evaluation of and Recommended Change to the Reliability Improvement Warranty (RIW) Guidelines," thesis, Air Force Institute of Technology, Wright Patterson Air Force Base, Ohio, 1979.

Joint Logistics Commanders Guide for the *Management of Multinational Programs,* Defense Systems Management College, Fort Belvoir, VA, July 1981.

Lenassi, John R., *Warranty Decision Process,* U.S. Army Materiel Systems Analysis Activity, Fort Lee, VA, 1982.

MANPRINT Primer, June 24, 1988.

Marshall, Clifford, *Incentivising Availability Warranties,* Office of Naval Research, Polytechnic Institute of New York, 1980.

Parkinson, David R., and Alan W. Schoolcroft, "An Evaluation of the Perceived Effectiveness of Reliability Improvement Warranties (RIW) Applied During the Air Force RIW Trial Period, thesis, Air Force Institute of Technology, Wright Patterson Air Force Base, Ohio, 1983.

Rannenberg, J.E., "Warranties in Defense Acquisition: the Concept, the Contest, and the Congress," thesis, Naval Postgraduate School Monterey, CA, 1984.

System Engineering Management Guide, published by the Defense Systems Management College.

MILITARY HANDBOOKS

MIL-HDBK-CALS	(Draft) *Computer-Aided Acquisition and Logistics Support Program Implementation Guide*
MIL-HDBK-157	*Transportability Criteria*
MIL-HDBK-245B	*Preparation of Statement of Work* (SOW)
MIL-HDBK-259	*Life Cycle Cost in Navy Acquisitions*
MIL-HDBK-472	*Maintainability Prediction*

MILITARY STANDARDS

MIL-STD-470	*Maintainability Program for Systems and Equipment*
MIL-STD-471	*Maintainability Demonstration/Validation*
MIL-STD-471A	*Maintainability Demonstration*
MIL-STD-480	*Configuration Control—Engineering Changes, Deviations, and Waivers*
MIL-STD-482	*Configuration Status Accounting, Data Elements, and Related Features*
MIL-STD-499A	*Engineering Management* (May 1, 1974)
MIL-STD-690	*Failure Rate Sampling Plans and Procedures*

MIL-STD-721	*Definitions of Terms for Reliability and Maintainability*
MIL-STD-785	*Reliability Program for Systems and Equipment Development and Production*
MIL-STD-1367	*Packaging, Handling, Storage, and Transportability Program Requirements for Systems and Equipments*
MIL-STD-1369A	*ILS Program Requirements*
MIL-STD-1388-1	*Logistic Support Analysis*
MIL-STD-1388-2	*Logistic Support Analysis Record*
MIL-STD-1472	*Human Engineering Design Criteria for Military Systems, Equipment and Facilities*
MIL-STD-1517	*Phased Provisioning*
MIL-STD-1561	*Provisioning Procedures, Uniform DoD*
MIL-STD-1629	*Procedures for performing a FMECA*

INDEX